Proceedings in Life Sciences

E. Petzinger, R. K. H. Kinne, H. Sies (Eds.)

Hepatic Transport of Organic Substances

With 118 Figures

Springer-Verlag Berlin Heidelberg New York
London Paris Tokyo

Professor Dr. ERNST PETZINGER
Institut für Pharmakologie und Toxikologie
Universität Gießen
Frankfurter Straße 107
6300 Gießen/FRG

Professor Dr. ROLF K. H. KINNE
Max-Planck-Institut für Systemphysiologie
Rheinlanddamm 201
4600 Dortmund 1/FRG

Professor Dr. HELMUT SIES
Institut für Physiologische Chemie I
Universität Düsseldorf
Moorenstraße 5
4000 Düsseldorf 1/FRG

ISBN-13:978-3-642-74249-1 e-ISBN-13:978-3-642-74247-7
DOI: 10.1007/978-3-642-74247-7

Library of Congress Cataloging-in-Publication Data.
Hepatic transport in organic substances / E. Petzinger, R. K. H.Kinne, H. Sies (ed.).
p. cm. -- (Proceedings in life sciences)
Proceedings of a symposium, held May 8-11, 1988 at Schloss Ringberg, Rottach Egern.
Includes index.
ISBN-13:978-3-642-74249-1 (U.S. : alk. paper)
1. Liver--Physiology-- Congresses. 2. Biological transport--Congresses.
3. Biomolecules--Metabolism--Congresses.
4. Organic compounds--Metabolism--Congresses.
I. Petzinger, E. (Ernst), 1950-. II. Kinne, Rolf K. H. III. Sies, H. (Helmut), 1942-. IV. Series.

[DNLM: 1. Liver--metabolism--congresses. WI 702 H5296 1988]
QP185.H47 1989 612'.35--dc 19 DNLM/DLC for Library of Congress 88-38907 CIP

Softcover reprint of the hardcover 1st edition 1989

2131/3020-543210 – Printed on acid-free paper

Preface

During recent years, the application of modern techniques has dramatically increased our insights into the cellular and molecular mechanisms comprising the chain of events involved in the transport of organic solutes into, out of, and across the hepatocyte. A major component facilitating this progress was the acceptance by the majority of hepatologists that general principles derived for membrane transport processes in epithelial cells such as kidney and intestine are also applicable to the hepatocyte despite its more complex morphology and the predominance of intracellular modification of organic compounds.

The Ringberg Symposium, whose proceedings are compiled in this volume, was thus aimed to foster this way of consideration even further by bringing together scientists whose work focusses on membrane events in general and transport events in particular. In doing so, this symposium can be considered as a continuation of the series of books initiated in 1981 by the publication of "Renal Transport of Organic Solutes" edited by R. Greger, F. Lang, and S. Silbernagl, followed in 1983 by "Intestinal Transport: Fundamental and Comparative Aspects" edited by M. Gilles-Baillieu and R. Gilles.

In planning the symposium, it became evident that time would not permit to include transport events which occur via vesicular uptake and transcytosis, although these transport processes are involved in a large number of hepatic functions.

The symposium, held from May 8-11, 1988 at Schloß Ringberg, Rottach Egern, coincided with the death of Dr. Hans Popper, eminent hepatologist and role model for generations of liver physiologists, clinicians, and pathologists. The participants and the organizing committee, therefore, decided to dedicate the symposium and this volume to him. In his letter, withdrawing his participation, Dr. Hans Popper expressed his eagerness to prepare the Summary Statements and "by doing so, to learn a great deal". We hope that this book confers some of this eagerness and interest onto its readers and makes the reading a fascinating and rewarding learning experience.

E. Petzinger
R.K.H. Kinne
H. Sies

List of Contributors

In Memoriam Hans Popper

We dedicate this volume to the memory of the eminent hepatologist Hans Popper, M.D. Ph.D., Gustave L. Levy Distinguished Service Professor, Mount Sinai School of Medicine, New York, who died shortly before this meeting to which he was invited as a Honorary Chairman.

Dr. Popper's intellect inspired and stimulated at least three generations of hepatologists and by many Dr. Popper is considered the founding father of hepatology as it is known today. His broad range of interest was linked to a sustained desire to transpose advances in modern biology, morphology, and physiology to medicine and pathobiology. During his whole scientific carreer he kept reaching out into new disciplines to learn more about the liver and its diseases. Dr. Popper played also a crucial role in catalyzing the development of hepatology on both sides of the Atlantic.

Dr. Popper had a unique fascinating personality and was an enthusiastic teacher and clinician. His efforts, like his life, serve as an inspiration to all of us.

Hans Popper 1903 - 1988

Acknowledgements

As in any endeavor a whole array of interfaces is necessary to complete the task successfully. In addition to financial support which was generously provided by the foundations and agencies mentioned below, moral support, scientific advice, and logistic support are indispensable. In this respect our enthusiastic colleagues and our secretarial staff, notably Daniela Mägdefessel and Christine Müller were instrumental. Without their effective, understanding, and thoughtful arrangements and the service of the staff of the Ringberg Castle, the conference would not have attained the ambiance necessary for the fruitful exchange of ideas in a relaxed and hospitable atmosphere. D. Mägdefessel, Ch. Müller, F. Dräger, and F. Sieland also carried the major burden in preparing the final version for the publication of the book which summarizes the papers and abstracts given at this meeting. Last but not least we would like to thank all the participants of the conference, notably the speakers, both the presenters and discussants, who made this meeting a stimulating voyage into an exciting scientific territory.

We would like to thank the Deutsche Forschungsgemeinschaft, the Max-Planck-Gesellschaft, the Paul-Martini-Stiftung and the Thyssen-Stiftung for their generous support.

Furthermore we acknowledge the financial assistance by Bayer AG, Leverkusen, Byk Gulden Lomberg Chemische Fabrik GmbH, Konstanz and Schering AG, Berlin.

Contents

Hepatic Transport of Organic Anions and Bile Acids

Part 1
General Aspects of Hepatic Transport

Glutamate-sodium Cotransport in the Kidney: An Example for the Plasticity of Transport Systems

Rolf K.H. Kinne
Max-Planck-Institut für Systemphysiologie
Rheinlanddamm 201
4600 Dortmund 1
F.R.G.

Introduction

Glutamate is an important metabolic intermediate as well as a neurotransmitter. Therefore, the question of its transport across plasma membranes has generated considerable interest. In the liver, the results concerning the ability to transport glutamate have been controversial. In liver slices (Hems et al., 1968) and in the perfused liver (Ross et al., 1967) the parenchymal cell membrane was found to be relatively impermeable to glutamate, whereas sodium-dependent glutamate transport was observed in rat hepatocytes in primary monolayer culture (Gebhardt and Mecke, 1983) and in a mixed preparation of plasma membrane vesicles from rat liver (Sips et al., 1982). The discrepancy between these different results was resolved recently by Ballatori et al. (1986) who demonstrated that the sodium-L-glutamate cotransport system was only present in the canalicular domains of liver plasma membranes. Since this domain is not exposed in slices or in the intact organ the strong polarity of the hepatocyte with regard to the distribution of transport systems (Kinne, 1987) could be made responsible for the divergence in the experimental results.

Hepatic Transport in Organic Substances
Ed. by E. Petzinger, R. K.-H. Kinne, H. Sies

In the kidney, electrophysiological studies on tubules perfused in vivo (Samarzija and Frömter, 1975; 1976) and experiments on isolated plasma membrane vesicles (Sacktor, 1981) showed consistently that in contrast to the liver, sodium-dependent L-glutamate transport systems are present both in the apical and in the basal pole of the proximal tubule cell. A discrepancy arose, however, with regard to the electrogenicity of the transport system. In the presence of K either intracellularly in vivo (Samarzija and Frömter, 1975; 1976) or inside brush border membranes (Burckhardt et al., 1980) the transport system was electrogenic whereas in the absence of K no or only a low electrogenicity could be demonstrated (Burckhardt et al., 1980; Schneider et al., 1980). Electrogenicity in the presence of trans K was also described for the sodium-L-glutamate cotransport system in liver plasma membrane vesicles (Sips et al., 1982; Ballatori et al., 1986).

This apparent unmasking of the electrogenicity by trans K^+ was even more puzzling since convincing evidence was provided that K^+ was countertransported by the sodium-L-glutamate cotransport system (Sacktor and Schneider, 1980; Sacktor et al., 1981; Koepsell et al., 1984). Further kinetic studies revealed a stoichiometry of two sodium, one glutamate, and one proton for the symport step, and of one potassium for the antiport step (Nelson et al., 1983; Fukuhara and Turner, 1985). These stoichiometries made clear that in the absence of intravesicular potassium - assuming an electroneutral carrier -the stoichiometry of glutamate translocation was two positive charges, whereas when simultaneous exchange occurred, the overall charge stoichiometry would <u>decrease</u> to +1. The experimental finding was, however, that the apparent electrogenicity of the initial rate of L-glutamate uptake <u>increased</u> in the presence of intravesicular K rather than decreased (Burckhardt et al., 1980).

The current contribution provides a model which can explain these results. It has been described in more detail elsewhere (Heinz et al., 1988). This model also demonstrates that the apparent electrogenicity exhibited by a particular transport system is a complex function of the experimental conditions employed. Thus conclusions about charge stoichiometry based on electrogenicity experiments should always be drawn with caution, a more reliable way seems to be the use of the thermodynamic properties of the transporter for this purpose (Turner, 1985).

Effect of Potassium on the Electrogenicity of the Sodium-L-Glutamate Cotransport System

As already mentioned in the introduction, potassium has a pronounced effect on the potential dependence of sodium-dependent L-glutamate transport. This is illustrated for rabbit renal brush borders in Fig. 1. Here different diffusion potentials generated by a sodium sulfate or a sodium nitrate gradient were imposed across the vesicle membrane and the initial rate of L-glutamate uptake in the presence or absence of intracellular potassium was measured. It is evident that potassium increases the potential sensitivity of the L-glutamate uptake significantly from a stimulation of 60% to a stimulation of 120%, whereas D-glucose uptake measured as a control remains unaffected.

As a more quantitative term one can introduce the relative electrogenicity which can be calculated by the equation

$$\frac{J_2 - J_1}{\epsilon_2 - \epsilon_1} \cdot \frac{1}{J_1}$$

where J_2 and J_1 represent the flux observed in the presence of the nitrate or sulfate gradient, respectively, and $\epsilon_2 - \epsilon_1$ the difference in membrane potential (here considered constant and

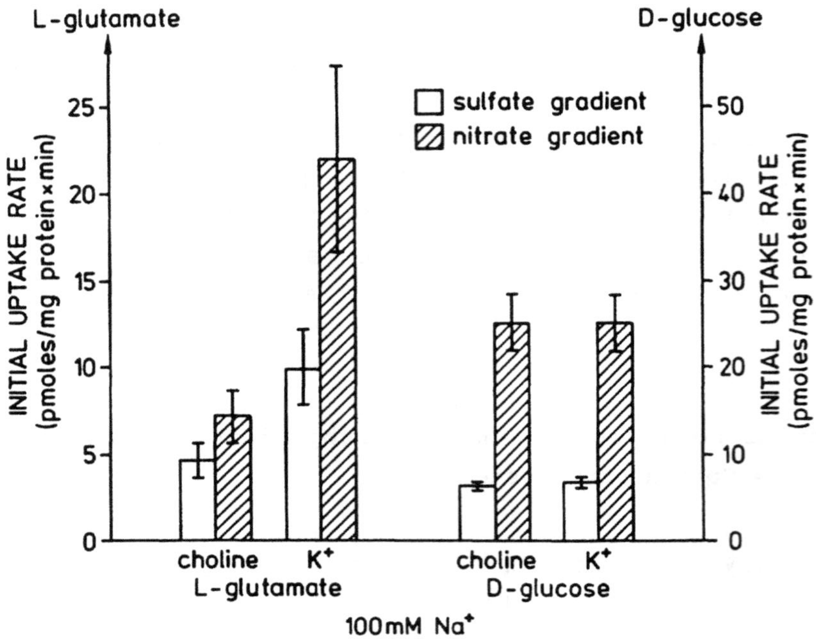

Figure 1: Effect of the electrical potential on the initial uptake rate of glutamate and glucose, respectively, in the absence and presence of intravesicular K^+. The electrical potential was changed by replacement of the sulfate gradient by a nitrate gradient. Each column represents the average of 11 independent rabbit kidney brush border membrane preparations. For glutamate the difference between NO_3 and SO_4 is not significant without potassium, but significant ($P < 0.05$) with potassium. For glucose the corresponding differences are highly significant, but there is no significant difference in the presence or absence of K^+. Reprinted from Heinz et al., 1988 with kind permission.

set at unity) (Heinz et al., 1988). The relative electrogenicity of L-glutamate in the absence of K thereby becomes 0.58 whereas it is 1.2 in the presence of K (see table 1).

Effect of Transportate Concentration

As shown in table 1 even in the presence of K the potential dependence of L-glutamate uptake is not constant but varies with the concentration of the transportates. In these

Table 1. Relative electrogenicity of sodium-dependent L-glutamate uptake (initial rate) into rabbit kidney brush border membranes: Effect of transportate

	20 mM Na	100 mM Na
without K	0.70	0.58
with 20 mM K intravesicular	2.11	1.21

Data have been recalculated from Heinz et al., 1988

experiments the sodium concentration - which due to the high stoichiometry has a very pronounced effect on the transport rate - has been altered. When decreasing the sodium concentration the relative electrogenicity of L-glutamate uptake almost doubles in the presence of intravesicular K, whereas only a small increase is observed in the absence of K.

A Model

In Fig. 2 a simplified model is presented in which the effect of K on the translocation steps of L-glutamate transport is depicted. In accordance with other authors (Nelson et al., 1983; Fukuhara and Turner, 1985) it is assumed that potassium facilitates the return of the unloaded carrier to the external face of the membrane by providing an additional translocation route via the XK^+ complex. This assumption explains the increased L-glutamate uptake in the presence of potassium. The occurrence of an additional translocation route has, however, also consequences for the relative electrogenicity of the initial L-glutamate uptake. In the presence of K, the charge translocating step where translocation of two sodium, one glutamate, and one proton takes place becomes rate limiting

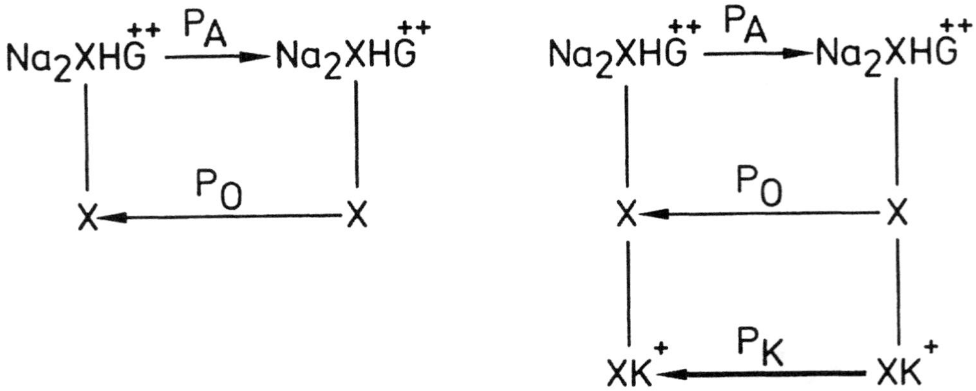

Figure 2: Simplified model of Na^+-linked glutamate transport without K^+ inside (left) and with K^+ inside (right). P_A, P_O, and P_K are the rate coefficient (probability coefficient) for the fully loaded, the empty, and the K^+-loaded carrier species, respectively. The translocator (X) is assumed to be neutral. Reprinted from Heinz et al., 1988 with kind permission.

and thus electrogenicity of the glutamate uptake is displayed. Consistent with this assumption the K effect is higher at low transportate concentrations (low sodium in table 1) and partially reversed when the transportate concentration is increased (compare 20 and 100 mM Na^+ in table 1). In the latter case the relative rate limitancy of sodium, glutamate, proton symport is lower than in the former case. This model thus predicts that the higher the rate limitancy of the glutamate translocating step (characterized by P_A in Fig. 2) the higher the relative electrogenicity and vice versa. Indeed, such a correlation could be found when, as shown in Fig. 3, relative rate limitancy and relative electrogenicity obtained in individual experiments were compared.

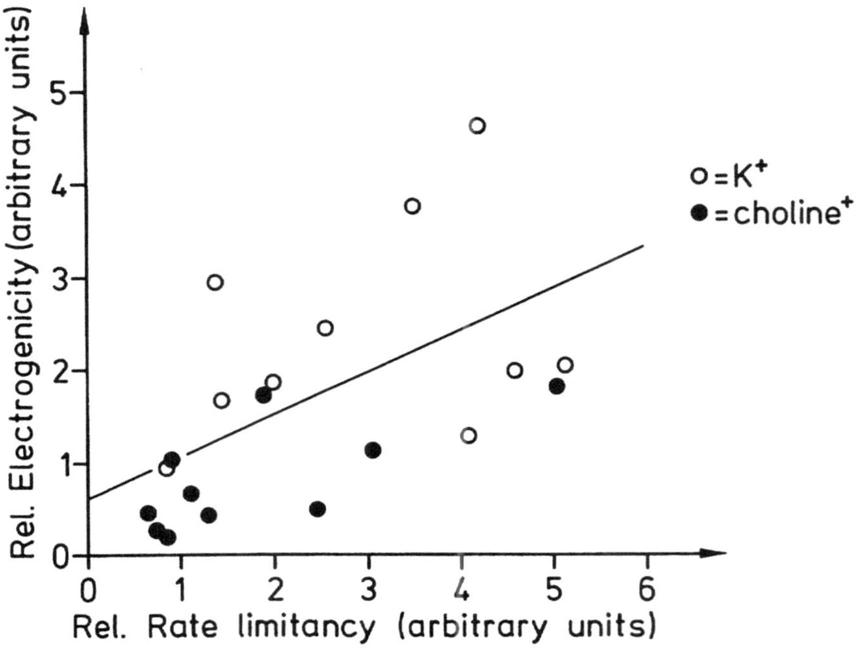

Figure 3: The relative electrogenicity is plotted versus the corresponding rate limitancy of altogether 20 independent experimental determinations. The dotted line is drawn by the method of least-squares. Its slope is significantly different from 0 (P = 0.01). Reprinted from Heinz et al., 1988 with kind permission.

Conclusion

The results presented above suggest that the relative electrogenicity displayed by transport systems under initial rate conditions may be a function of the relative rate limitancy of that step in the transport cycle where charge translocation occurs. Changes in the relative rate limitancy can produce results which apparently are at odds with assumed charge stoichiometries. Knowledge of the potential dependence of a transport system can thus be only of limited value with regard to the actual number of charges being tranlocated.

References

Ballatori N, Moseley RH, Boyer JL (1986) Sodium gradient-dependent L-glutamate transport is localized to the canalicular domain of liver plasma membranes. Studies in rat liver sinusoidal and canalicular membrane vesicles. J Biol Chem 261:6216-6221

Burckhardt G, Kinne R, Stange G, Murer H (1980) The effects of potassium and membrane potential on sodium-dependent glutamic acid uptake. Biochim Biophys Acta 599:191-201

Fukuhara Y, Turner J (1985) Cation dependence of renal outer cortical brush border membrane L-glutamate transport. Am J Physiol 248:F869-F875

Gebhardt R, Mecke D (1983) Glutamate uptake by cultured rat hepatocytes is mediated by hormonally inducible, sodium-dependent transport systems. FEBS Lett 161:275-278

Heinz E, Sommerfeld DL, Kinne RKH (1988) Electrogenicity of sodium/L-glutamate cotransport in rabbit renal brush-border membranes: a reevaluation. Biochim Biophys Acta 937:300-308

Hems R, Stubbs M, Krebs HA (1968) Restricted permeability of rat liver for glutamate and succinate. Biochem J 107:807-815

Kinne RKH (1987) Modulation of membrane transport in epithelia: lessons for the liver. In: Reutter W, Popper H, Arias IM, Heinrich PC, Keppler D, Landmann L (eds) Modulation of liver cell expression, Falk Symposium 43. MTP Press Ltd, Lancaster/Boston/The Hague/Dordrecht, p 95-106

Koepsell H, Korn K, Ferguson D, Menuhr H, Ollig D, Haase W (1984) Reconstitution and partial purification of several Na^+ cotransport systems from renal brush-border membranes. Properties of the L-glutamate transporter in proteoliposomes. J Biol Chem 259:6548-6558

Nelson PJ, Dean GE, Aronson PS, Rudnick G (1983) Hydrogen ion cotransport by the renal brush border glutamate transporter. Biochemistry 22:5459-5463

Ross BD, Hems R, Krebs HA (1967) The rate of gluconeogenesis from various precursors in the perfused rat liver. Biochem J 102:942-951

Sacktor B (1981) L-Glutamate transport in renal plasma membrane vesicles. Mol Cell Biochem 39:239-251

Sacktor B, Lepor N, Schneider EG (1981) Stimulation of the efflux of L-glutamate from renal brush-border membrane vesicles by extravesicular potassium. Biosci Rep 1:709-713

Sacktor B, Schneider EG (1980) The singular effect of an internal K^+ gradient (Kl^+ K^+) on the Na^+ gradient (Na^+ Na^+)-dependent transport of L-glutamate in renal brush border membrane vesicles. Int J Biochem 12:229-234

Samarzija I, Frömter E (1975) Electrical studies on amino acid transport across brushborder membrane of rat proximal tubule in vivo. Pflügers Arch 359:R119

Samarzija I, Frömter E (1976) Renal transport of glutamate and aspartate. Evidence for Na-dependent uptake from the peritubular surface into proximal tubular cells. Pflügers Arch 365:R15

Schneider EG, Hammerman MR, Sacktor B (1980) Sodium gradient-dependent L-glutamate transport in renal brush border membrane vesicles. Evidence for an electroneutral mechanism. J Biol Chem 255:7650-7656

Sips HJ, De Graaf PA, Van Dam K (1982) Transport of L-aspartate and L-glutamate in plasma-membrane vesicles from rat liver. Eur J Biochem 122:259-264

Turner RJ (1985) Stoichiometry of cotransport systems. Ann NY Acad Sci 456:10-25

Energetic Coupling in Secondary Active Transport

E. Heinz
Max-Planck-Institut für Systemphysiologie,
Rheinlanddamm 201
4600 Dortmund 1, FRG

The transport of most organic and inorganic solutes across biological membranes is secondary active, i.e. driven by the electrochemical potential difference of certain ion species, mostly by symport with Na^+ or H^+, and in some cases also by antiport with K^+. This requires the energetic coupling between the flow of the transported substrate and the parallel (or anti-parallel) flow of the "driver" ion. The molecular mechanism of this coupling is not precisely known, but the underlying principles presumably are: amplification of transport-effective pathways and/or obstruction of leakage pathways. As instrumental for this purpose two effects of the driver ion on the translocator have been postulated: one on the affinity of the latter for the substrate (affinity effect) and/or one on the velocity of the translocator (velocity effect). The first one should promote the formation, and the second one the translocation of the substrate-loaded translocator species (Heinz et al., 1972; Heinz, 1978). Each of these effects by itself should be able to bring about coupling, but so far it seems that both effects operate simultaneously.

Hepatic Transport in Organic Substances
Ed. by E. Petzinger, R. K.-H. Kinne, H. Sies

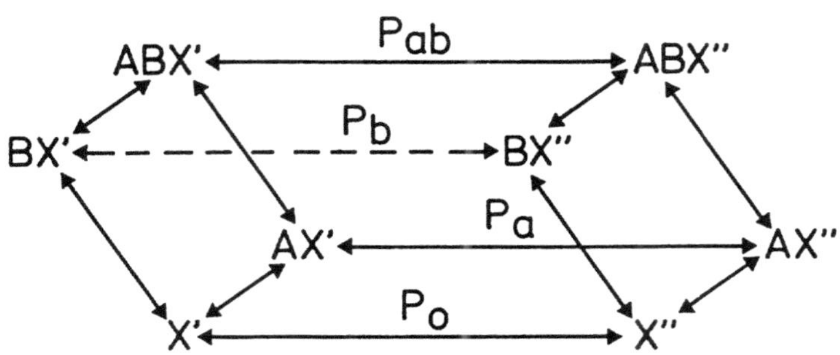

Figure 1: General case of secondary active transport. Assumed is quasi equilibrium between translocator species (X) and the ligands A and B in the adjacent bulk solutions. Asymmetry and electrogenicity are neglected as irrelevant in the present context.

These two effects should be reflected by the relative magnitude of the corresponding velocity and affinity parameters of the system. For a most general system of secondary active transport (Fig. 1), in which the transport of a substrate A is driven by the electrochemical potential difference of an ion B, by symport or antiport, the basic requirement for energetic coupling can most simply be expressed by the following ratio (Heinz et al., 1972):

$$R_c = \frac{r_{ab} \cdot P_o \cdot P_{ab}}{P_a \cdot P_b} \tag{1}$$

r_{ab} is the coefficient of cooperativity between solute A and B with respect to binding to the translocator, X. P_{ab}, P_o, P_a, and P_b denote the velocity coefficients of the four translocator species involved: the fully loaded one, the empty one, and the two partially loaded ones with A and B, respectively.

For symport R_c must be greater than unity, and for antiport smaller than unity. An affinity effect comes about, whenever the cooperativity coefficient, r_{ab}, differs from unity, and a velocity effect, whenever the ratio $P_{ab} \cdot P_o / P_a \cdot P_b$ differs from unity. These relationships are of general validity: they are not affected by asymmetry or electrogenicity.

As to the velocity effect, there are various possibilities to meet the above requirement, the simplest one being be that all rate coefficients are about equal with the exception of one or two "critical" ones that by being different would afford a velocity effect.

The velocity parameters P_a and P_o would appear not to be critical, as they refer to pathways not carrying the driver ion. This does not follow from equ. (1), but becomes more plausible, if we consider a system that combines both symport and antiport, such as glutamate transport in various systems to be discussed below. There is good reason to assume that in such transport, which is energized by both a parallel Na gradient and an opposing K gradient, the two modes involve the same translocator system (Fig 2). The corresponding ratio can be derived for each mode from equ. (1) by inserting the appropriate parameters, as follows:

$$R_{sym} = \frac{r_{NaG} \cdot P_{NaG} \cdot P_o}{P_{Na} \cdot P_G} \qquad R_{ant} = \frac{r_{KG} \cdot P_{KG} \cdot P_o}{P_K \cdot P_G} \tag{2}$$

The indices G, Na, and K refer to the translocator ligands glutamate, sodium, and potassium, respectively. We can combine these ratios to obtain the overall ratio

$$R_{comb} = \frac{R_{sym}}{R_{ant}} = \frac{r_{NaG} \cdot P_{NaG} \cdot P_K}{r_{KG} \cdot P_{KG} \cdot P_{Na}} \tag{3}$$

which must be well above unity in order to warrant effective coupling. The parameters P_G and P_0 cancel as they are common to both modes, but with different functions: P_G is a leakage coefficient for the symport mode, but essential for the antiport mode, whereas for P_0 the inverse is true.

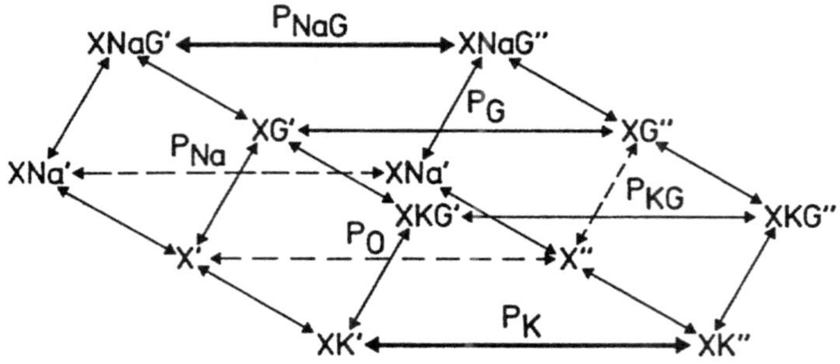

Figure 2: Combination of Na symport and K antiport within the same system, 1:1 ratio. Assumptions as in Fig. 1. The fat arrows represent transport-effective pathways, the thin ones leakage pathways.

Experimentally, the relative magnitude of P_{Na} and P_K can be tested by trans effects of Na^+ and K^+, respectively, on the substrate uptake under <u>zero-trans</u> conditions.

As to Na^+, a marked transinhibition of the initial rate of substrate uptake has been reported for numerous systems of Na-linked symport.

To the extent that the experimental conditions have been sufficiently close to zero-trans, i.e. that the average activity of substrate on the transside during the experimental period can be neglected, and that the addition of the transion does not appreciably change the p.d., these results show that the Na-loaded translocator is less mobile than the empty one, i.e. that

$$P_{Na} < P_0.$$

As to the transeffects of K^+, the opposite has been found: In all known systems of antiport with this ion, such as glutamate transport in renal and intestinal brush border vesicles (Schneider and Sacktor, 1980; Burckhardt et al., 1980; Berteloot, 1984) or hydroxy tryptamine transport in blood platelets (Nelson and Rudnick, 1979), the initial uptake rate is strongly stimulated by K^+ on the transside. On the other hand, glucose transport in renal brush border vesicles, which does not have K antiport, does not show this transstimulation (Centelles et al., 1988). With the reservations mentioned for Na, the transstimulation by K^+ can be taken to indicate that the K-loaded translocator is m o r e mobile than the empty one, i.e. that

$$P_K > P_o.$$

These transeffects appear to reveal distinct velocity effects that could account for energetic coupling in symport and antiport, respectively.

As to affinity effects, they may be tested for by equilibrium exchange, which should not depend on P_{Na} and P_K, but should instead be stimulated by the driver ion that cooperates positively with the substrate, e.g. by Na^+, and inhibited by the one that cooperates negatively with the substrate, e.g. by K^+.

The few equilibrium exchange experiments carried out so far agree with these predictions: Systems with Na-linked symport, such as glucose transport in renal brush border vesicles, are indeed stimulated by Na^+ (Lin et al., 1988), whereas in a system known to have a K^+ antiport, such as glutamate transport in the same tissue, equilibrium exchange is inhibited by K^+ (Heinz et al., 1988).

These results, however, while being consistent with the expected cooperativity effects, do not prove these. They merely tell us that in symport $r_{NaG} \cdot P_{NaG} > P_G$, and in antiport, $r_{KG} \cdot P_{KG} > P_G$. Accordingly, the same results would be expected without cooperativity effects, namely in the presence of additional velocity effects in that Na accelerated, and K^+ retarded, the substrate-bearing translocator, i.e. if $P_{NaG} > P_G > P_{KG}$. The separation between cooperativity and velocity effects is, however, possible, based on Michaelis-Menten kinetics of equilibrium exchange (s. Appendix). It can be derived for symport that the maximum rate must increase with increasing Na, whenever the ternary translocator complex moves faster than the binary, substrate-bearing one, i.e. if $P_{ab} > P_b$. Under the same conditions the Michaelis constant, K_m, should decrease, whenever there is positive cooperativity between substrate and driver ion, i.e. if $r_{ab} > 1$. Experimentally, it has been found for Na-linked glucose transport in renal brush border membrane that with increasing Na concentration the maximum rate remains rather unchanged, whereas K_m decreases (Centelles et al., 1988). These results argue against a second velocity effect, but in favor of positive cooperativity, and hence of an affinity effect to be present at least in this system. They are in line with the finding that in renal brush border vesicles Na increases the affinity of the glucose binding site of the translocator, for phlorizin (Aaronson, 1978; Silverman, 1981; Turner, 1981) and for glucose (Kinne, 1976).

In antiport, the situation should be analogous but inverse: Increasing the concentration of the antiporting ion should depress the maximum exchange rate, whenever the ternary complex moves more slowly than the binary, substrate-bearing one, whereas the Michaelis constant should rise, if there is negative cooperativity. With glutamate transport in renal brush border membranes, it has been found that the addition of K^+, up to about 10 mM, leaves the maximum rate rather unchanged, but does raise the Michaelis constant.

The observations available so far are not sufficient to be generalized and ought to be confirmed for many other systems. They, nonetheless, point towards a rather simple unifiying hypothesis, namely that coupling in each symport and antiport involves a combination of only two effects: an affinity effect, through cooperativity, and a velocity effect, through acceleration or deceleration, respectively, of the substrate-free translocator.

It is interesting to view these results in terms of the underlying principles of coupling which are mentioned in the beginning. In symport the affinity effect obviously serves to expand a transport-effective step, namely the translocation of the ternary complex. The velocity does, contrary to expectation, not directly concern substrate translocation, but serves to restrict a leakage step, namely "slipping" of the (binary) Na complex. In antiport the reverse is true: Here, the (positive) velocity effect serves to expand a transport-effective pathway, namely the translocation of the (binary) substrate complex, whereas the (negative) affinity effect serves to restrict a leakage pathway, namely the translocation of the (ternary) complex of the substrate with K^+. The end effect is the same in each mode: to provide efficient coupling between the substrate flow on the one hand, and each of the flows of the driver ions on the other hand.

As can be shown by computer simulation of the simple model of symport in Fig. 1, the affinity effect on the initial net flux of the substrate (zero-trans) seems to be more powerful than the velocity effect, if each acts alone. But, if both effects are present simultaneously, they reinforce each other so that the joint effect may greatly exceed the sum of the two components (Fig. 3)

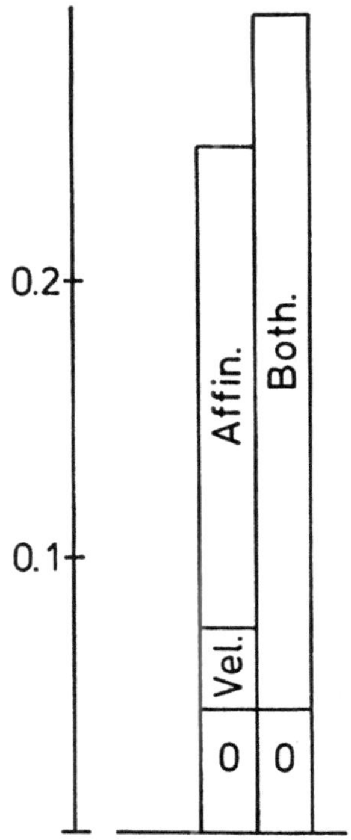

Figure 3: Velocity and affinity effects of transport rate at zero-trans for symport, computer simulation. Left column: velocity and affinity effect, each acting alone. Right column: Both acting simultaneously. Assumed substrate concentration 0.01 x K_G, concentration of the driver ion K_{Na}, cooperativity coefficient, r_{NaG} assumed to be 10, ratio P_{Na}/P_o to be 0.01.

Appendix

For the simple model (Fig. 1) the general equation for equilibrium exchange is

$$\overset{e.e.}{\tau_a} = \frac{(P_a + P_{ab}\, r_{ab}\, \beta)\cdot \alpha}{1 + \beta + (1 + r_{ab}\, \beta)\, \alpha}$$

$$\max \overset{e.e.}{\tau_a} = \frac{P_a + P_{ab}\, r_{ab}\, \beta}{1 + r_{ab}\, \beta}$$

$$\overset{e.e.}{K_m} = \frac{1 + \beta}{1 + r_{ab}\, \beta}$$

$\tau_a^{e.e.}$ = equilibrium exchange of substrate G

$$\alpha = \frac{[A]}{K_a} \qquad \beta = \frac{[B]}{K_b}$$

The other symbols are the same as in equ. (1).

It is seen that $^{max}\tau_a^{e.e.}$ is independent of ß, whenever $P_a = P_{ab}$; whereas K_m depends on ß whenever r_{ab} is different from unity.

References

Aronson PS (1978) Energy-dependence of phlorizine binding to isolated renal microvillus membranes. J Membr Biol 42:81-98

Berteloot A (1984) Characteristics of glutamic acid transport by rabbit intestinal brush-border membrane vesicles. Biochim Biophys Acta 775:129-140

Burckhardt G, Kinne R, Stange G, Murer H (1980) The effects of potassium and membrane potential on sodium-dependent glutamic acid uptake. Biochim Biophys Acta 599:191-201

Centelles T, Heinz E, Kinne R (in preparation)

Heinz E (1978) Mechanics and Energetics of Biological Transport. Springer, Heidelberg New York London

Heinz E, Geck P, Wilbrandt W (1972) Coupling in secondary active transport. Activation of transport by co-transport and/or counter-transport with the fluxes of other solutes. Biochim Biophys Acta 255:442-461

Heinz E, Sommerfeld D, Kinne R (1988) Electrogenicity of Na-L-glutamate cotransport in rabbit renal brush border membranes: A reevaluation. Biochim Biophys Acta 937:300-308

Kinne R (1976) Properties of the glucose transport system in the renal brush border membrane. Current Top 8:209-267

Lin J-T, Xu ZJ, Lovelace C, Windhager EE, Heinz E (submitted) Effect of Ca on Na-D-glucose cotransport across isolated renal brush border membranes.

Nelson PT, Rudnick G (1979) Coupling between platelet 5 hydroxytryptamine and potassium transport. J Biol Chem 254:10084-10089

Schneider EG, Sacktor B (1980) Sodium gradient-dependent L-glutamate transport in renal brush border membrane vesicles. J Biol Chem 255:7645-7649

Turner RT, Silverman M (1981) Interaction of phlorizin with the renal brush-border membrane D-glucose transporter: Stoichiometry and order of binding. J Membr Biol 58:43-45

The Passage of Molecules into Bile: Paracellular and Transcytotic Pathways

WGM Hardison and PJ Lowe.
Department of Medicine (111-D)
VA Medical Center
3350 La Jolla Village Drive and
University of California
San Diego, CA 92161.

Early observations that entry of certain inert solutes into bile was restricted gave rise to the concept of bile/plasma permeability barrier (Forker, 1968). The situation was analogous to that in the capillary which had received scrutiny by such eminent physiologists as Landis, Pappenheimer and Zweifach, to name a few. The capillary wall was viewed as containing static pores with degree of solute filtration depending upon hydrostatic pressure, fluid viscosity, surface area, solute concentration and size, shape and physical properties of the solute (Pappenheimer, 1953). The capillary itself was passive. This basic theory was later modified slightly by Grotte (1956) who postulated a small number of large pores to account for the finite permeability of macromolecules.

Such a theory was directly applicable to biliary capillary permeability and remains with us today (Forker, 1967; Wheeler et al, 1968). Such a theory has weaknesses, however. It largely ignores the transcellular contribution to biliary clearance and allows for changes in permeability only by rather destructive unphysiologic perturbations such as high levels of bile acids (Layden and Boyer, 1977; Reichen and Le, 1983), cytoskeletal poisons (Elias et al, 1980), high

Hepatic Transport in Organic Substances
Ed. by E. Petzinger, R. K.-H. Kinne, H. Sies

biliary pressure (Toyota et al, 1984) and cellular reorganization (Yee and Revel, 1978). Such a theory seems indeed "static" when one considers the advances made in our understanding of epithelial/endothelial permeability in other tissues. Since the work of Majno (1969) such permeability has become progressively viewed as a kinetic hormonally regulated phenomenon (Crone 1986; Svensjö and Grega, 1986). The major site of regulation appears to be the tight junction; the major mechanism is probably alteration of cytoskeleton components mediated by changes in intracellular calcium (Cereijido et al, 1981; Olesen and Crone, 1986; Keller and Mooseker, 1982; Burgess 1982; Madara 1983; Madara et al, 1986; Meza et al 1980; Hirokawa et al, 1983). It is unlikely that the liver is alone in having its permeability adequately described by the static pore theory. Biliary permeability is probably regulated and recognition of this fact would be important in fashioning our concepts of biliary permeability. It is difficult to assign a physiologic role to a system the functions of which we do not fully appreciate.

Although our knowledge of epithelial paracellular permeability has recently improved, we still know little about transcytosis largely because the methods for studying it are cumbersome. Nonetheless quantitative estimates of transcytosis across the hepatocytes have been derived for ligands with liver cell surface receptors (St. Hilare et al, 1983; Schiff et al, 1984; Kloppel et al, 1986; Rank and Wilson, 1983). Few data are available on transcytosis of solutes taken up by bulk-phase endocytosis. Although this pathway is generally thought not to be a major one for small inert solutes, recent data suggest its contribution is greater than previously thought (Lake et al, 1985; Lorenzini et al, 1986). Until easier methods are developed to quantify transcytosis, we will have no appreciation of its role in elaboration of bile and no knowledge of possible regulators of this process. The only factor so far identified which might affect this process is elevated biliary pressure (Rank and Wilson, 1983). We need more information about it.

For several years we have been using a method which allows us to distinguish transcellular (TC) from paracellular (PC) biliary access in the isolated perfused rat liver. The probe is horseradish peroxidase (HRP), a stable glycoprotein of about 40kD with a hydrated diameter of 5.34nm and a pI of about 7.5. One mg of HRP is injected as a bolus into the perfusate of a rat liver perfused _in situ_ with non-recirculating Krebs-Ringer bicarbonate buffer pH 7.4. Flow (3.5-4.5ml/gm liver/min) is sufficient to insure adequate oxygenation. Taurodehydrocholate, 22μM, to assure adequate bile flow and nitroprusside, 20μM, to inhibit vasospasm are included in the perfusate. After 30 min of perfusion, HRP is injected and bile is collected every 2 minutes for 10 minutes and then every 5 minutes for 30 more minutes. Bile flow is calculated and biliary HRP assayed (Galatti and Pracht, 1985). Usually ^{14}C sucrose, 10 μCi/l, is included in the perfusate as a second permeability probe.

The use of HRP has virtues and limitations. Advantages include stability, low cost, and a simple and extremely sensitive assay. HRP is a neutral molecule with no receptor on the liver surface membrane so it is taken up by bulk-phase endocytosis. It is nontoxic and interacts with neither the cell surface nor albumin. The extremely small fraction of an injected bolus appearing in bile (<.001%) is compensated for by the extremely sensitive assay which can accurately quantify <50pg. The disadvantages are two. There is receptor-mediated uptake of HRP by Kupffer cells so that figures for total hepatic extraction of HRP do not reflect hepatocyte activity. Second, HRP is heavily glycosylated and we have not been able to accomplish nondestructive iodination to high enough specific activity to use it as a tool to quantify biliary output. We are limited therefore to assessment of enzyme activity in bile and cannot quantify appearance of inactive catabolites. Sizing on a gel permeation column indicates that all active enzyme in bile is of the same molecular size as the parent compound. When HRP is administered to the perfused liver it appears in bile in two peaks: an early peak maximal at about 2 min which represents paracellular access and a much broader late peak maximal at about 12-15 min which represents transcellular access (Lowe et al, 1985). Validation of assignment of the early peak to paracellular access includes prompt appearance, co-appearance with inulin when the latter is injected as a bolus with HRP, co-appearance with ^{14}C sucrose (fig. 1a), and a good correlation between the amount of HRP entering bile under the first peak and the bile to plasma (B/P) ^{14}C sucrose ratio (fig 1b). That transcellular access accounted for the second peak seemed likely because the HRP peak appeared in bile coincident with polymeric IgA known to traverse a transcellular pathway into bile, and was depressed by colchicine (Lowe et al, 1985).

To avoid arbitrary assignment of areas to the first and second peaks and to improve quantification of the two pathways, we defined peak shapes closely by making bile collections every 20 seconds. We found on fitting these curves to a normal, a square-root, and a log-normal distribution, that the latter distribution fit extremely well. We still were unable, however, to use this fit directly since our samples were collected routinely over longer intervals (2 min and 5 min). We therefore considered these samples (units of pg HRP/sample) to be the integral of the log-normal curve over the time of collection. This method of modeling gave an accurate fit of the 20-sec data when the latter were pooled into 2- and 5-minute collection periods. Currently we compute the best-fit curve for the data by an iterative least squares program from which the areas of the peaks are derived.

The technique has yielded provocative data for both transcellular and paracellular access of HRP into bile. We have done least with the transcellular pathway but have accumulated much data passively since each time we examine the first peak, we inevitably define the second peak. We have found that the first peak is sensitive to perfusate calcium concentration through a wide range but that the second peak drops precipitously only at very low perfusate concentrations (fig 2a). We confirmed the work of Rank and

(a)

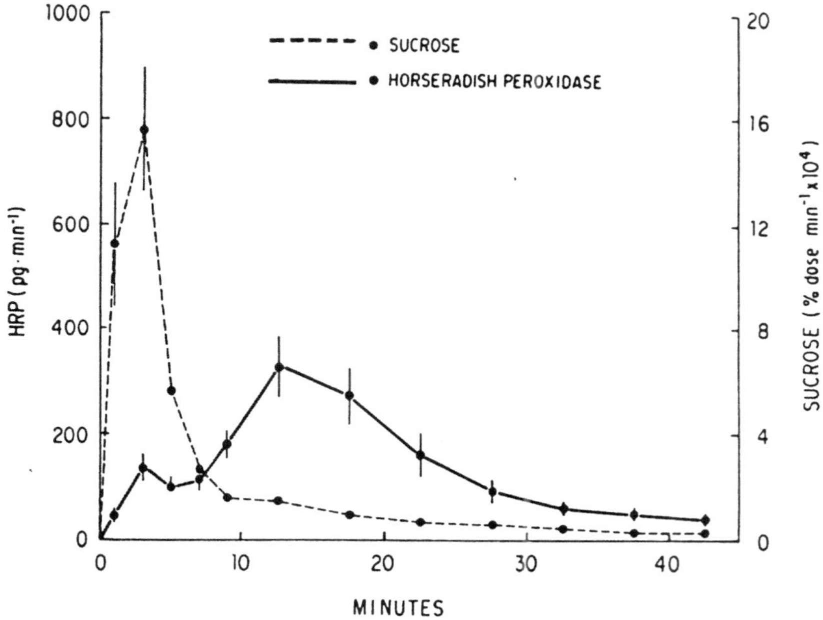

(b)

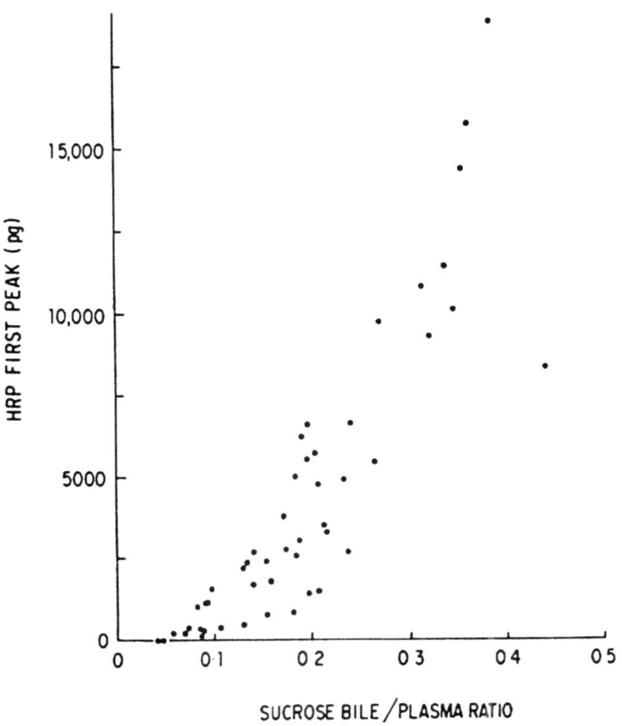

Fig 1:

(a) The appearance of first-peak HRP in bile coincides with the appearance of bolus sucrose.

(b) Sucrose bile: plasma ratio increases with increasing HRP first peak.

Fig 2

(a)

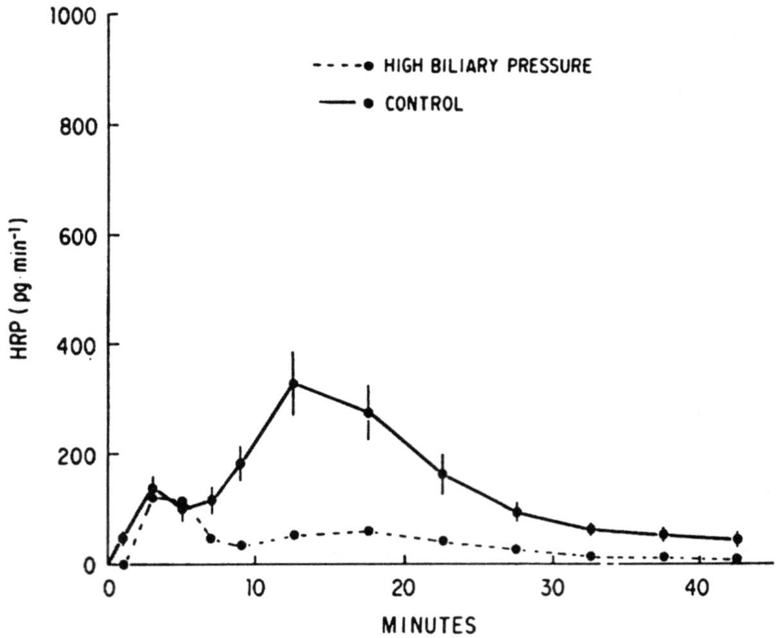
HIGH BILIARY PRESSURE
CONTROL
HRP (pg·min⁻¹)
1000
800
600
400
200
0
0
10
20
30
40
MINUTES

(b)

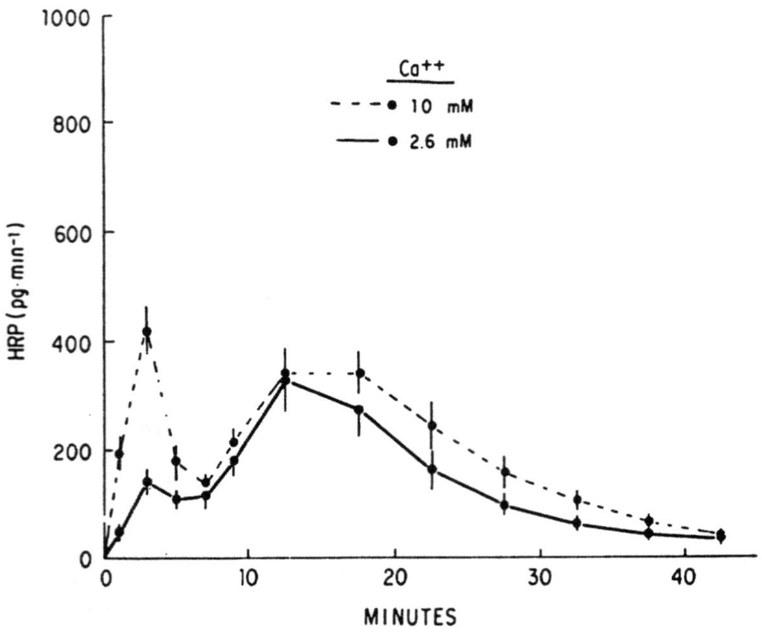
Ca++
10 mM
2.6 mM
HRP (pg·min⁻¹)
1000
800
600
400
200
0
0
10
20
30
40
MINUTES

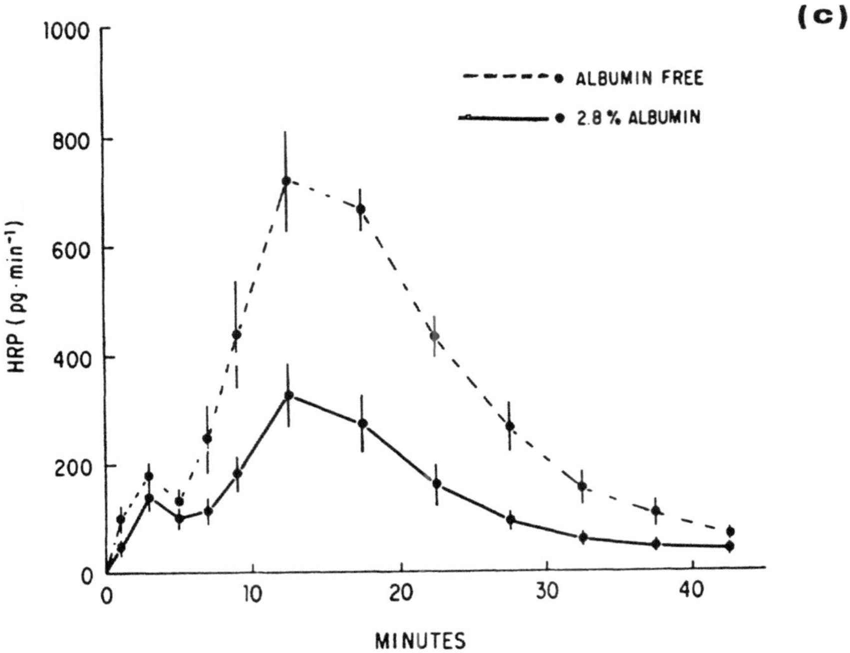

Fig 2:

(a) Moderate lowering of perfusate Ca^{++} augments first-peak HRP and affects second-peak HRP less.
(b) Elevated biliary pressure (10 cm H_2O) severely diminishes second-peak HRP.
(c) Inclusion of 2.8% albumin in the perfusate reduces second-peak HRP.

Wilson (1983) and Kloppel et al (1986) that elevated biliary pressure depresses transcytosis (fig 2b). We confirmed in the liver the work of Clough and Michel (1981) in frog capillary endothelium that albumin depresses endo- and transcytosis (fig 2c), suggesting this process is common to a wide variety of endo- and epithelia.

These data are interesting in their own right and they also demonstrate the power of the technique. We have, however, been primarily interested in the paracellular pathway. Stimulated by work suggesting that tight junctions were dynamically regulatable structures, we examined the effect of vasopressin (VP) on the appearance of HRP in bile. We found that VP acutely and reversibly augmented the area of the first peak of HRP but not the second, and that the B/P ratio of ^{14}C sucrose was similarly affected. Since VP presumably exerts its actions initially by increasing intracellular calcium concentration, we studied other agents with specific receptors on the hepatocyte surface which acted in the same way: angiotensin II (AII) and epinephrine. Each acutely increased both ^{14}C sucrose B/P ratio and first peak HRP area in bile. Moreover the effects of each of these latter compounds could be eliminated by specific antagonists, [sar^{1}, thr^{8}]-AII for AII, and prazosin for epinephrine. Table I shows the maximally effective dose of agonist, the pg of HRP excreted under the first peak and the corresponding B/P ratio. The agent dibutyryl cyclic AMP (DBcAMP), had no effect on first-peak HRP suggesting that hormones with cAMP transduction pathways do not mediate these effects.

TABLE 1

	dose (M)	1st-peak HRP(pg)	^{14}C Sucrose B/P
control	---	223$\pm$ 60	0.09
VP	10^{-8}	4729$\pm$1365	0.35
AII	10^{-9}	2973$\pm$ 749	0.24
epinephrine	10^{-6}	1694$\pm$ 260	0.24
DBcAMP	10^{-4}	110$\pm$ 55	0.09

The technique also allowed us to determine the site of paracellular leakage. One minute after injection of a large dose of HRP (25mg), the liver was sectioned. In control livers only about 2% of canaliculi contained stainable HRP whereas after VP treatment all canaliculi contained stainable HRP with some of these canaliculi showing HRP streaming across tight junctions. Although bile ductules showed intercellular HRP staining, none was traversing tight junctions nor had entered the lumen of the ductule. The data confirm earlier suggestions (Elias et al, 1980) that the site of enhanced paracellular permeability is the hepatocyte tight junction.

These observations allow a unique approach to several important questions with regard to hepatocyte tight junctions and biliary physiology in general. Because the technique clearly separates transcellular from paracellular pathways it can be effectively used to re-examine the effect of molecular charge on transcellular as well as paracellular permeability. The use of HRP is especially appropriate since this neutral protein (pI about 7.5) can be easily derivatized to strongly cationic (p>9.5) and strongly anionic (pI about 3.5) isoenzymes with little size change. When these isoenzymes are studied, paracellular discrimination against anionic HRP is evident confirming earlier data using smaller neutral and anionic probes (Bradley and Herz, 1978). In addition the same discrimination was manifest for transcellular access. Work with primary liver cultures suggests this discrimination is not at the level of uptake or intracellular catabolism.

The question of the specificity of ^{14}C sucrose as a paracellular permeability probe can be readdressed with new precision. When sucrose entry into bile is modeled as we have HRP, not only is transcellular access found to be substantial (as already suggested by Lake et al, 1985; Jaeschke et al, 1987) but the proportion of sucrose entering by each pathway is variable depending upon conditions. As was true for HRP, VP increased paracellular access of sucrose. More interestingly, the fasting animal has a more restricted paracellular but a less restricted transcellular access of ^{14}C sucrose than the fed animal. In that circumstances, transcellular access exceeds 50% of the total. It appears that inert molecules the size of sucrose are not very discerning paracellular probes.

The role of junctional permeability in bile formation may also be addressed. When VP is given to augment paracellular permeability, no change in bile flow or taurodehydrocholate output occurs. It appears that junctional permeability has no influence on bile formation. It seems likely that in the hepatobiliary system fluid flux is relatively unrestricted and is determined by osmotic gradients rather than by the opening or closing of anatomic portals. In view of the fact that organic anions (bile acids) are the major osmotic determinants of bile flow, the marked discrimination against anion flux noted earlier would protect bile flow from all but major alterations in junctional permeability. One might predict, however, that biliary cations would be more susceptible to changes in junctional permeability. Using acetyl procainamide ethylbromide (APAEB) a cation excreted in bile (Hwang and Schanker, 1973), we have found that VP administration caused a fall in APAEB output. These data suggest that the junctional complex is instrumental in allowing selective resorption of biliary cation.

These studies not only illustrate the dynamic qualities of biliary permeability, but also suggest that the techniques described can provide answers to a number of unresolved questions in biliary physiology.

References

Bradley SE, Herz R (1978) Permselectivity of biliary canalicular membranes in rats: clearance probe analysis. Am J Physiol 235:E570-E576

Burgess DR (1982) Reactivation of intestinal epithelial cell brush border motility: ATP-dependent contraction via a terminal web contractile ring. J Cell Biol 95:853-863

Cereijido M, Meza I, Martinez-Palomo A (1981) Occluding junctions in cultured epithelial monolayers. Am J Physiol 240:C96-C102

Clough G, Michel CC (1981) The role of vesicles in the transport of ferritin through frog endothelium. J Physiol 315:127-142

Crone C (1986) Modulation of solute permeability in microvascular endothelium. Fed Proc 45:77-83

Elias E, Hruban Z, Wade JB, Boyer JL (1980) Phalloidin-induced cholestasis: a microfilament-mediated change in junctional complex permeability. Proc Nat Acad Sci 77:2229-2233

Forker EL (1967) Two sites of bile formation as determined by mannitol and erythritol clearance in the guinea pig. J Clin Invest 46:1189-1195

Forker EL (1968) Bile formation in guinea pigs: analysis with inert solutes of graded molecular radius. Am J Physiol 215:56-62

Galatti VH, Pracht I (1985) Horseradish peroxidase: kinetic studies and optimization of the peroxidase activity determination with the substrates H_2O_2 and 3,3',5,5' tetramethylbenzidine. J Clin Chem Biochem 23:453-460

Grotte G (1956) Passage of dextran molecules across the blood-lymph barrier. Acta Chir Scand Suppl 211:1-84

Hirokawa N, Keller III TCS, Chasan R, Mooseker M (1983) Mechanism of brush border contractility studied by the quick-freeze, deep-etch method. J Cell Biol 96:1325-1336

Hwang SW, Schanker LS (1973) Hepatic uptake and biliary excretion of N-acetyl procaine amide ethobromide in the rat. Am J Physiol 225:1437-1443

Jaeschke H, Krell H, Pfaff E (1987) Quantitative estimation of transcellular and paracellular pathways of biliary sucrose in isolated perfused rat liver. Biochem J 241:635

Keller III TCS, Mooseker MS (1982) Ca^{++}-calmodulin-dependent phosphorylation of myosin, and its role in brush border contraction in vitro. J Cell Biol 95:943-959

Kloppel TM, Brown WR, Reichen J (1986) Mechanisms of secretion of proteins into bile: studies in the perfused rat liver. Hepatology 6:587-594

Lake JR, Licko V, Van Dyke RW, Scharschmidt BF (1985) Biliary secretion of fluid-phase markers by the isolated perfused rat liver. Role of transcellular vesicular transport. J Clin Invest 76:676-684

Layden TJ, Boyer JL (1977) Taurolithocholate-induced cholestasis: taurocholate, but not dehydrocholate, reverses cholestasis and bile canalicular membrane injury. Gastroenterology 73:120-128

Lorenzini I, Sakisaka S, Meier PJ, Boyer JL (1986) Demonstration of a transcellular vesicle pathway for biliary excretion of inulin in rat liver. Gastroenterology 91:1278-1288

Lowe PJ, Kan KS, Barnwell SG, Sharma RK, Coleman R (1985) Transcytosis and paracellular movements of horseradish peroxidase across liver parenchymal tissue from blood to bile. Effects of α-naphthylisothiocyanate and colchicine. Biochem J 229:529-537

Madara JL (1983) Increases in guinea pig small intestinal transepithelial resistance by osmotic loads are accompanied by rapid alterations in absorptive-cell tight junction structure. J Cell Biol 97:125-136

Madara JL, Barenberg D, Carlson S (1986) Effects of cytochalasin D on occluding junctions of intestinal absorptive cells: further evidence that cytoskeleton may influence paracellular permeability and junctional charge selectivity. J Cell Biol 102:2125-2136

Majno G, Shea SM, Leventhal M (1969) Endothelial contraction induced by histamine type mediators: an electron microscopic study. J Cell Biol 42:647-672

Meza I, Ibarra G, Sabanero M, Martinez-Palombo A, Cereijido M (1980) Occluding junctions and cytoskeletal components in a cultured transporting epithelium. J Cell Biol 87:746-754

Olesen S-P, Crone C (1986) Substances that rapidly augment ionic conductance of endothelium in cerebral venules. Acta Physiol Scand 137:233-241

Pappenheimer JR (1953) Passage of molecules through capillary walls. Physiol Rev 33:387-423.

Rank J, Wilson ID (1983) Changes in IgA following varying degrees of biliary obstruction in the rat. Hepatology 3:241-247

Reichen J, Le M (1983) Taurocholate, but not taurodehydrocholate, increases biliary permeability to sucrose. Am J Physiol 245:G651

Schiff JM, Fisher MM, Underdown BJ (1984) Receptor-mediated biliary transport of IgA and asialoglycoprotein: sorting and missorting of ligands revealed by two radiolabeling methods. J Cell Biol 98:79-89

St. Hilare RJ, Hradek GT, Jones AL (1983) Hepatic sequestration and biliary secretion of epidermal growth factor: evidence for a high-capacity uptake system. Proc Natl Acad Sci 80:3797-3801

Svensjö E, Grega GJ (1986) Evidence for endothelial cell-mediated regulation of macromolecular permeability by postcapillary venules. Fed Proc 45:89-95

Toyota N, Miyai K, Hardison WGM (1984) The effect of biliary pressure vs. high bile acid flux on permeability of hepatocellular tight junction. Lab Invest 50:536-542

Wheeler HO, Ross ED, Bradley SE (1968) Canalicular bile production in dogs. Am J Physiol 214:866-874

Yee AG, Revel J-P (1978) Loss and appearance of gap junctions in regenerating liver. J Cell Biol 78:554-564

Increase in Paracellular Permeability as a Pathophysiological Principle of Drug-Induced Intrahepatic Cholestasis in Rats

H. Krell
Institut für Toxikologie der Universität Tübingen
Wilhelmstraße 56
D-7400 Tübingen

Since the early studies on estrogen cholestasis (Forker, 1969), alterations in the permeability between canalicular and sinusoidal spaces have been regarded as a pathogenetic factor in drug-induced intrahepatic cholestasis. The general relevance of this alteration, however, was not clearly worked out. Isolated perfused livers pretreated with different drugs are a valuable tool to verify this concept. Thus, it was demonstrated that paracellular permeability to inert solutes such as sucrose and inulin was increased upon treating rats with α-naphthylisothiocyanate (ANIT, Krell et al., 1982) and estradiolvalerate (EV, Jaeschke et al., 1987a). Since it was observed that inert solutes were transported via transcellular pathways as well, quantitative determination was required of both transcellular and paracellular pathways of inert solutes into bile. This can be achieved by analyzing biliary off-kinetics of radioactive permeability markers after omission of the markers from the perfusion medium (Jaeschke et al., 1987b).

Using this method together with conventional flow analysis (Forker, 1969), the cholestatic actions of additional drugs were studied. Upon treatment of rats with 1,3-bis(2-chloroethyl)-1-nitrosourea (BCNU) for 24 h with a single dose of 36 mg/kg, the paracellular portion of biliary [^{14}C] sucrose

Hepatic Transport in Organic Substances
Ed. by E. Petzinger, R. K.-H. Kinne, H. Sies

clearance was increased from 0.09±0.03 to 0.31±0.09 μl/min/g liver wt. (n=4)

Insertion of the paracellular portion of [^{14}C] sucrose clearance into the conventional clearance versus flow relationship (Strasberg et al., 1979) indicated increased permeability by an increase of the diffusion permeability coefficient K from 0.15 to 0.39 and a decrease of the reflection factor σ from 0.96 to 0.71.

Pretreatment of rats with endotoxin (0.3 mg/kg) increased the paracellular portion of [^{14}C] sucrose clearance to 0.33± 0.12, that of [^{14}C] inulin clearance from 0.038±0.012 to 0.211±0.085 μl/min/g liver wt. (n=4).

By applying the same method to the biliary off-kinetics of [^{32}P] orthophosphate, a differential picture of paracellular permeability can be drawn. While the permeability of [^{32}P] orthophosphate was markedly increased in the cholestasis induced by ANIT and BCNU, only minor changes were observed in EV-cholestasis. No increase in [^{32}P] orthophosphate was detectable upon treatment of rats with endotoxin.

In conclusion, paracellular permeability between sinusoidal and biliary spaces of liver was increased under a variety of cholestatic conditions in rats. However, there are marked differences with respect to extent and specificity of the alterations.

References

Forker EL (1969) The effect of estrogen on bile formation in the rat. J Clin Invest 48:645-663

Krell H, Hoeke H, Pfaff E (1982) Development of intrahepatic cholestasis by alphanaphthylsothiocyanate in rats. Gastroenterology 82:507-514

Jaeschke H, Treummer E, Krell H (1987a) Increase in biliary permeability subsequent to intrahepatic cholestasis by estradiol valerate in rats. Gastroenterology 93:533-538

Jaeschke H, Krell H, Pfaff E (1987b) Quantitative estimation of transcellular and paracellular pathways of biliary sucrose in isolated perfused rat liver. Biochem J 241:635-640

Strasberg SM, Petrunka CN, Ilson RG, Paloheimo JE (1979) Characteristics of inert solute clearance by the monkey liver. Gastroenterology 76:259-266

The Hepatocyte's Plasma Membrane Domains. Interrelations with the Endocytic Compartment

W. Howard Evans and Carlos Enrich[1],
National Institute for Medical Research,
Mill Hill,
London NW7 1AA,
United Kingdom

Introduction

The plasma membrane of the hepatocyte consists of three major functionally and biochemically characterised domains (Evans, 1980; 1981). The isolation of membrane vesicles or sheets from the sinusoidal, lateral and canalicular domains has allowed many aspects of cell surface physiology to be investigated, for example, the properties of transport mechanisms and details of the channels operational at the canalicular and sinusoidal poles of the cell have been identified (Berk et al., 1986; Meier,1986). The polarised organisation of the surface membrane of the hepatocyte raises basic questions regarding the nature of the sorting mechanisms directing newly synthesised proteins to different domains and the need to circumvent the barrier posed by tight junctions to lateral diffusion of proteins between the apical and baso-lateral domains (Evans et al., 1980; Bartles et al.,1988). The movement of ions, organic solutes, metabolites etc. from the blood to the bile across the interior of the hepatocyte raises questions as to the mechanisms, routes and the nature of the membrane

1 Permanent address. Departmento de Biologia Celular, Facultad de Medicina, Universidad de Barcelona, Barcelona 08028, Spain

Hepatic Transport in Organic Substances
Ed. by E. Petzinger, R. K.-H. Kinne, H. Sies

networks involved. Furthermore, it has become evident that the endocytic compartment, a recently characterised tubulo-vesicular membrane-bound organelle, plays a key role in regulating ligand uptake mechanisms and it provides a vesicular pathway for trans-epithelial transport, for example of polymeric IgA (Schiff et al., 1986). Clearly the endocytic compartment is positioned to play a general role in regulating transepithelial transport of ions and organic solutes. The present chapter describes molecular relationships between the plasma membrane domains of the hepatocyte and proposes that the endocytic compartment is an important intermediary in transport of materials between the blood and bile.

Plasma Membrane Domains in Hepatocytes

The regionalisation of the hepatocyte's plasma membrane has been intensively explored. A number of ecto-enzymes are present at high specific activities at the bile canalicular plasma membrane, but, surprisingly, the functions of many of these enzymes, e.g. 5'-nucleotidase, leucylnaphthylamidase etc. still remain to be elucidated. Enzymes facilitating ion transport also show a polarised distribution, with a highly active Ca^{2+}-ATPase located at the bile canalicular face pumping Ca^{2+} into bile, and a Na^{+}-K^{+}-ATPase being functional at the sinusoidal/lateral faces (Meier, 1986). Mechanisms facilitating amino acid, bile acid, bilirubin, glucuronide, fatty acid uptake are located at the sinusoidal face. A wide range of receptors to which hormones, neurotransmitters, and metabolites bind, with the accompanying signal transduction and effector mechanisms are also located at the sinusoidal and lateral domains of the plasma membrane. Extensive and rapid recycling of receptors occurs between the sinusoidal plasma membrane and the membranes comprising the endocytic compartment.

Lateral plasma membranes are heavier than the sinusoidal and canalicular membranes, and they are probably the easiest of the hepatocyte's plasma membranes to isolate in high purity. These membranes, recovered as sheets and vesicles, are characterised by the presence of junctional proteins, e.g. those of desmosomes and gap junctions as well as the various cadherins including cell-CAM. Analysis of the distribution of extracellular matrix components indicates that laminin beta chains are associated mainly with lateral plasma membrane fractions (Enrich, unpublished observations), whereas fibronectin is associated with the membranes recovered from all three plasma membrane domains. However, fibronectin isoforms with a mol. wt. of 180kD, in addition to the conventional 210-220kD isoforms were detected in lateral membranes (Enrich et al., 1988).

Fig. 1 shows a two-dimensional comparison of the large number of proteins located at the three major domains of the hepatocyte's plasma membrane. Although many polypeptides of varying size and charge are common to all three domains, some appear to be confined to one or other domains. By the use of specific reagents, e.g. antibodies to asialoglycoprotein receptors, lectins etc., a dissection of this complexity is now possible. A further example of this analytical approach is shown in Fig. 2 in which the polypeptides in the three plasma membrane domains and in endosomes that bind ^{125}I-calmodulin specifically were examined.

The Bile Canalicular Plasma Membrane: Biochemical Complexity

The plasma membrane domain encircling the biliary lumen has long been regarded as performing a mainly excretory function (de Duve and Wattiaux, 1966). However, extensive analysis of bile canalicular vesicles points to a complexity suggesting possible other funtions. Indeed, Fig. 2 emphasises that a far wider range of calmodulin-binding proteins are present in bile canalicular membranes in contrast to the sinusoidal and lateral domains. This may reflect the presence of a well developed actin-myosin contractile apparatus that is necessary for the pumping of bile into the biliary spaces.

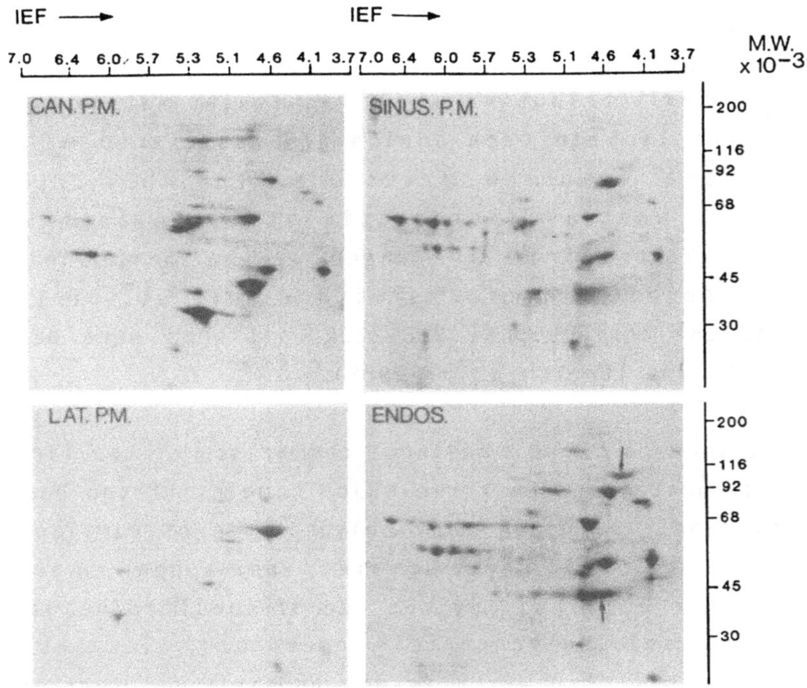

Fig.1 Two-dimensional comparison of rat liver canalicular (CAN.PM), sinusoidal (SINUS.PM) lateral (LAT.PM) plasma membranes with a liver endosomal fraction (ENDOS). Membranes were prepared by techniques referred to in the text, resolved by the O'Farrell (1970) method and the polypeptides stained with Coomassie Blue. Arrows in endosomal fraction indicate positions of 115kD calmodulin binding polypeptide and the asialoglycoprotein receptor. IEF isolectric focussing across the indicated pH range. Molecular weight (M.W.) scale is indicated.

Studies on the distribution of D-*myo*inositol 1,4,5-trisphosphate and 1,3,4,5-tetrakisphosphate-5-phosphatase in liver plasma membrane domains showed clearly that these enzymes were preferentially localised in bile canalicular vesicles (Shears et al., 1988). Furthermore, these enzyme activities involved in inositol metabolism were latent in the canalicular vesicles recovered in a right-side-out configuration, indicating a cytoplasmic location on the membrane. This contrasted with the extracellular location on the canicular membrane of 5'-

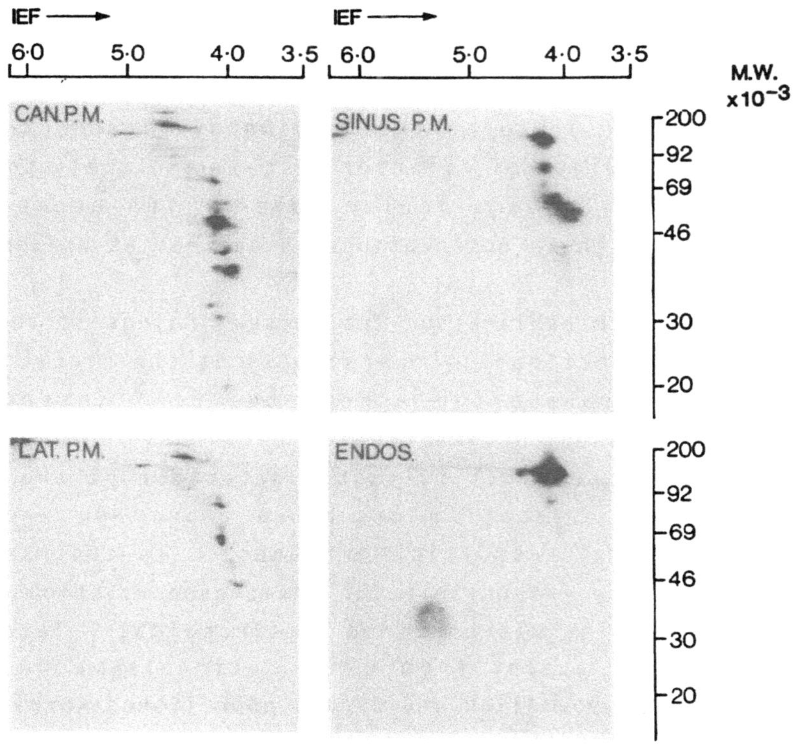

Fig.2 Two-dimensional comparison of calmodulin-binding polypeptides of canalicular, sinusoidal and lateral plasma membranes with endosomal membranes. Details are explained in Fig.1 and in the text. Calmodulin-binding polypeptides after electrophoretic transfer to nitrocellulose sheets, were identified on the basis of specific binding of ^{125}I-calmodulin.

nucleotidase, alkaline phosphodiesterase, etc. In view of the role of phosphatases specifically hydrolysing inositol phosphates in receptor signalling at the sinusoidal pole, where the hepatocyte is bathed by blood-borne hormones and neurotransmitters, the high levels of these enzymes at the bile canalicular plasma membrane is puzzling.

Many of the receptors located at the cell surface are coupled to effector systems by guanyl nucleotide binding proteins (G-proteins). In a survey of the distribution in plasma membrane subfractions of the alpha and beta subunits of the G-protein using antibodies and toxin catalysed ADP-ribosylation, it was shown that these proteins, in addition to their location in sinusoidal plasma membranes, were also found at high levels in the bile canalicular membrane (Ali and Evans, unpublished work). Since trans-membrane signalling in response to the binding of ligands to receptors involving modulation of effector systems is likely to be a minor event at the bile canalicular pole of the hepatocyte, the significance of these observations is unclear at present.

One possible explanation for these findings is that a number of molecules functionally operational at the sinusoidal plasma membrane are transported across the cell interior by the underlying trans-epithelial flow of materials from blood to bile. Supporting this possibility is the detection of the enzymes of phosphoinositol metabolism described above and G-proteins in 'early' and 'late' endocytic membranes. In contrast, protein kinase C activity responsible for the phosphorylation of membrane and other proteins was confined to sinusoidal / lateral plasma membranes being absent from canalicular plasma membranes and endosomes. (Kouyoumdjian and Evans, unpublished work). Thus there is a highly specific distribution of receptors, proteins and enzymes featuring in signalling and transport between the plasma membrane domains and the endocytic membranes.

The Endocytic Compartment

The endocytic compartment is a tubulo-vesicular membrane network located in the cytoplasm of animal cells (Geuze et al., 1984). A major role credited to this organelle is the reception from coated invaginating regions of the plasma membrane ("coated pits") of receptor ligand complexes. After transfer into the endocytic membrane networks, many receptor-ligand complexes undergo dissociation at the pH of 5.5 - 6.0 present in the compartment, allowing them to be independently sorted, a prelude to their onward dispatch to lysosomes or to the sinusoidal or canalicular domains of the liver plasma membrane (Fig. 3). Additional roles played by the endocytic compartment may include

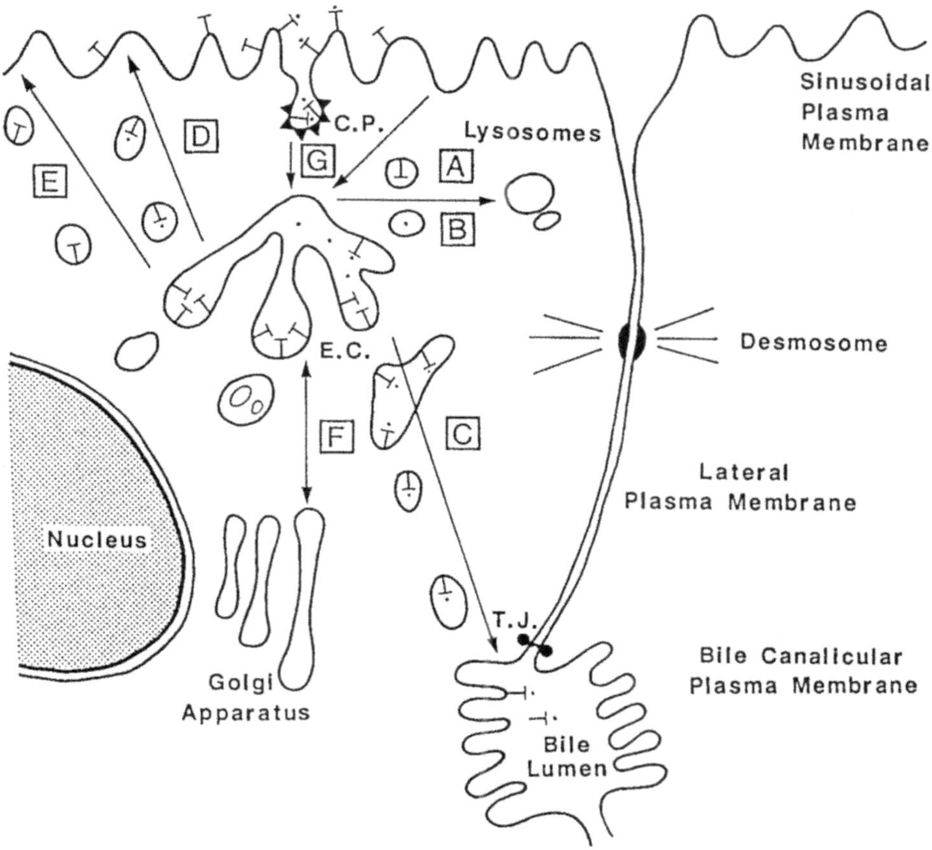

Fig.3 Intracellular routes followed by receptor (T) ligands (•) and complexes (Ṫ) in the hepatocyte, illustrating the possible intermediary role of the endocytic compartment (E.C.) in facilitating transepithelial transport. Routes followed by receptors and ligands after transfer from coated pits (CP) or non coated plasma membrane regions (route G) involve onward transfer to the bile canaliculus (route C), Golgi apparatus (route F - a minor route), to lysosomes (routes A,B) and to the sinusoidal plasma membrane (routes D and E) T.J., tight junction. For other details see text.

signalling by the internalised receptors and attached ligands, the restricted proteolytic processing of the internalised ligands and the modulation of components of the extracellular matrix (Enrich and Evans, 1987; Evans and Enrich, 1988). Geographical areas of the compartment that correlate with the kinetic position in endosome subfractions after internalisation of circulating ligands are also discerned. Ligand-receptor complexes are first observed in vacuolar structures located mainly underneath the plasma membrane; later they are transferred to multi-vesicular bodies located deeper in the cytoplasm, often in the juxta-nuclear environs. In a polarised cell, exemplified by the hepatocyte in Fig. 3, it is assumed that the "late" endocytic structures are located in the vicinity of the bile canaliculi and these may be the carriers of materials transported from blood to bile.

The isolation by a variety of subcellular fractionation techniques of endocytic vesicles (endosomes) has now allowed many of the biochemical features of these membranes to be deduced (Evans and Enrich, 1988). Endocytic membranes are empoverished in the marker enzymes characteristic of other subcellular membranes and organelles, and they have been purified on the basis of the enrichement, up to two-hundred fold, of endocytosed iodinated ligands - e.g. insulin, asialotransferrin, etc. (Debanne et al., 1982). They are generally recovered in the 'light' region of density gradients and up to three vesicle-containing endosome subfractions, characterised on the basis of the kinetic position of the internalised ligands have been described. (Belcher et al., 1987; Evans and Flint, 1986; Quintart et al., 1984; Kennedy and Cooper, 1988).

Although endogenous marker constituents are proving difficult to identify, positive markers have now been deduced for hepatic endosomes. It was shown, for example, that a monensin-activated Mg^{2+}–ATP-ase was enriched up to sixty-fold relative to the tissue homogenate in hepatic endosomes, and when the activity is taken in conjunction with the presence of undegraded ligands inside the endocytic vesicles, this fulfills the biochemical marker requirements (Saermark et al., 1984). A calmodulin-binding 115kD polypeptide (Fig.2) was also shown to emerge as a candidate as an endosome-specific component, although the precise function of this polypeptide in the endocytic compartment remains to be established (Enrich et al., 1988). Immunological approaches are also being brought to bear on the problem of identifying and isolating endosome-specific components. Antibodies to the calmodulin-binding 115kD polypeptide as well as the various receptors concentrated in the endocytic compartment, e.g. the asialoglycoprotein receptor, have been generated, characterised and shown by electron microscopy to locate to vacuolar endocytic structures (Enrich and Evans, unpublished work).

Analysis of the distribution in subcellular fractions of plasma membrane ectoenzymes, G proteins and enzymes featuring in inositol phosphate metabolism has shown that these components are also present in early and late endosomes. Indeed, all these components were shown to be present at higher specific activities in 'late' than in 'early' endosomes suggesting a mechanism in which late endosomes participate in the transport of specific components present at high levels in the bile canalicular plasma membrane. (Fig.3).

The availability of endosomes corresponding to early and late regions of the endocytic compartment together with membranes originating from the three plasma membrane domains should prove useful in delineating the routes followed by organic solutes across the hepatocyte and especially in determining whether and to what extent these newly discovered membrane networks are involved in facilitating trans-epithelial transport.

References

Bartles JR, Feracci HM, Steiger B, Hubbard AL (1987) Biogenesis of rat hepatocyte plasma membrane _in vivo_: Comparison of the pathways taken by apical and basolateral proteins using subcellular fractionation. J Cell Biol 105:1241-1251

Belcher JD, Hamilton RL, Brady SE, Hornick CA, Jaekle S, Schneider WJ, Havel RJ (1987). Isolation and characterisation of three endosomal fractions from the liver of oestradiol-treated rats. Proc Natl Acad Sci USA 84:6785-6789

Berk PD, Potter BJ, Stremmel W (1986) The sinusoidal surface of the hepatocyte: a dynamic plasma membrane specialised for high volume molecular transit. In: Reutter W, Popper H, Arias IM, Heinrich PC, Keppler D, Landmann L (eds) Modulation of liver cell expression. Falk Symp 43 MTP Press, Lancaster Boston-The Hague-Dordrecht, p 107-125

Debanne MT, Evans WH, Flint N, Regoeczi E (1982) Receptor-rich intracellular membrane vesicles transporting asialotransferrin and insulin in liver. Nature 298:398-400

Duve C de, Wattiaux R (1966) Functions of lysosomes. Ann Rev Physiol 28:435-492

Enrich C, Bachs O, Evans WH (1988) A 115kD calmodulin-binding protein is located in rat liver endosome fractions. Biochem J in press

Enrich C, Evans WH (1987) Evidence for the role of the hepatic endocytic compartment in the modulation of the extracellular matrix. Exptl Cell Res 173:99-108

Enrich C, Evans WH, Gahmberg CG (1988) Fibronectin isoforms in plasma membrane domains of normal and regenerating rat liver. FEBS Letters 228:135-138

Evans WH (1980) A biochemical dissection of the functional polarity of the plasma membrane of the hepatocyte. Biochim Biophys Acta 604:27-64

Evans WH (1981) Membrane traffic at the hepatocytes sinusoidal and canlicular surface domains. Hepatology 1:452-457

Evans WH, Enrich C (1988) In: Hepatic endosome subfractions: preparation, biochemical properties and functional roles. Morré DJ, Howell K, Cook GMW, Evans WH (eds) Cell Free Analysis of Membrane Traffic. Alan R Liss, New York

Evans WH, Flint N (1985) Subfractionation of hepatic endosomes in Nycodenz gradients and by free-flow electrophoresis. Separation of ligand-transporting and receptor enriched membranes. Biochem J 232:25-32

Evans WH, Flint NA, Vischer P (1980) Biogenesis of hepatocyte plasma-membrane domains. Biochem J 192:903-910

Geuze HJ, Slot JW, Strous GJAM, Peppard J, Figura K, Hasilik A, Schwartz AL (1984) Intracellular receptor sorting during endocytosis: Comparitive immunoelectron microscopy of multiple receptors in rat liver. Cell 37:195-204

Kennedy G, Cooper C, (1988) The time-dependent distribution of ^{125}I-asialo-orosmucoid horseradish peroxidase and ^{131}I-immunoglobulin A among three endosomal subfractions isolated from rat liver. Biochem J 252:739-752

Meier PJ (1986) Transport processes at the canalicular surface of rat hepatocytes. In: Reutter W, Popper H, Arias IM, Heinrich PC, Keppler D, Landmann L (eds) Modulation of liver cell expression. Falk Symp 43 MTP Press, Lancaster—Boston—The Hague-Dordrecht, p 127-141

O'Farrell PH (1975) High resolution two-dimensional electrophoresis of proteins J Biol Chem 250:4007-4021

Quintart J, Courtoy P, Baudhuin P (1984) Receptor-mediated endocytosis in rat liver: Purification and enzymic characterisation of low density organelles involved in the uptake of galactose-exposing proteins. J Cell Biol 98:877-884

Saermark T, Flint N, Evans WH (1985) Hepatic endosome fractions contain an ATP-driven proton pump. Biochem J 225:51-58

Schiff JM, Fisher MM, Jones AL, Underdown BJ (1986) Human IgA as a heterovalent ligand: switching from the asialoglycoprotein receptor to secretory component during transport across the rat hepatocyte. J Cell Biol 102:920-931

Shears S, Evans WH, Kirk CJ, Michell RH (1988) Preferential localisation of rat liver D-myoinositol 1,4,5-trisphosphate/1,3,4,5-tetrakisphospate 5-phophatase in bile canalicular plasma membrane and 'late' endosomal vesicles. Biochem J in press

Hepatic Spaces and Transport in the Perfused Liver

J. Reichen
Department of Clinical Pharmacology
University of Berne
Murtenstrasse 35
CH-3010 Berne
SWITZERLAND

The liver, in contrast to other organs such as the kidney, is able to extract highly protein-bound compounds. This is related to the particular structure of the liver "capillaries", the sinusoids. The endothelial lining cell carries sieve-plates with fenestrations permitting exchange of high molecular weight compounds including proteins between the intravascular (sinusoidal) and extravascular space, the space of Disse. The characteristics of these sieve-plates have been assessed by morphometric means by different authors (Wisse et al., 1983; Horn et al., 1986). The average diameter of these fenestrations is 105 and 175 nm, when assessed by scanning and transmission elecron microscopy, respectively; it decreases slightly along the acinus, while the number of fenstrations per square micron increases from 9 to 13 (Wisse et al., 1985). The number and/or diameter of these fenestrations is subject to regulation by different factors (Table 1) including pressure (Nopanitaya et al., 1976, Fraser et al., 1980), alcohol (Mak and Lieber, 1984; Horn et al., 1987), carbon tetrachloride (Okazaki et al., 1973), hypoxia, serotonin, norepinephrine, and cytochalasin B (Stefan et al. 1987).

Hepatic Transport in Organic Substances
Ed. by E. Petzinger, R. K.-H. Kinne, H. Sies

Table 1.

FACTORS INVOLVED IN REGULATION OF SINUSOIDAL FENESTRATIONS

I. SIZE

Increase:	Decrease:
- Alcohol	- Serotonin
- Pressure	- Norepinephrine
- Hypoxia	
- CCl4	

II. NUMBER

Increase:	Decrease:
- Cytochalasin B	- Alcohol

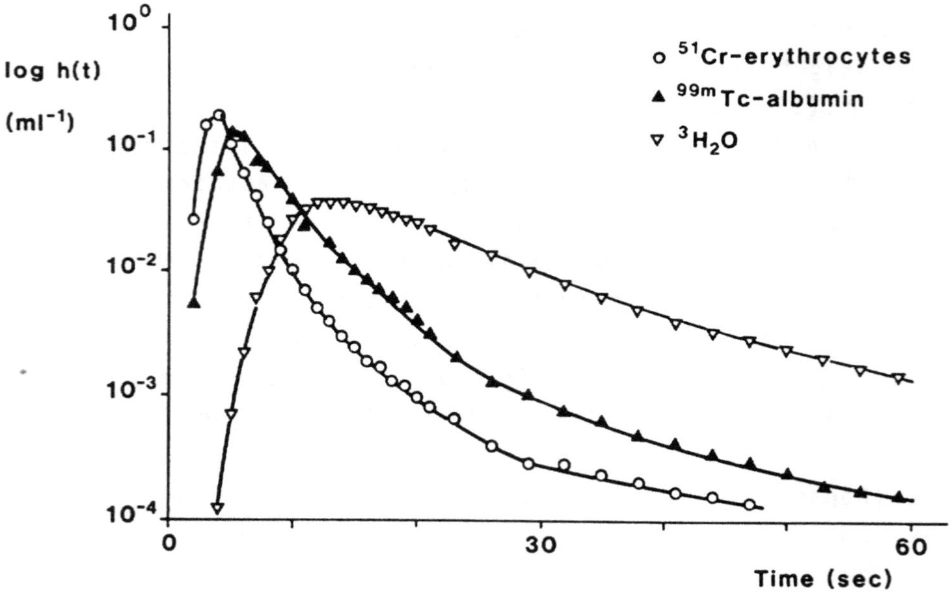

Figure 1: Multiple indicator dilution curve of erythrocytes, albumin, and water in the in situ perfused rat liver showing the typical flow-limited pattern in hepatic venous effluent: The erythrocytes appear earlier and peak higher than albumin, since the latter distributes into the space of Disse. The water curve, distributing over the whole water space of the liver, is still more delayed and dampened.

Goresky was the first to characterize these particular exchange characteristics in dog liver by a multiple indicator dilution technique (Goresky, 1963). In organs with tight capillaries exchange between the intra- and extravascular space is limited by diffusion, while in the liver exchange is limited by flow (Goresky et al., 1970). The characteristics of this flow-limited exchange are demonstrated in Fig. 1.

Erythrocytes appear earlier and peak higher than albumin; this is due to unhindered distribution of albumin into its space of distribution, which includes sinusoidal plasma space and the space of Disse. Water appears even later and is dampened more, since it distributes over the whole water space of the liver. The indicator dilution curves shown in Fig. 1 were obtained in the in situ perfused rat liver using techniques described before (Reichen and Paumgartner, 1976; Reichen and Le, 1986). The major advantage of the perfused rat liver is the lack of recirculation of indicator, which avoids potentially erroneous extrapolations and a better appreciation of the distribution of transit times.

In a first set of experiments we investigated the effects of norepinephrine on hepatic spaces and single pass extraction of different drugs. The following substances were used: 1 μg of ^{14}C-taurocholate, a high extraction compound entering the hepatocyte by active transport, and two drugs entering the hepatocyte passively, namely 1 mg of d-propranolol (high extraction) and 2.5 mg of aminoantipyrine (low extraction). Norepinephrine induced a dose-dependent increase in hepatic vascular resistance; the maximal effect was reached at an infusion rate of 0.5 mg/min, corresponding to a concentration of about 2×10^{-5} M. This pharmacological concentration decreased intra- and extravascular spaces to about the same extent (Fig. 2). A similar decrease in erythrocyte has been observed in dog liver by Cousineau et al. (1983); however, these authors found no effect on extravascular sucrose space. This divergence is probably due to the fact that we studied a pharmacological norepinephrine concentration in contrast to

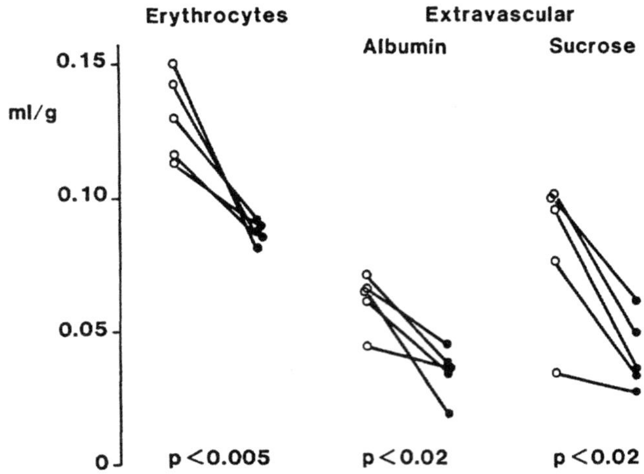

Figure 2: Effect of norepinephrine at maximal dose on hepatic spaces. Both intra- and extravesicular space are significantly decreased. The decrease is similar in magnitude for all three indicators studied suggesting an alteration in perfusion distribution rather than an effect on sinusoidal fenestrae.

the studies cited above which used norepinephrine at physiologic concentrations. In the present experiments norepinephrine reduced intra- and extravascular spaces to a similar extent. This is compatible with a reduction in the number and/or length of sinusoids being perfused.

The effect of norepinephrine on extraction of the model compounds is shown in Fig. 3. The excretion of the passively transported compounds aminopyrine and propranolol, but not that of the actively transported taurocholate, was significantly affected by the maximally effective dose of norepinephrine. It has to be pointed out that only a tracer amount of taurocholate (1 μg) was studied in the experiments shown in Fig. 3; when a higher dose of taurocholate was given, a decrease in extraction was observed in line with published data on the effects of norepinephrine on bile salt transport (Ballet et al., 1987). This dose dependence could either be

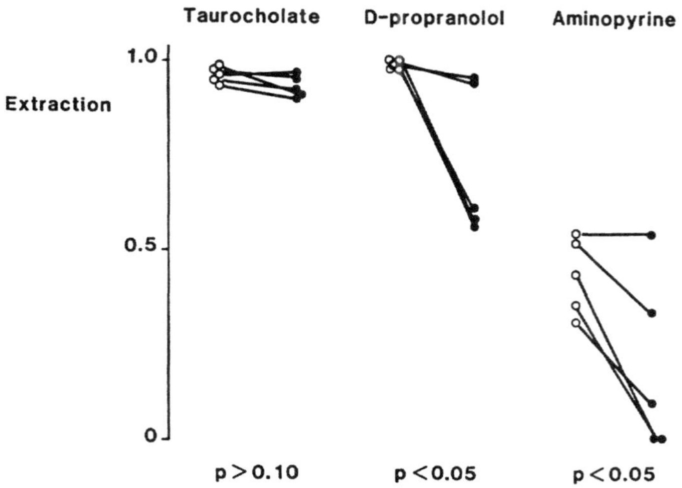

Figure 3: Effect of norepinephrine on hepatic extraction efficiency of three model compounds. Transport of taurocholate, which is actively taken up, is not affected, while extraction of the two passively transported compounds is significantly diminished.

due to an effect of norepinephrine on the taurocholate transport system or to a decrease in the length of sinusoids being perfused. In view of the finding of a parallel reduction of intra- and extravascular spaces induced by norepinephrine we interprete our data in the sense that norepinephrine leads to a preferential distribution of flow into shorter sinusoids. However, our results do not support a major effect of norepinephrine on size and/or number of fenestrations, since the extravascular albumin space per erythrocyte space was not affected. By which effect norepinephrine affects the population of sinusoids being perfused remains to be determined.

Popper was the first to demonstrate sinusoidal capillarization in cirrhotic liver (Popper et al., 1952; Schaffner and Popper, 1963). It took over two decades until the functional importance of this phenomenon was recognized. In cirrhotic human as well as rat liver a change from the flow-limited to a diffusion-limited pattern has been demonstrated and associated with

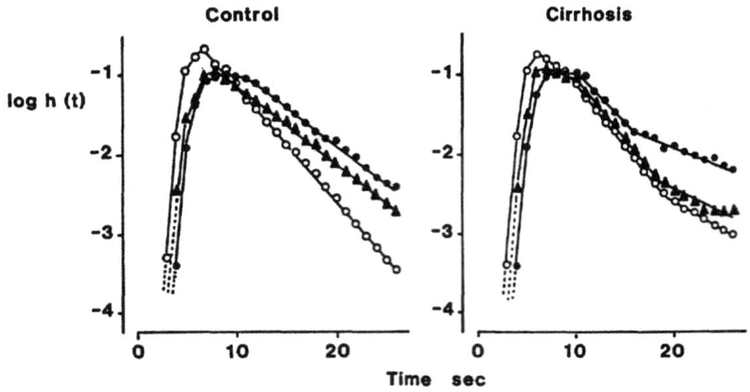

Figure 4: Sinusoidal capillarization in cirrhotic rat liver. The control liver shows the normal, flow-limited behaviour (see also Fig. 1), while the cirrhotic liver shows a diffusion-limited pattern: the albumin curve is almost superimposed upon the erythrocyte curve, while the sucrose curve shows a bimodal distribution due to capillary exchange at the newly formed diffusional barrier at the sinusoidal wall.

impairment of the hepatic clearance function (Huet et al., 1982; Varin and Huet, 1985; Reichen and Le, 1986). Fig. 4 shows an example of this sinusoidal capillarization. In contrast to the normal, flow-limited pattern seen in the control rat, the albumin curve in the cirrhotic liver is almost superimposed upon the erythrocyte curve, while the sucrose curve shows a biphasic behaviour due to diffusional exchange at the capillarized sinusoid. Accordingly, the extravascular albumin space is markedly reduced.

In the following we are going to examine the determinants of hepatic extraction of the same three model compounds studied above, namely of taurocholate, propranolol, and of aminopyrine. It has long been known that in contradiction to pharmacokinetic theory (Rowland et al., 1973; Wilkinson and Shand, 1975) the clearance of low and of high extraction compounds is diminished to a similar extent in liver cirrhosis (Branch et al., 1976). Different explanations have been forwarded to resolve this apparent contradiction, the most popular being the so-called intact cell hypothesis of liver disease (Wood et al., 1979). The cirrhosis model used in these investigations, chronic exposure of rats to phenobarbital and CCl_4, presumably fits the criteria of the intact cell hypothesis (Reichen et al., 1987). In addition to determination of the first-pass hepatic extraction of the different compounds we have determined intrahepatic shunting by a microsphere technique, the hepatic erythrocyte, albumin and sucrose space by the multiple indicator dilution technique and hepatocellular volume by morphometric techniques, as previously described (Reichen and Le, 1986; Reichen et al., 1987; Gross et al., 1987). In addition, we have measured the intrinsic clearance of aminopyrine by determining the enzyme kinetics of aminopyrine N-demethylase, the rate limiting step in aminopyrine excretion in hepatic microsomes, as previously described (Reichen et al., 1987).

The extraction efficiencies of all three model compounds was markedly reduced in cirrhotic as compared to control livers (Fig. 5). Intrahepatic shunting was increased in cirrhotic livers in accordance with previous results from our own and other laboratories (Varin and Huet, 1985; Reichen et al., 1986; Reichen and Le, 1986). In the present experiments it ranged from 0.5 to 9.8% . Morphometrically determined hepatocellular volume was reduced in 50% of the animals to a similar extent, as in previously published studies from our laboratories (Reichen et al., 1987). The same held true for intrinsic aminopyrine clearance (Reichen et al., 1987). The thorough characterization of these cirrhotic livers permitted analysis by stepwise logistic regression analysis to define the determinants of the different transport functions. Taurocholate extraction efficiency correlated, among others, with hepatocellular volume ($r=0.846$; $p < 0.001$), extravascular albumin space ($r=0.821$; $p < 0.001$) and inversely with intrahepatic shunting ($r=-0.826$). It did not correlate with hepatic blood flow, however. A Cox model selected only extravascular albumin space and hepatocellular volume as determinants of taurocholate extraction ($TCE=8.34.EVA+0.05.HCV-0.12$; $r=0.913$,

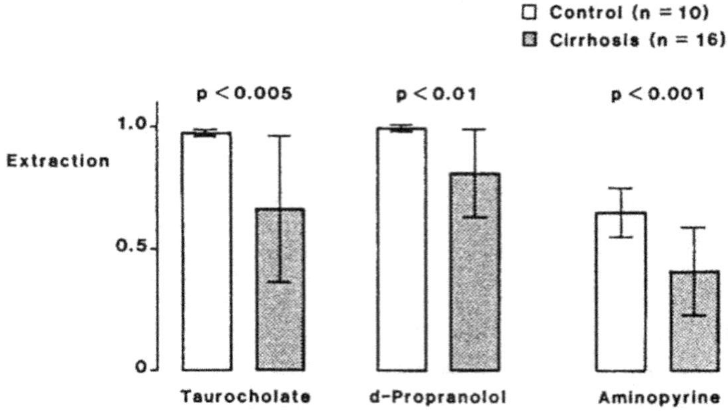

Figure 5: Extraction efficiency of three model compounds in control and cirrhotic rats. Extraction of all compounds, whether transported actively (taurocholate) or passively (propranolol and aminopyrine), were significantly reduced in cirrhotic rat liver.

$p < 0.0001$). These results demonstrate that the reduced capillary exchange is a major, but not the sole, determinant of bile salt handling in cirrhotic rat liver. When the determinants of serum bile acid levels in vivo are considered, we have recently shown that intrahepatic shunt fraction is the cnly determinant.

A similar result was obtained when the determinants of propranolol extraction were considered (Fig. 6). Similarly to the actively transported taurocholate, extraction efficiency of the passively transported propranolol correlated with hepatocellular volume, extravascular albumin space, and intrahepatic shunting, but not with hepatic blood flow. Stepwise logistic regression analysis defined hepatocellular volume and extravascular albumin space as the sole determinants of propranolol extraction ($y=0.34+3.74.EVA+0.03.HCV$; $r=0.856$, $p < 0.0005$). Thus, another high extraction model compound exhibits a major dependence on its accessability to the extravascular space.

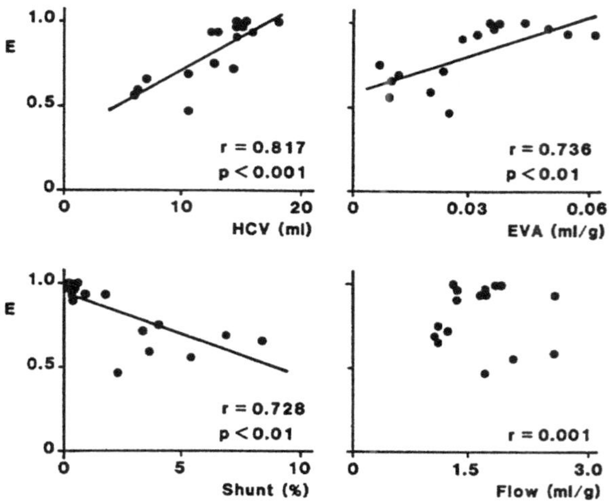

Figure 6: Determinants of propranolol extraction in cirrhotic rat liver. Stepwise logistic regression analysis identified extravascular albumin space (EVA) and hepatocellular volume (HCV) as the sole determinants of propranolol extraction, the regression equation being E=0.34+3.74.EVA+0.03.HCV (r=0.856; $p < 0.0005$).

The determinants of the extraction of the enzyme-limited compound, aminopyrine, showed a slightly different pattern (Fig. 7): the best single correlation was observed with its intrinsic clearance, followed by hepatocellular volume, shunt fraction, and extravascular albumin space. As expected, portal flow did not correlate with aminopyrine extraction at all (data not shown). Stepwise logistic regression analysis picked intrinsic clearance as the sole determinant of aminopyrine extraction. This is in line with the conception that hepatic clearance of low extraction ("enzyme-limited" regime) depends mostly on the amount of enzyme which limits their hepatic clearance (Rowland et al., 1973; Wilkinson and Shand, 1975).

We conclude from these data that sinusoidal capillarization as measured by the accessability of albumin to the extravascular space is the major determinant of the hepatic transport of high extraction ("flow-limited") compounds. In this analysis

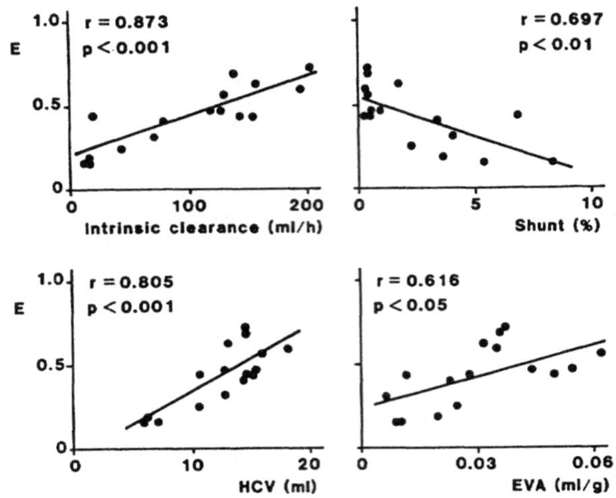

Figure 7: Determinants of aminopyrine extraction in cirrhotic rat liver. Stepwise logistic regression analysis identified intrinsic clearance as the only determinant of aminopyrine extraction ($y=0.19+0.0024.x$; $r=0.873$, $p < 0.0001$) in line with the concept that aminopyrine clearance follows an enzyme-limited regime.

hepatocellular volume is the second most important determinant of clearance of these high extraction compounds. The extraction of enzyme-limited compounds such as aminopyrine, by contrast, is determined mainly by their intrinsic clearance as predicted by classical pharmacokinetic theory. Nevertheless, microvascular exchange as assessed by the accessability of the extravascular space to albumin still shows a statistically significant correlation with aminopyrine extraction. Moreover, when we investigated aminopyrine disposition in vivo, it was the major determinant of aminopyrine clearance.

These data support the notion that microvascular exchange is an important determinant of hepatic clearance. This is further supported by our recent demonstration (Reichen and Le, 1986) that verapamil, a calcium antagonist, is able to improve microvascular exchange in cirrhotic rat liver (Fig. 8). This improvement was associated with an amelioration of hepatic

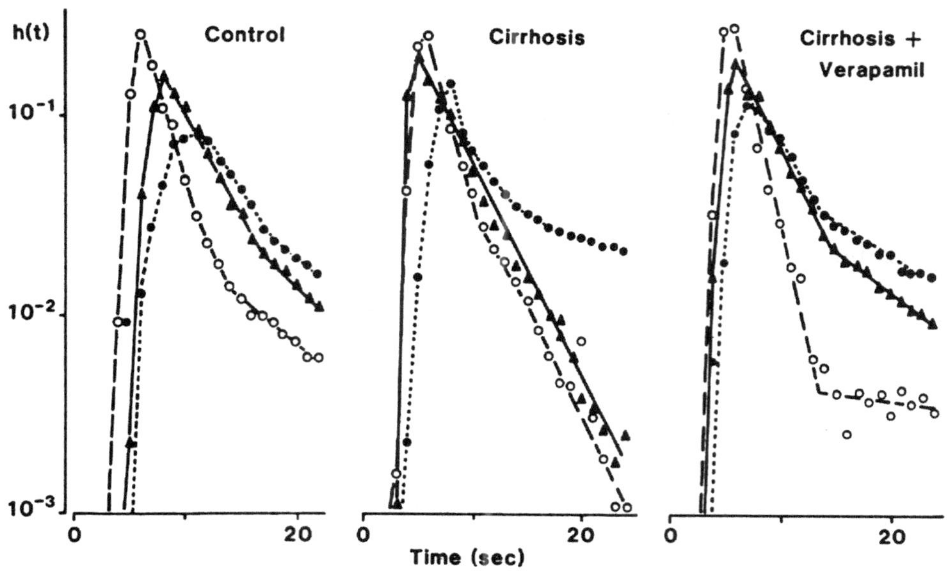

Figure 8: Effect of acutely administered verapamil on microvascular exchange in cirrhotic liver. The sinusoidal capillarization (see also Fig. 4) is markedly ameliorated after verapamil, as evidenced by the increased distribution of albumin into an extravascular space at the longer transit times. Reprinted from the Journal of Clinical Investigation 78:448 (1986) with permission from the Publisher.

antipyrine clearance. Administered chronically, the same phenomenon associated with a marked improvement in microsomal function could be observed (Reichen et al., 1986).

The distribution of perfused sinusoids and their microvascular exchange characteristics can adequately be assessed by the multiple indicator dilution technique. Such studies can be performed in both acute and chronic type experiments, as exemplified by the foregoing sketchy demonstrations. Disturbed hepatic microvascular exchange is a major determinant of hepatic clearance function in models of chronic liver disease.

Acknowledgements

The author was the recipient of a Research Career Development Award of the Swiss National Foundation for Scientific Research (SNF 3.731). Supported by SNF grants 3.823 and 3.986. I wish to thank Ms. M. Kappeler for the artwork and Ms. R. Steiner for secretarial assistance.

References

Ballet F, Chretien Y, Rey C, Poupon R (1987) Norepinephrine: A potential modulator of the hepatic transport of taurocholate. A study in the isolated perfused rat liver. J Pharmacol Exp Ther 240:303-307

Branch RA, James JA, Read AE (1976) The clearance of antipyrine and indocyanine green in normal subjects and in patients with chronic liver disease. Clin Pharmac Ther 20:81-89

Cousineau D, Goresky CA, Rose CP (1983) Blood flow and norepinephrine effects on liver vascular and extravascular volumes. Am J Physiol 244:H495-H504

Fraser P, Bowler LM, Day WA, Dobbs B, Johnson HD, Lee D (1980) High perfusion pressure damages the sieving ability of sinusoidal endothelium in rat liver. Br J Exp Pathol 61:222-228

Goresky CA (1963) A linear method for determining liver sinusoidal and extravascular volume. Am J Physiol 204:626-640

Goresky CA, Ziegler WH, Bach GG (1970) Capillary exchange modelling. Barrier-limited and flow-limited distribution. Circ Res 27:739-764

Gross JB, Reichen J, Zeltner T, Zimmermann A (1987) The evolution of changes in quantitative liver function tests in a rat model of cirrhosis. Hepatology 7:457-463

Horn T, Christoffersen P, Henriksen JH (1987) Alcoholic liver injury: Defenestration in noncirrhotic livers - A scanning electron microscopic study. Hepatology 7:77-82

Horn T, Henriksen JH, Christoffersen P (1986) The sinusoidal lining cells in "normal" human liver. A scanning electron microscopic investigation. Liver 6:98-110

Huet PM, Goresky CA, Villeneuve JP, Marleau D, Lough JO (1982) Assessment of liver microcirculation in human cirrhosis. J Clin Invest 70:1234-1244

Mak KM, Lieber CS (1984) Alterations in endothelial fenestrations in liver sinusoids of baboons fed alcohol: A scanning electron microscopic study. Hepatology 4:386-391

Nopanitaya W, Lamb JC, Grisham JW, Carson JL (1976) Effect of hepatic venous outflow obstruction on pores and fenestrations in sinusoidal endothelium. Br J Exp Path 57:604-609

Okazaki I, Tsuchiya M, Kamegaya K (1973) Capillarization of hepatic sinusoids in carbon tetrachloride-induced hepatic fibrosis. Bibl Anat 12:476-483

Popper H, Eliás H, Petty DE (1952) Vascular pattern of cirrhotic liver. Am J Clin Pathol 22:717-729

Reichen J, Arts B, Schafroth U, Zimmermann A, Zeltner TB, Zysset T (1987) Aminopyrine N-demethylation by rats with liver cirrhosis. Evidence for the intact cell hypothesis. A morphometric-functional study. Gastroenterology 93:719-726

Reichen J, Hirlinger A, Ha HR, Saegesser H (1986) Chronic verapamil administration lowers portal pressure and improves hepatic function in rats with liver cirrhosis. J Hepatol 3:49-58

Reichen J, Le M (1986) Verapamil favourably influences hepatic microvascular exchange and function in rats with cirrhosis of the liver. J Clin Invest 78:448-455

Reichen J, Paumgartner G (1976) Uptake of bile acids by the perfused rat liver. Am J Physiol 231:734-742

Rowland M, Benet LZ, Graham GG (1973) Clearance concepts in pharmacokinetics. J Pharmacokin Biopharm 1:123-136

Schaffner F, Popper H (1963) Capillarization of hepatic sinusoids in man. Gastroenterology 44:239-242

Stefan AM, Gendrault JL, Kirn A (1987) Increase in the number of fenestrae in mouse endothelial liver cells by altering the cytoskeleton with cytochalasin B. Hepatology 7:1230-1238

Varin F, Huet PM (1985) Hepatic microcirculation in the perfused cirrhotic rat liver. J Clin Invest 76:1904-1912

Wilkinson GR, Shand DG (1975) A physiologic approach to hepatic drug clearance. Clin Pharm Ther 18:377-390

Wisse E, De Zanger RB, Jacobs R, McCuskey RS (1983) Scanning electron microscope observations on the structure of the portal veins, sinusoids and central veins in rat liver. Scan Electron Microsc. 3:1441-1452

Wisse E, De Zanger B, Charels K, Van Der Smissen P, McCuskey RS (1985) The liver sieve: Considerations concerning the structure and function of endothelial fenestrae, the sinusoidal wall and the space of Disse. Hepatology 5:683-692

Wood AJJ, Villeneuve JP, Branch RA, Rogers LW, Shand DG (1979) Intact hepatocyte theory of impaired drug metabolism in experimental cirrhosis in the rat. Gastroenterology 76:1358-1362

New Methodologic Approaches for the Study of Transport in the Liver

J.L. Boyer
J. Graf
A. Gautam
Department of Medicine
and Liver Center
Yale School of Medicine
333 Cedar Street
New Haven, CT USA 06510

Introduction

Knowledge of the mechanisms of hepatic transport systems has increased rapidly in the last few years, as new techniques have become available to study this complex physiologic process. Traditionally, investigators have relied on the bile fistula animal, isolated perfused liver preparations, and isolated hepatocyte suspensions to probe these transport systems. While each model has provided significant information, certain disadvantages are inherent in each of the models that limit interpretation. For example, the bile fistula animal has proven useful for characterizing both stimulants and inhibitors of bile secretion. However, problems arise from the inability to monitor changes in hepatic circulation or to distinguish parenchymal cell from bile duct functions. For this reason, the isolated perfused liver has been a much used technique for studies of hepatic uptake and biliary secretion and has the advantage over the bile fistula animal that hepatic hemodynamics (flow and perfusion pressure) can be monitored in a controlled fashion. Since the blood supply to the bile ducts is usually excluded in these preparations, transport processes across the bile duct epithelia are minimized, although probably not entirely

Hepatic Transport in Organic Substances
Ed. by E. Petzinger, R. K.-H. Kinne, H. Sies

eliminated even in species like the rat where only a small component of net secretion appears to be bile duct-related. The major disadvantage of the isolated perfused liver is the difficulty in localizing transport functions to specific plasma membrane or intracellular domains as well as the effects of lobular gradients on transport phenomena in periportal and pericentral zones of the liver. The latter problem is largely overcome for studies of hepatic uptake by using isolated hepatocyte suspensions which effectively eliminate the lobular gradient. This preparation provides a population of liver cells that are easily divided into multiple samples of similar composition. However, to the extent that there may be regional differences in transport between periportal and perivenous areas, techniques will be required for separating hepatocytes from separate zones in the lobule. Loss of a functional excretory domain is another characteristic of hepatocyte suspensions that restricts their use in transport studies of the liver. Single cells in isolation lose polarity of apical domain transporters and certain intracellular organelles including the Golgi apparatus (Wanson et al., 1977; Gebhardt, 1986).

Other investigators utilize hepatocytes in monolayer cultures in order to prolong the lifespan of the isolated cells which deteriorate over a few hours when maintained in suspension. A number of different culture systems have been described and are widely used. While certain functions of hepatocytes may also deteriorate in cell culture, the use of biomatrix (collagen, Matrigel) or co-cultures with other epithelial cells helps to maintain normal function and prolong viability (Rojkind et al., 1980; Gugen-Guillouzo et al., 1983; Bissell et al., 1987). Monolayer cell cultures of hepatocytes are used for the study of mechanisms of hepatic uptake as well as intracellular metabolism. However, like isolated hepatocyte suspensions, monolayer cultures of hepatocytes initially lose their excretory polarity which must be reconstituted once the cells reach confluence and reestablish junctional contacts and bile canalicular lumens (Wanson et al., 1977). Yet even when morphologic evidence of an excretory system is reestablished, normal excretory functions may not always be restored (Gebhardt, 1986).

Over the past several years, technical advances in the disciplines of membrane biology and cellular and molecular physiology have provided newer techniques that have begun to increase our understanding of the transport processes that involve the excretory domain of the liver cell. These advances include: (1) the ability to separate and purify plasma membranes from both the sinusoidal and canalicular domains (Inoue et al., 1983; Meier et al., 1984), a development that has greatly facilitated characterization and localization of a number of receptors, carriers, enzymes, and proteins to these respective domains (see Table); and (2) the ability to test the physiologic significance of these transporters in an intact cell system that retains polarity of its cellular membrane domains (the isolated hepatocyte couplet, Graf et al., 1984; Gautam et al., 1987).

Knowledge of the presence and location of a number of membrane transporters has led to the development of a working model of hepatic bile formation that can now be tested more directly (Boyer, 1986; Moseley and Boyer, 1985; Meier, 1987). The model holds that the sodium pump actively maintains an inwardly directed sodium and outwardly directed potassium gradient through the expenditure of cellular energy. Potassium is also conducted out of the cell through channels and this outward flux of potassium is a major determinant of the intracellular negative electrical potential difference. The inwardly directed sodium gradient and the intracellular negative electrical potential both serve as driving forces for the uptake of solutes from the blood into the liver. For example, alanine transport into the hepatocyte is coupled to sodium and is also driven by the negative intracellular potential. An electrogenic transport process is also thought to facilitate taurocholate uptake into the hepatocyte, an initial step in the process of generating bile acid-dependent bile flow (Edmondson et al., 1985). Based on studies in membrane

Polarity of Hepatocyte Transport Mechanisms

	Sinusoidal Domain	Canalicular Domain
Enzymes:		
Na^+,K^+-ATPase; Ca^{++}-ATPase	+	-
leucine aminopeptidase	-	+
alkaline phosphatase	-	+
alkaline phosphodiesterase	-	+
γ-glutamyl transpeptidase	-	+
Mg^{++}-ATPase	-	+
Receptors:		
LDL; secretory components	+	-
asialoglycoprotein	+	-
transferrin	+	-
Proteins:		
100-110K	-	+
Carriers:		
Na^+-coupled bile acid transporter (48K)	+	-
bile acid transporter (potential driven) (100K)	-	+
multispecific organic anion transporter (54K)	+	?
Na^+-coupled alanine transporter	+	+
Na^+,K^+-coupled glutamate transporter	-	+
GSH	+	+
GSSG	-	+
cysteine, glycine	+	+
Antiports and Symports:		
Na^+/H^+ antiport	+	-
HCO_3^-/Cl^-	-	+
$SO_4^=/OH^-$ antiport	+	-
$SO_4^=/HCO_3^-$	-	+
Na^+,HCO_3^- symport	+	?
Channels:		
K^+	+	+
Cl^-	+	+

vesicles, the negative intracellular potential difference provides the major driving force for the transcanalicular membrane transport of taurocholate (Inoue et al., 1984; Meier et al., 1984). As taurocholate accumulates in the canalicular lumen, osmotic pressure rises and water and small ions move passively into the canaliculus to stimulate the flow of bile. Bile acid-independent flow (BAIF) is thought to occur from the accumulation of other osmotically active solutes in the canalicular lumen, although evidence for this process is imcomplete (Boyer, 1986). For example, sodium/proton exchange on the sinusoidal membrane may lead to the intracellular accumulation of bicarbonate equivalents above their electrical and chemical equilibrium. Bicarbonate could then exchange with chloride at the canalicular domain and if chloride is recycled via a canalicular chloride channel (and these anions are retained within the canalicular lumen by the tight junctions), bile acid-independent flow would occur (Moseley and Boyer, 1985; Moseley et al., 1986). Other candidates for the bile acid-independent secretory process include SO_4/HCO_3 exchange (an example of tertiary active transport) (Meier, 1987) and electric potential driven canalicular excretion of reduced and oxidized glutathione (Inoue et al., 1983; Akerboom et al., 1984). It is likely that several transport processes account for the generation of BAIF.

Isolated Hepatocyte Couplets

The isolated hepatocyte couplet model is providing a new opportunity to assess the physiologic significance of many of the transporters defined (see Table). This cell system represents a primary secretory unit, devoid of influence of blood flow and bile duct function, yet it retains a highly polarized cell ultrastructure in short-term culture that elaborates bile into a closed canalicular space between the two attached cells (Oshio and Phillips, 1981; Boyer et al., 1985). Couplets are prepared by collagenase perfusion of rat or mouse liver and represent two hepatocytes that continue to be attached at their adjacent tight junctions. When some of these couplets are placed on glass coverslips and are maintained in a short-term culture, their junctional contracts reseal spontaneously and secretion is elaborated into a closed lumen between two adjacent cells. In couplets where a complete tight junctional barrier is restored, the apical domain gradually expands with time in culture so that by three to four hours, a lumen of 2-5 microns or more in diameter develops. Morphometric studies indicate that the surface area of the remaining canalicular domain increases over time, largely by a process of membrane recruitment from the hemicanalicular domains that no longer remain in contact with adjacent hepatocytes (Gautam et al., 1987). Thus, the canalicular membrane is reorganized to the remaining excretory pole of the cell, a process that can be blocked by the microfilament poison, cytochalasin D. Unlike isolated hepatocyte suspensions and some monolayer cultures, ultrastructural studies indicate that the Golgi apparatus and lysosomes remain adjacent to the canalicular domain. Cytoskeletal elements including actin and tubulin are also increased at the excretory pole of the cell (Nickola and Frimmer, 1986; Sakisaka et al., 1988). With time in culture, the canalicular lumen of the couplet expands as secretion is elaborated and then periodically contracts and collapses, emptying its contents paracellularly into the surrounding media. Oshio and Phillips, 1981 first demonstrated this process utilizing time lapse cinephotomicrography and concluded that the cytoskeleton both maintained canalicular ultrastructure as well as provided a contractile force that facilitated movement of bile in the intact liver. However, in the isolated couplet, the process of contraction disrupts the junctional contacts which may either reseal and undergo a second cycle of secretion or fail to reseal with subsequent loss of secretory polarity (Gautam et al., 1987).

Functional Properties of Hepatocyte Couplets

Several simple functional studies can be performed in selected couplets that indicate that this isolated cell system retains secretory properties that are quite similar to those in the intact liver. First, the addition of bile acids (taurocholate) to the media results in a prompt accumulation of fluid in the canalicular lumen (Gautam et al., 1986). Secondly, the addition of fluorescein-diacetate to the media results in rapid hepatic uptake and the appearance of fluorescence within the cells as esterases hydrolyze the ester bond and release fluorescein (Graf et al., 1984). Fluorescein subsequently accumulates within the canalicular lumen resulting in an increase in fluorescence intensity in couplets that retain their secretory polarity. Thirdly, when horseradish peroxidase (HRP) is added to the media, and electron micrographs are prepared from preparations of couplets at timed intervals, HRP containing vesicles move from the sinusoid to the remaining pericanalicular zone in 10 to 20 minutes, a time frame that is comparable to the phenomenon of transcytotic vesicle transport in the intact liver (Sakisaka et al., 1988). Thus, the hepatocyte couplet not only rapidly transports organic anions that generate secretion but endocytosis continues and vesicles are transported to the apical domain in a manner that is consistent with an intact secretory process. Finally, the expanding canalicular lumen provides a unique opportunity to directly examine events at the apical domain of the bile secretory unit, both by allowing insertion of microelectrodes as well as permitting optical imaging of secreted fluorescent compounds (Graf et al., 1984).

There are also significant limitations to the model. Studies are limited to short-term experiments lasting up to six to eight hours since cell viability deteriorates rapidly in these non-confluent cultures. Secretion is elaborated into a closed rather than an open luminal space, resulting in periodic collapse of the space which may terminate the experiment. In culture, the cells are extremely sensitive to photo damage and must be examined with high resolution imaging systems utilizing an inverted microscope and perfusion chambers that maintain temperature at 37°C and allow for rapid changes of media. Despite these shortcomings, the isolated hepatocyte couplets represents a unique primary bile secretory unit in which the determinants of the bile secretory process can be examined directly.

Applications of Hepatocyte Couplets for the Study of Bile Secretory Function

1. Quantitative assessment of canalicular bile secretion.

One of the unique aspects of this cell culture system is the ability to directly assess agents that stimulate or inhibit canalicular bile formation, without confounding effects of blood flow of bile ducts. The process can be evaluated qualitatively but can also be quantitated by utilizing optical sectioning techniques (Gautam et al., 1986). With this technique, changes in canalicular space volume with time can be obtained by video-enhanced microscopic optical planimetry utilizing high resolution Nomarski optics. By obtaining multiple optical sections, one micron apart, from the top to the bottom of the canalicular lumen, cross sectional areas and perimeters can be measured and used to calculate canalicular volumes and surface areas. Utilizing this technique, taurocholic and ursodeoxycholic acid stimulate canalicular secretion to comparable degrees. There is a close correlation between the magnitude of the bile acid-stimulated choleretic response and the surface area of the canalicular membrane in a given hepatocyte couplet and no hypercholeretic effect is seen with ursodeoxycholic acid. These findings suggest either that ursodeoxycholic acid may stimulate secretion at sites that are distal to the canaliculus in the intact animal or that ursodeoxycholic acid is distributed to a greater number of hepatocytes within the hepatic lobule than taurocholic acid. Quantitative studies of canalicular secretion in isolated hepatocyte couplets can also be performed to examine the role of bicarbonate and hormones in the generation of bile acid-independent secretion. Preliminary results suggest that agents that activate adenylcyclase influence the production of canalicular bile (Nathanson and Gautam, 1988). Thus, by utilizing isolated hepatocyte couplets and quantitative measurements of bile secretion, a clearer understanding of the mechanisms of choleretic agents should be obtained.

2. Assessment of Bile Canalicular Contractility

Utilizing time lapse cinephotomicrography, Phillips and colleagues (1981) have examined bile canalicular membrane motility. Regular periodic contractions have been observed for periods up to sixteen hours that result in closure of the canalicular lumen. Canalicular contractions can be stimulated by the addition of taurocholate or the microinjection of calcium (Miyairi et al., 1984; Watanabe and Phillips, 1984). These studies suggest that taurocholate may induce secretion in part by an effect on the canalicular contractile apparatus, a process that would also appear to be calcium-dependent.

3. Microinjection

Watanabe et al. (1984) have used micro pipettes prepared from glass capillaries to inject calcium into the cell cytoplasm. By examining hepatocyte triplets, they have demonstrated that the calcium induced signal can be transmitted from one cell to the next leading to a coordinated subsequent contraction of the adjacent canalicular lumen.

Graf et al. (1984) have used similar micro pipettes to directly enter the canalicular lumen. This approach has allowed direct measurement of the intracanalicular luminal electrical potential.

4. Fluorescence Microscopy

A wide variety of fluoresent compounds can be utilized for the study of hepatic transport functions in this cell preparation. As previously mentioned, cellular uptake and luminal secretion of fluorescein anions can be readily demonstrated (Graf et al., 1984). Fluorescent bile acid derivatives can also be used to study the determinants of bile acid uptake and canalicular secretion in this cell system

(Weinman et al., 1988). Hepatocytes rapidly take up pH-sensitive dyes such as carboxyfluorescein-acetate and bis(carboxyethyl)carboxy-fluorescein (BCECF). These compounds have been utilized to study regulation of intracellular pH and are currently being utilized to examine intracanalicular pH. Dye injection has also been used to study cell coupling through intracellular communications by microinjection of Lucifer Yellow (Spray et al., 1986). Future applications include the measurement of intracellular calcium with calcium-sensitive dyes, the study of vesicle transport utilizing fluorescent dextrans and the movement of membrane lipids with fluorescent derivatives of these compounds.

5. Electrophysiology

The isolated hepatocyte couplet system has been particularly useful for applying classical methods of electrophysiology to analyze transport processes associated with bile formation (Graf et al., 1984; 1987). Specific electrophysiologic applications include:

1. The determination of transepithelial potential profiles (Graf et al., 1984).
2. The measurement of electrical resistances across sinusoidal, and canalicular membranes as well as the paracellular pathway (Graf et al., 1987).
3. Measurements of intracellular ion activities and an analysis of partial ion conductances of the sinusoidal membrane (Graf et al., 1987).
4. An analysis of the conductive properties of intercellular communications (Spray et al., 1986).
5. Studies of pH_i regulation by sodium proton exchange (Henderson et al., 1987)
6. Analysis of Na^+,K^+-ATPase activity (Graf et al., 1987).

Measurements of intracellular potential (-30 to -40 mV) are in good agreement with studies in the intact liver. In addition, measurements of ion concentrations with ion selective electrodes indicate that large electrochemical gradients exist for the entry of sodium across the luminal and basolateral cell membrane and for the exit of potassium from the cell. Chloride anions appear to be in electrochemical equilibrium as has been observed by other investigators (Graf et al., 1987). For the first time, a canalicular luminal potential (-5 mV) has been directly determined (Graf et al., 1984). The magnitude of this potential supports earlier assumptions that the potential is due to a Donnan distribution of inorganic electrolytes across the paracellular pathway as a result of the accumulation of negatively charged organic anions.

Membrane resistances can be measured with 2 microelectrodes (1 in the cell and another in the lumen or bath) by determining the voltage change that results when current is passed across specific membrane domains ($V = I \times R$). These studies indicate that the resistance of the basolateral membrane (0.15 GΩ) is similar to that obtained by indirect measurements from intact tissue. A high luminal membrane resistance of 0.7 GΩ and a low transepithelial resistance of approximately 20 MΩ indicate that the paracellular pathway exhibits a high specific conductance as has been classically observed in leaky epithelia.

Partial ionic conductances (g_i) of the basolateral cell membrane have been obtained by measuring changes in the intracellular voltage during variations of individual ion concentrations in the bathing media. These findings indicate that $g_k > g_{Cl} > g_{Na}$ and that g_k = 2.3 nS, g_{Na} = 0.4 nS, and g_{Cl} = 1.7 nS. Further studies have established that potassium conductance is the major determinant of the intracellular electrical potential, findings confirmed by whole cell recordings utilizing Patch Clamp techniques (Henderson et al., 1988). Others have examined the conductance of intercellular communications and have demonstrated regulation of the gap junctions with cyclic nucleotides and a reduction in conductance with low intracellular pH (Spray et al., 1986; Saez et al., 1986).

Intracellular ion activities of sodium and potassium have been measured with ion-sensitive electrodes and are within the range observed in intact liver. Maintenance of these ion gradients requires the continued utilization of ATP by membrane Na^+, K^+-ATPase. This pump activity can be inhibited either by introducing ouabain or by removing potassium from the external medium (Graf et al., 1987). The latter maneuver results in the depletion of intracellular potassium and a gain of sodium. When potassium is readmitted to the external medium, there is instantaneous activation of the Na^+, K^+-ATPase pump, resulting in hyperpolarization of the basolateral cell membrane and a small negative deflection in the canalicular luminal potential. This latter observation indicates that the pump is predominantly if not exclusively located at the basolateral cell membrane, since if the pump were active at the canalicular pole, a positive deflection in the luminal potential should have been observed. These findings confirm recent immunocytochemical data indicating that Na^+,K^+-ATPase catalytic activity is a property of the basolateral membrane in hepatocytes (Stzul et al., 1987).

More recent studies have examined the role of the electrical potential as a determinant for both the hepatic uptake and canalicular excretion of bile acids. By combining electrophysiologic and fluorescent techniques, it has been possible to vary the electrical potential within the hepatocyte by first impaling the cell with microelectrodes and then injecting current. When couplets are exposed to a fluorescent derivative of taurocholate (7-Beta-nitrobenzoxadiazol taurocholate), changes in the electrical potential from -40 to +20 meV had no significant effect on the initial uptake rates of this fluorescent derivative of taurocholic acid. In contrast, when canalicular lumen accumulation of the fluorescent bile acid was examined, increases or decreases in the negative intracellular potential induced by altering K^+ gradients produced parallel changes in the rate of canalicular bile acid fluorescence intensity (Weinman et al., 1988). Furthermore, in a separate study, measurements of changes in the canalicular volume were obtained in the presence of 5 μM taurocholic acid, and significant increases in canalicular secretion were observed when the negative intracellular potential was increased.

Comparable changes in intracellular potential produced no significant changes in luminal volume in the absence of bile acid loading (Weinman et al., 1987). Together, these studies provide the first direct physiologic confirmation of the potential driven canalicular bile acid transport system that was previously demonstrated in isolated membrane vesicle system (Inoue et al., 1984; Meier et al., 1984).

Conclusions

Isolated hepatocyte couplets are providing a unique opportunity to assess the physiologic significance of several membrane transport systems that have been implicated in the generation of hepatic bile secretion by more indirect techniques. To date, these studies have confirmed the critical role of potassium conductance and the maintenance of intracellular ion gradients in developing the transmembrane electrical potential. No role for active chloride transport has been found and the electrical potential has been demonstrated to be a major driving force for canalicular bile acid excretion. The "leakiness" and low electrical resistance of the tight junctions suggest that it may be difficult to generate secretion by the active extrusion of small ions. However, definitive studies have yet to be made, and the role of other organic anion transport systems in the generation of hepatocyte bile secretion remain to be assessed. In addition to clarifying mechanisms of water and electrolyte secretion, hepatocyte couplets should also be useful for the evaluation of mechanisms of protein and lipid excretion into bile.

References

Akerboom T, Inoue M, Sies H, Kinne R, Arias IM (1984) Biliary transport of glutathione disulfide studied with isolated rat-liver canalicular membrane vesicles. Eur J Biochem 141: 211-215

Bissel DM, Arenson DM, Maher JJ, Roll FJ (1987) Support of cultured hepatocytes by a Laminin-rich gel - Evidence for a functionally significant subendothelial matrix in normal rat liver. J Clin Invest 79:801-812

Boyer JL, Ng OC, Gautam A (1985) Formation of canalicular spaces in isolated rat hepatocyte couplets. Trans Assoc Am Phys 98:21-29

Boyer JL (1986) Mechanisms of bile secretion and hepatic transport. In: Andreoli TE, Hoffman JF, Fanestil DD, Schultz SG (eds) Physiology of membrane disorders. Plenum Publishing Corp., New York, p 609-636

Edmondson JW, Miller BA, Lumeng L (1985) Effect of glucagon on hepatic taurocholate uptake: relationship to membrane potential. Am J Physiol 12:G427-G433

Gautam A, Scaramuzza D, Boyer JL (1986) Quantitative assessment of primary canalicular secretion in isolated rat hepatocyte couplets (IRHC) by optical planimetry. Gastroenterology 90:1727

Gautam A, Ng OC, Boyer JL (1987) Isolated rat hepatocyte couplets in short-term culture: structural characteristics and plasma membrane reorganization. Hepatology 7:216-223

Gebhardt R (1986) Use of isolated and cultured hepatocytes in studies on bile formation. In: Guillouzo A, Guguen-Guillouzo C (eds) Research in isolated and cultured hepatocytes. John Libbey Ltd., London, p 353-376

Graf J, Gautam A, Boyer JL (1984) Isolated rat hepatocytes couplets: A primary secretory unit for electrophysiologic studies of bile secretory function. Proc Natl Acad Sci USA 81:6516-6520

Graf J, Henderson RM, Krumpholz B, Boyer JL (1987) Cell membrane and transepithelial voltages and resistances in isolated rat hepatocyte couplets. J Membrane Biol 95:241-254

Guguen-Guillouzo C, Clement B, Baffet G, Beaumont C, Morel-Chaney E, Glaise D, Guillouzo A (1983) Maintenance and reversibility of active albumin secretion by adult rat hepatocytes co-cultured with another liver epithelial cell type. Exp Cell Res 143:47-54

Henderson RM, Graf J, Boyer JL (1987) Na-H exchange regulates intracellular pH in isolated rat hepatocyte couplets. Am J Physiol 252:G109-G113

Henderson RM, Graf J, Boyer JL (1988) Inward rectifying potassium channels in isolated hepatocytes. Clin Res 36:558A

Inoue M, Kinne R, Arias IM (1983) The mechanism of biliary secretion of reduced glutathione. Eur J Biochem 134:467-471

Inoue M, Kinne R, Tran T, Biempica L, Arias IM (1983) Rat liver canalicular membrane vesicles. Isolation and topological characterization. J Biol Chem 258:5183-5188

Inoue M, Kinne R, Tran T, Arias IM (1984) Taurocholate transport by rat liver canalicular membrane vesicles. J Clin Invest 73:659-663

Meier PJ, St Meier-Abt A, Barrett C, Boyer JL (1984) Mechanisms of taurocholate transport in canalciular and basolateral rat liver plasma membrane vesicles - evidence for an electrogenic canalicular organic anion carrier. J Biol Chem 259:10614-10622

Meier PJ, Valantinas J, Hugentobler G, Rahm I (1985) Bicarbonate sulfate exchange in canalicular rat liver plasma membrane vesicles. Am J Physiol 253:G461-G468

Meier PJ (1987) Transport processes at the canalicular surface of rat hepatocytes. In: Reutter W, Popper H, Arias IM, Heinrich PC, Keppler D, Landman L (eds) Modulation of liver expression. MTP Press Ltd., Lancaster, p 127-141

Miyairi M, Oshio C, Watanabe S, Smith CR, Yousef IM, Phillips MJ (1984) Taurocholate accelerates bile canalicular contractions in isolated rat hepatocytes. Gastroenterology 87:788-792.

Moseley RH, Boyer JL (1985) Mechanisms of electrolyte transport in the liver and their functional significance. Sem Liver Dis 5:122-135

Moseley RH, Meier PJ, Aronson PS, Boyer JL (1986) Na-H exchange in rat liver basolateral but not canalicular plasma membrane vesicles. Am J Physiol 250:G35-G43

Nathanson MH, Gautam A (1988) Activation of endogenous adenylate cyclase (AC) stimulates choleresis in isolated rat hepatocyte couplets (IRHC). Gastroenterology 94:A619

Nickola I, Frimmer M (1986) Preservation of cellular polarity in isolated hepatocytes - Visualization of cytoskeletal structures by indirect immunofluorescence and fluorescent staining with tetramethylrhodaminyl-phalloidin. Cell Tissue Res 243:437-440

Oshio C, Phillips MJ (1981) Contractility of the bile canaliculi: implications for liver function. Science 212:1041-1042

Rojkind M, Gatmaitan Z, Machensen S, Giambrone MA, Ponce P, Reid LM (1980) Connective tissue biomatrix: its isolation and utilization for long-term cultures of normal rat hepatocytes. J Cell Biol 87:255-263

Saez JC, Spray DC, Nairn AC, Hertzberg E, Greengard P, Bennett MVL (1986) cAMP increases junctional conductance and stimulates phosphorylation of the 27-KDa principle gap junction polypeptide. Proc Natl Acad Sci USA 83:2473-2477

Sakisaka S, Ng OC, Boyer JL (1988) A tubulo-vesicular transcytotic pathway in isolated rat hepatocyte couplets in culture - effect of colchicine and taurocholate. Gastroenterology (in press)

Spray DC, Ginzberg RD, Morales EA, Gaitmaitin Z, Arias IM (1986) Electrophysiological properties of gap junctions between dissociated pairs of rat hepatocytes. J Cell Biol 103:135-144

Stzul ES, Biemesderfer D, Caplan MJ, Kashgarian M, Boyer JL (1987) Localization of Na^+,K^+-ATPase α-subunit to the sinusoidal and lateral but not canalicular membranes of rat hepatocytes. J Cell Biol 104:1239-1248

Wanson J-C, Drochmans P, Mosselmans R, Ronveaux M-F (1977) Adult rat hepatocytes in primary monolayer culture - Ultrastructural characteristics of intercellular contacts and cell membrane differentiations. J Cell Biol 74:858-877
Watanabe S, Phillips MJ (1984) Calcium causes active contraction of bile canaliculi: direct evidence from microinjection studies. Proc Natl Acad Sci USA 81:6164-6168
Weinman SA, Scaramuzza DM, Boyer JL (1987) Direct evidence that electrical potential difference drives taurocholate-dependent fluid secretion across the canalicular membrane of isolated rat hepatocyte couplets. Hepatology 7:1105
Weinman S, Thom H, Kurz G, Boyer JL (1988) Uptake and secretion of 7-nitrobenzoxadiazol taurocholate (NBDTC) by the isolated rat hepatocyte couplet: effects of membrane potential. FASEB J 2:A1725

Transport Processes in Mammalian Gallbladder and Bile Duct

K.-U. Petersen
Institut für Pharmakologie
RWTH Aachen
Schneebergweg
D-5100 Aachen
Federal Republic of Germany

Introduction

Pioneered by Diamond's early work (Diamond, 1968), gallbladder epithelial transport has attracted considerable attention, both as a model and a transporting epithelium in its own right. By comparison, knowledge of transport across the bile duct system is scarce. It is the scope of this brief review to summarize current understanding of mammalian gallbladder fluid and electrolyte transport and, on this background, examine how far these concepts fit with data on bile duct function. Two main conditions will be considered: the normal, absorptive state and the secretory one that can be induced by cyclic AMP (cAMP). Broader reviews of gallbladder properties have been provided recently (Rose, 1987; Reuss, 1988). Short summaries of bile duct transport are parts of recent reviews on biliary function (Boyer, 1986; Erlinger, 1987).

Gallbladder: Absorbing Cells

Available data on resistances, voltages and ionic composition of absorbing gallbladder cells (rabbit and guinea-pig) are summarized in Table 1. Its low transmural resistance categorizes mammalian gallbladder as typically leaky. In tissues of this type (Frömter, 1972), the paracellular route allows large passive so-

Hepatic Transport in Organic Substances
Ed. by E. Petzinger, R. K.-H. Kinne, H. Sies

lute flows and, by this, efficiently short-circuits the two limiting cell membranes. Intracellular Na activities are below and those of K and Cl above the values expected for passive distribution. The luminal cell membrane is mainly K-selective (Henin and Cremaschi, 1975; van Os and Slegers, 1975; Wehner and Petersen, 1984). In addition, it possesses a minor permeability for Na (Cremaschi and Meyer, 1982; Wehner and Petersen, 1984), while the evidence in favor of anion permeabilities (only guinea-pig, Cremaschi and Meyer, 1982) is not convincing (Stewart et al., 1988). Also the basolateral membrane appears to be K-selective (Henin and Cremaschi, 1975; van Os and Slegers, 1975); a Na permeability is not detectable (Gunther-Smith, 1986). On the other hand, additional ion permeabilities must be present unless reported mem-

Table 1. Properties of gallbladder epithelial cells

	Guinea-pig	Rabbit
R_t (Ωcm^2)	110 (f); 140 (n)	25 (a); 28 (c); 30 (f)
R_a (Ωcm^2)	460 (a); 960 (i)	540 (a); 390 (e)
R_b (Ωcm^2)	480 (a); 280 (i)	560 (a); 160 (e)
R_a/R_b	1.0 (a); 2.3 (d); 2.4 (l); 7.0 (i); 7.9 (j); 8.6 (m)	1.0 (a); 1.0 (g); 2.5 (e)
V_{mc} (mV)	-44 (a); -48 (d,l); -50 (a); -53 (i,m)	-49 (c); -56 (a) -59 (g); -64 (e) -69 (k); -72 (h)
aNa_i (mM)	36 (b)	15 (h); 46 (b)
aK_i (mM)	87 (b)	53 (b); 73 (e)
aCl_i (mM)		35 (c)

All data in the presence of HCO_3. R_t, transepithelial resistance; R_a, apical membrane resistance; R_b, basolateral membrane resistance; V_{mc}, apical membrane voltage; aX_i, intracellular activity of ion X. Since transepithelial voltage is close to zero, basolateral membrane potential largely equals V_{mc}.

Reference key: a, Cremaschi and Meyer, 1982; b, Cremaschi et al., 1983b; c, Duffey et al., 1978; d, Gunther-Smith, 1984; e, Gunther-Smith and Schultz, 1982; f, Heintze et al., 1981; g, Henin and Cremaschi, 1975; h, Moran et al., 1986; i, Petersen and Wehner, 1987; j, Stewart et al., 1988; k, van Os and Slegers, 1975; l, Wehner and Petersen, 1984; m, Wehner and Petersen, unpublished; n, Winterhager et al., 1986

brane voltages (Table 1) are artifactially low. This is because K equilibrium potential is about -80 mV which, in the absence of other conductive pathways, would be equal to the zero-current voltage of the basolateral membrane and, at high values of R_a/R_b (Table 1) also be close to membrane voltages.

Diamond (1968) was first to describe isotonic fluid absorption accomplished by electroneutral absorption of NaCl. A mutual dependency of Na and Cl absorption (Diamond, 1968), first explained by direct coupling at the apical membrane (Frizzell et al., 1975; Henin and Cremaschi, 1975), in all likelihood reflects parallel operation of Na/H and Cl/HCO_3 exchangers (Fig. 1). Intracellular

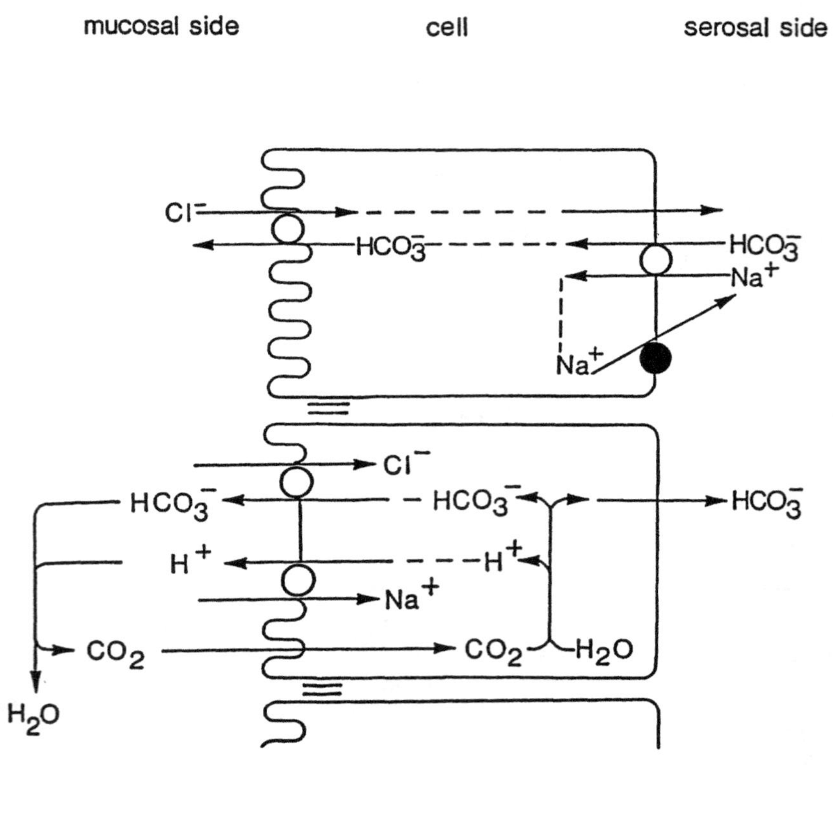

Fig.1: Transcellular ion transport in gallbladder epithelium. Upper cell: secretory HCO_3 flux. Filled circle: Na,K-ATPase. Lower cell: Absorptive HCO_3 flux and apical membrane entry of Na and Cl. From Winterhager et al. (1986), with permission.

supply of H and HCO_3 for secretion is maintained by hydration of CO_2 (Petersen et al., 1981). The double ion exchange also explains stimulation of NaCl absorption by luminal HCO_3 (Cremaschi et al., 1979; Heintze et al., 1981). According to evidence in Necturus gallbladder, the two exchangers are not directly coupled (Reuss, 1988). Additional entry mechanisms (true NaCl co-transport) have been proposed to account for the higher rates of NaCl absorption in rabbit gallbladder (Cremaschi et al., 1983a).

Beside Cl, also HCO_3 is absorbed in rabbit gallbladder (Diamond, 1968). This is most easily explained by the notion that only part of the HCO_3 formed in the cell (Fig. 1) is exchanged for luminal Cl. Consequently, Na/H exchange and luminal acidification predominate. A similar flux of HCO_3 has been demonstrated in guinea-pig gallbladder (Winterhager et al., 1986), although, provided the availability of serosal HCO_3, it even alkalinizes its lumen. This is due to additional HCO_3 influx at the basolateral membrane (probably cotransport with Na, Fig. 1) resulting in excess operation of the anionic countertransport and, finally, transepithelial Cl/HCO_3 exchange (Heintze et al., 1979).

Gallbladder: Secreting Cells (Effects of cAMP)

In guinea-pig gallbladder epithelium, rises in the intracellular concentration of cAMP can be achieved by a variety of agents (Wood and Svanvik, 1983) including prostaglandin E_1 (Petersen et al., 1982). As a consequence, the apical membrane becomes highly permeable to Cl and HCO_3 (Stewart et al., 1988). These changes lead to substantial decreases in apical membrane voltage and resistance. As to Cl, they are identical with cAMP effects known from Necturus (Petersen and Reuss, 1983) while the rise in HCO_3 permeability seems to be special to the guinea-pig. In the rabbit, PGE_1 has no effect on apical membrane properties (Petersen et al., 1987).

In rabbit gallbladder, cAMP reduces NaCl absorption, a finding interpreted to mean inhibition of coupled NaCl entry at the apical membrane (Frizzell et al., 1975). Replacement of the cotransport model by the double ion exchange (Fig. 1) necessitated a new analysis. In Necturus gallbladder, cAMP blocks Na/H exchange (Reuss and Petersen, 1985). A similar action in guinea-pig gallbladder (Petersen et al., 1985) can account for the amiloride-like inhibition of the absorptive HCO_3 flux observed in both guinea-pig and rabbit (Petersen et al., 1987; Stewart et al., 1988). As a possible contribution to transport inhibition, a reduction of Cl/HCO_3 exchange must be considered (Reuss, 1987).

In guinea-pig, cAMP-induced anion permeabilities re-direct transport of Cl and HCO_3 (Fig. 2). Part of HCO_3 secretion stays dependent on luminal Cl and sensitive to SITS and is therefore thought to proceed by way of apical Cl/HCO_3 exchange. Cl, in turn, recycles into the lumen by electrodiffusion. The remaining HCO_3 exits via the HCO_3 permeability. In both cases, conductive anion secretion ensues. Net HCO_3 flux accounts for the short-circuit current induced by PGE_1 (Stewart et al., 1988). Under open-circuit conditions, the lumen-negative voltage (~8 mV) leads to paracellular

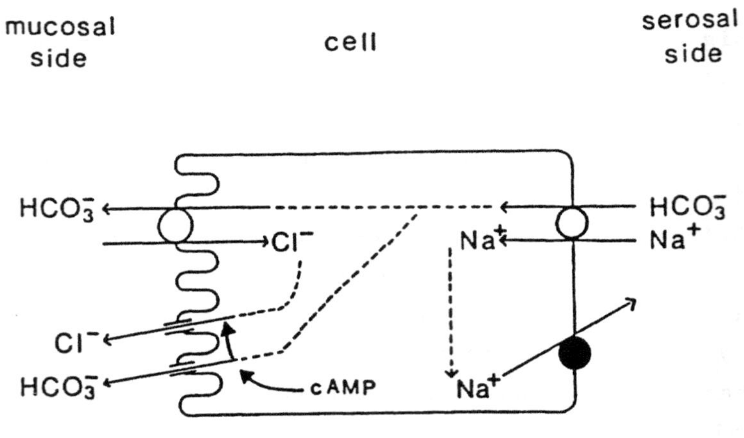

Fig.2: Electrogenic HCO_3 secretion in gallbladder epithelium. Filled circle: Na,K-ATPase. See also Heintze et al., 1984

flow of Na and K. In addition, transcellular K secretion can be predicted from the PGE_1-induced membrane depolarization. Transport of HCO_3, Na, and K accounts for the fluid secretion elicited by PGE_1 or cAMP (Heintze et al., 1979; Petersen et al., 1982).

Short-chain fatty acids can substitute for HCO_3 in most cases. Luminal butyrate stimulates NaCl absorption, probably by replacing the HCO_3/CO_2 shuttle with the system butyrate/butyric acid (Petersen et al., 1981; cf. Fig. 1). Serosal butyrate seems to enter the cell by Na-dependent mechanisms and leave it in exchange for luminal Cl. In the presence of cAMP, co-existence of apical Cl/butyrate exchange and Cl permeability gives rise to electrogenic butyrate secretion (Petersen and Macherey, 1987).

A reversal of fluid absorption into secretion by cAMP-related agents occurs also in vivo. In the cat gallbladder, vasoactive intestinal peptide converted net absorption of fluid, Na, K, and HCO_3 into secretion (Jansson et al., 1978), consistent with electrogenic HCO_3 secretion. In experimental cholecystitis, production of prostaglandins is elevated; the ensuing fluid secretion may contribute to biliary pain (Thornell, 1982).

Transport in the Bile Duct/Ductules

Information on bile ductal/ductular transport is less direct since it mostly derives from whole-liver studies. The picture is complicated by species differences and a large variety of experimental designs. Therefore, this overview attempts to combine some of the loose ends into a working hypothesis rather than to summarize current knowledge.

Like gallbladder, bile duct epithelium is easily permeable to anorganic ions (see Boyer, 1986) and, therefore, would be classified as leaky. Circumstantial evidence suggests that it is able to reabsorb water (see Erlinger, 1987). By contrast, isolated common bile ducts of dog (Nahrwold and Shariatzedeh, 1971) and rabbit (Chenderovitch, 1972) spontaneously secrete a fluid containing Na, K, HCO_3, and Cl. In the latter species, omission of one of the electrolytes reduced fluid and, to different extents, ion secretion. A lumen-negative potential difference of 5.6 mV was recorded that was HCO_3-dependent and decreased upon removal of either Na or Cl. Since reported ion fluxes do not balance and flux determinations were incomplete, straightforward modelling is impossible. However, it is interesting to note that these data can be explained in terms of (cAMP-regulated) electrogenic secretion of HCO_3 and/or Cl by mechanisms similar to those illustrated in Fig. 2.

The choleretic effects of secretin (which can be localized to the bile duct system; see Erlinger, 1987) are likely to be mediated by cyclic AMP (Levine and Hall, 1976). In dog isolated liver, secretin-induced choleresis goes along with a rise in bile HCO_3 concentration and lumen-negative voltage. The latter was taken to indicate active anion secretion (London et al., 1967). In the guinea-pig, liver perfusion with a nominally HCO_3-free solution reduced bile flow (Fig. 3) and Na excretion (Rutishauser, 1985). Conversely, Li-for-Na substitution inhibited bile flow (Fig. 3), associated with a fall of HCO_3 excretion. These effects were referred to alterations of ductal/ductular transport (Rutishauser, 1985). In the cat, secretin-like effects of prostaglandins E_1 and E_2 (Krarup et al., 1976) point to cAMP-mediated effects even though the authors disfavored this possibility.

For the model of Fig. 2 to be correct, nominal removal of HCO_3 should abolish fluid secretion in the isolated bile duct. Actually, inhibition was only partial (Chenderovitch, 1972). Likewise, acetazolamide lowered HCO_3 output to a greater extent than that of fluid in man (Waitman et al., 1969). This could be understood if the blood side entry mechanism, in the absence of HCO_3, also accepted Cl, as postulated for duodenal mucosa (Flemström et al., 1982) and guinea-pig gallbladder (Stewart et al., 1988). In support of this view, Na removal lowers bile flow also in the nominal absence of HCO_3/CO_2 (Fig. 3), concomitant with a decrease in

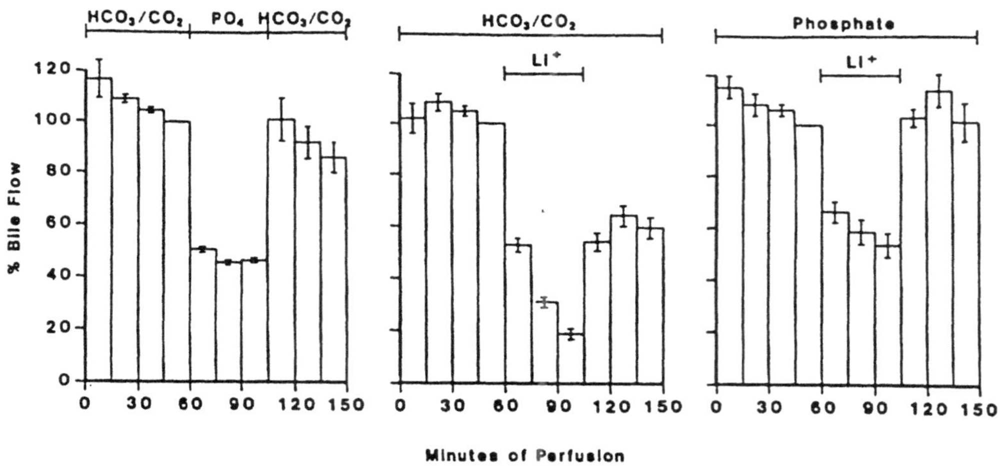

Fig.3: Effects of ion substitutions (phosphate buffer plus Cl for HCO_3/CO_2 and/or Li for Na) on guinea-pig bile flow. From Rutishaüser (1985), with permission.

Cl concentration that is lacking in the presence of HCO_3/CO_2 (Rutishauser, 1985). Secretion of Cl is also evident from the observation that secretin raises Cl concentration in dog bile despite a considerable increase in the lumen-negative potential difference (London et al., 1967).

Thus, bile duct and gallbladder secretion may turn out to follow similar pathways. Absorption and secretion might be conditioned by environmental demands or even co-existing at different sites of the bile collecting system. Clearly, more direct approaches such as isolated duct perfusion are called for before unequivocal definitions of bile duct transport mechanisms are possible.

Acknowledgements: Work from the author's laboratory was supported by grants from the Deutsche Forschungsgemeinschaft (SFB 160 "Eigenschaften biologischer Membranen", Projekt C2, and Pe 280/2-1).

References

Boyer JL (1986) Mechanisms of bile secretion and hepatic transport. In: Andreoli TE, Hoffman JF, Fanestil DD, Schultz SG (eds) Physiology of membrane disorders. Plenum Medical Book Company, New York London, p 609

Chenderovitch J (1972) Secretory function of the rabbit common bile duct. Am J Physiol 223:695-706

Cremaschi D, Meyer G (1982) Amiloride-sensitive sodium channels in rabbit and guinea-pig gallbladder. J Physiol (Lond) 326: 21-34

Cremaschi D, Henin S, Meyer G (1979) Stimulation by HCO_3^- of Na^+ transport in rabbit gallbladder. J Membrane Biol 47:145-170

Cremaschi D, Meyer G, Bermano S, Marcati M (1983a) Different sodium chloride cotransport systems in the apical membrane of rabbit gallbladder epithelial cells. J Membrane Biol 73:227-235

Cremaschi D, Meyer G, Rossetti C (1983b) Bicarbonate effects, electromotive forces and potassium effluxes in rabbit and guinea-pig gallbladder. J Physiol (Lond) 335:51-64

Diamond JM (1968) Transport mechanisms in the gallbladder. In: Heidel W, Code CF (eds) Handbook of Physiology: Section 6, Alimentary Canal, American Physiological Society, Washington DC, Vol V, p 2451

Duffey ME, Turnheim K, Frizzell RA, Schultz SG (1978) Intracellular chloride activities in rabbit gallbladder: Direct evidence for the role of the sodium gradient in energizing "uphill" chloride transport. J Membrane Biol 42:229-245

Erlinger S (1987) Physiology of bile secretion and enterohepatic circulation. In: Johnson LR (ed) Physiology of the gastrointestinal tract, 2nd edition. Raven Press, New York, p 1557

Flemström G, Heylings JR, Garner A (1982) Gastric and duodenal HCO_3^- transport in vitro: Effects of hormones and local transmitters. Am J Physiol 242:G100-G110

Frizzell RA, Dugas MC, Schultz SG (1975) Sodium chloride transport by rabbit gallbladder. Direct evidence for a coupled NaCl influx process. J Gen Physiol 65:830-834

Frömter E (1972) The route of passive ion movement through the epithelium of Necturus gallbladder. J Membrane Biol 8:259-301

Gunther-Smith PJ (1986) Serosal membrane conductances and the effect of reducing bath Na in guinea pig gallbladder. J Gen Physiol 88:26a

Gunther-Smith PJ, Schultz SG (1982) Potassium transport and intracellular potassium activities in rabbit gallbladder. J Membrane Biol 65:41-47

Heintze K, Petersen KU, Olles P, Saverymuttu SH, Wood JR (1979) Effects of bicarbonate on fluid and electrolyte transport by the guinea-pig gallbladder: A bicarbonate-chloride exchange. J Membrane Biol 45:43-59

Heintze K, Petersen KU, Wood JR (1981) Effects of bicarbonate on fluid and electrolyte transport by guinea-pig and rabbit gallbladder: Stimulation of absorption. J Membrane Biol 62: 175-181

Heintze K, Stewart CP, Winterhager JM, Petersen KU (1984) Sodium- and chloride-dependent bicarbonate secretion by guinea-pig gallbladder. In: Skadhauge E, Heintze K (eds) Intestinal absorption and secretion. MTP Press, Lancaster, p 353

Henin S, Cremaschi D (1975) Transcellular ion route in rabbit gallbladder. Electrical properties of the epithelial cells. Pflügers Arch 355:125-139

Jansson, R, Steen G, Svanvik J (1978) Effects of intravenous vasoactive intestinal peptide (VIP) on gallbladder function in the cat. Gastroenterology 75:47-50

Krarup N, Larsen JA, Munck A (1976) Secretin-like choleretic effect of prostaglandins E_1 and E_2 in cats. J Physiol (Lond) 254:813-820

Levine RA, Hall RC (1976) Cyclic AMP in secretin choleresis. Evidence for a regulatory role in man and baboons but not in dogs. Gastroenterology 70:537-544

London CD, Diamond JM, Brooks FP (1968) Electrical potential differences in the biliary tree. Biochim Biophys Acta 150: 509-517

Moran WM, Hudson RL, Schultz SG (1986) Transcellular sodium transport and intracellular sodium activities in rabbit gallbladder. Am J Physiol 251:G155-G159

Nahrwold DL, Shariatzedeh AN (1971) Role of the common bile duct in formation of bile and in gastrin-induced choleresis. Surgery 70:147-153

Petersen KU, Macherey HJ (1987) Electrogenic butyrate secretion by guinea-pig gallbladder epithelium: Mechanism and inhibition by blockers of anion exchange and Cl channels. Naunyn-Schmiedeberg's Arch Pharmacol 335:R45

Petersen KU, Reuss L (1983) Cyclic AMP-induced chloride permeability in the apical membrane of Necturus gallbladder. J Gen Physiol 81:705-729

Petersen KU, Wehner F (1987) Predictable transcellular K secretion in guinea-pig gallbladder epithelium. Z Gastroenterol 25:389

Petersen KU, Wood JR, Schulze G, Heintze K (1981) Stimulation of gallbladder fluid and electrolyte absorption by butyrate. J Membrane Biol 62: 183-193

Petersen KU, Osswald H, Heintze K (1982) Asymmetric release of cyclic AMP from guinea-pig gallbladder. Naunyn-Schmiedeberg's Arch Pharmacol 318:358-362

Petersen KU, Wehner F, Winterhager JM (1985) Na/H exchange at the apical membrane of guinea-pig gallbladder epithelium: Properties and inhibition by cyclic AMP. Pflügers Arch 405: S115-S120

Petersen KU, Wehner F, Winterhager JM (1987) Effects of cyclic AMP on transcellular HCO_3 transport in rabbit gallbladder epithelium. Pflügers Arch 408:R34

Reuss L (1987) Cyclic AMP inhibits Cl^-/HCO_3^- exchange at the apical membrane of Necturus gallbladder epithelium. J Gen Physiol 90:173-196

Reuss L (1988) Salt and water transport by gallbladder epithelium. In: Schultz SG (sect ed), Frizzell RA, Field M (vol eds) Handbook of Physiology: Intestinal absorption and secretion, Vol IV. American Physiological Society, 1988 (in press)

Reuss L, Petersen KU (1985) Cyclic AMP inhibits Na^+/H^+ exchange at the apical membrane of Necturus gallbladder epithelium. J Gen Physiol 85:409-429

Rose RC (1987) Absorptive functions of the gallblader. In: Johnson LR (ed) Physiology of the gastrointestinal tract, Second edition. Raven Press, New York, p 1455

Rutishauser SCB (1985) The sodium and bicarbonate dependence of bile secretion in the guinea-pig. Comp Biochem Physiol 82A: 317-322

Stewart CP, Winterhager JM, Heintze K, Petersen KU (1988) Electrogenic bicarbonate secretion by guinea-pig gallbladder epithelium: Apical membrane exit. Am J Physiol (submitted)

Thornell E (1982) Mechanisms in the development of acute cholecystitis and biliary pain. Scand J Gastroenterol 17 (Suppl 76)

van Os CH, Slegers JFG (1975) The electrical potential profile of gallbladder epithelium. J Membrane Biol 24:341-363

Waitman AM, Dyck WP, Jawitz HD (1969) Effect of secretin and acetazolamide on the volume and electrolyte composition of hepatic bile in man. Gastroenterology 56:286-294

Wehner F, Petersen KU (1984) Apparent ion permeabilities in the apical membrane of guinea-pig gallbladder epithelium. Gastroenterol Clin Biol 8:874

Winterhager JM, Stewart CP, Heintze K, Petersen KU (1986) Electroneutral secretion of bicarbonate by guinea-pig gallbladder epithelium. Am J Physiol 250: C617-C628

Wood JR, Svanvik J (1983) Gall-bladder water and electrolyte transport and its regulation. Gut 24:579-593

Transcellular Butyrate Transport in Gallbladder Epithelium

K.-U. Petersen, F. Wehner, J.M. Winterhager
Institut Pharmakologie
RWTH Aachen
Schneebergweg
5100 Aachen, FRG

In guinea-pig gallbladder epithelium, HCO_3 plays a dual role: first, it stimulates apical membrane uptake of Na and Cl by supplying (via the HCO_3/CO_2 shuttle) counter ions to the double exchange Na/H and Cl/HCO_3; second, it can be secreted into the lumen by electroneutral and (in the presence of cAMP) rheogenic mechanisms. There is evidence to suggest that butyrate can substitute for HCO_3 in most of these transports.

Like HCO_3, luminal butyrate stimulates NaCl absorption; its influx into the epithelium, measured as short-term uptake of ^{14}C-butyrate, is inhibited by prostaglandin E_1 (PGE_1) and ouabain. This effect is consistent with cAMP-sensitive, Na,K-ATPase-dependent Na/H exchange leading to cell entry of undissociated butyric acid. Influx of fatty acid is directly evident from cellular acidification in Necturus gallbladder exposed to luminal propionate. At steady-state, rates of butyrate uptake exceed its absorptive flux (J_{ms}^{Bu}) by 4 $\mu mol/cm^2$ h, the value by which butyrate enhances absorption of Na and Cl. This difference must be accounted for by backflux into the lumen, for instance by Cl/butyrate exchange. Net absorption of butyrate is only 0.3 $\mu mol/cm^2$ h. As deduced from the ability to stimulate Na and Cl absorption, the anion exchanger

Hepatic Transport in Organic Substances
Ed. by E. Petzinger, R. K.-H. Kinne, H. Sies

seems to accept also branched-chain fatty acids. This ability is lowered by methylation or dimethylation, particularly at sites more distal of the carboxylic group.

The following experiments support the existence of Cl/butyrate exchange at the apical membrane: Luminal Cl raises the secretory flux of butyrate (J_{sm}^{Bu}) from ~0.5 to ~1.3 μmol $cm^{-2} \cdot h^{-1}$; likewise, serosal butyrate increases the corresponding flux of Cl, J_{ms}^{Cl}, from ~1.6 to ~3.2 μmol $cm^{-2} \cdot h^{-1}$. Under conditions optimized for butyrate secretion (butyrate in the serosal, Cl in the mucosal bath), PGE_1 induces a short-circuit current of ~0.8 μmol $cm^{-2} \cdot h^{-1}$ without changing J_{sm}^{Bu} (~1.3 μmol $cm^{-2} \cdot h^{-1}$). This effect is prevented by removal of Cl or butyrate and dependent on serosal Na; it is abolished by serosal ouabain, luminal SITS (10^{-3} M), and the Cl channel blocker, 5-nitro-2-(phenylpropylamino-)benzoic acid (3 x 10^{-7} M, luminal side). Microelectrode experiments yield no evidence in favor of a cAMP-induced butyrate permeability at the apical membrane. Thus, electrogenic butyrate secretion is best explained by Na-dependent entry at the serosal membrane followed by apical Cl/butyrate exchange plus conductive Cl recycling into the lumen. In the rabbit, luminal butyrate is handled similarly as in the guinea-pig (involvement in the double ion exchange, partial absorption).

Accessibility of Primary Bile in Liver Cell Culture

F. Wehner, E. Petzinger, H. Acker, J. Hentschel, R.K.H. Kinne
Max-Planck-Institut für Systemphysiologie
Rheinlanddamm 201
4600 Dortmund 1, FRG

For the study of the biophysical and biochemical mechanisms involved in bile secretion, it is a prerequisite to know the composition of canalicular bile. Here, we describe a primary cell culture system, which allows the sampling of primary bile as well as the electrophysiological analysis of bile secretion.

When placed on gas permeable membranes (Petriperm[R]), isolated rat hepatocytes become flat, epithelial-like cells and form an almost completely confluent monolayer. Monolayer formation starts after 4 h and is complete within 24 h.
Within Day 1 in culture, a network of canaliculi is formed around the cells, which are bordered by tight junctions. This canalicular system represents a separate compartment, into which primary bile is secreted because:

(1) When cell cultures were exposed to the non-fluorescent 6-carboxyfluorescein diacetate ester, strong fluorescence was observed inside the canaliculi, indicating that carboxyfluorescein acetate is transported into the liver cell, deesterified, and secreted into the canalicular space as carboxyfluorescein.

Hepatic Transport in Organic Substances
Ed. by E. Petzinger, R. K.-H. Kinne, H. Sies

(2) It is possible to collect primary bile selectively from the dilated, bulb-like canaliculi by use of an empty micropipette (0.1 - 0.2 μm tip diameter).

(3) Measurements of electrical potentials revealed that there exist differences between the canalicular space (V_{can}) and the intracellular space (V_{cell}) or the incubation medium, respectively. The absolute magnitude of the electrical potentials, thereby depends critically on the composition of the solutions used for the filling of the microelectrodes:

	3 M KCl	0.5 M KCl	1 M Mg-acetate
V_{cell}	-44.3 ± 6.2 mV n = 15	-34.8 ± 1.3 mV n = 9	-32.6 ± 3.4 mV n = 14
V_{can}	-30.8 ± 7.4 mV n = 11	-16.9 ± 1.9 mV n = 10	-24.2 ± 4.5 mV n = 14

We conclude that primary cultures of hepatocyte monolayers grown on gas permeable membranes are a suitable model for the analysis of primary bile composition and of electrogenic mechanisms involved in canalicular bile formation. Among the electrode filling solutions used 0.5 M KCl seems to be the less interfering one.

References

Petzinger E, Föllmann W, Acker H, Hentschel J, Zierold K, Kinne RKH (1988) Primary liver cell cultures grown on gas permeable membrane as source for the collection of primary bile. In vitro (in press)

Polarity of the Hepatocyte Surface Membrane: Influence of Lipids on Protein Function

F.R. Simon
Department of Medicine
Division of Gastroenterology
University of Colorado School of Medicine and
Veterans Administration Hospital
4200 East Ninth Avenue
Denver, Colorado 80262, USA

Introduction

The surface membrane of epithelial cells, including the liver, is organized into two major domains, apical (or bile canalicular) and the basolateral (or sinusoidal) (Simons and Fuller, 1985). These surface membrane domains are responsible for vectorial transport, secretion and uptake (or absorption), and are accomplished by localizing a distinct set of components to one of the two membrane surfaces. In addition to having distinct proteins such as enzymes, receptors, and carriers, localized to one pole, the lipid composition and fluidity of the apical and basolateral domains are different (van Meer and Simons, 1988; Schachter, 1984).

A schematic model for the organization of proteins in the liver surface membrane is shown in Fig. 1. Numerous studies have localized distinct bile acid transport systems and enzyme activities to one domain or the other (Evans, 1980; Meier et al., 1984). Furthermore, several groups have established that membrane lipid fluidity is higher in the sinusoidal domain compared to the bile canaliculus (Schachter, 1984). However, in contrast to kidney proximal tubule and small intestinal

Hepatic Transport in Organic Substances
Ed. by E. Petzinger, R. K.-H. Kinne, H. Sies

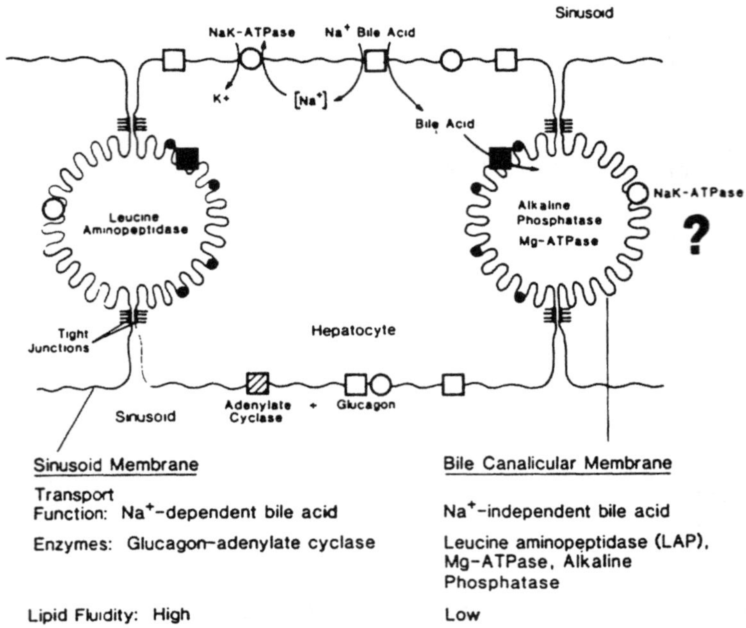

Figure 1: Hepatocyte polarity: A schematic model showing the distribution of proteins and lipids in the hepatocyte. NaK-ATPase localization to the bile canaliculus is controversial

epithelial cells where the location of NaK-ATPase is well established to the basolateral domain, its location in the hepatocyte is controversial (Leffert et al., 1985). The aims of the present studies were three-fold: (1) to present evidence supporting the surface polarization of lipids, (2) to suggest that lipid fluidity may be involved in regulating the apparent location of NaK-ATPase, and (3) to demonstrate the role of abnormalities in lipid composition and fluidity in the pathogenesis of cholestasis.

Methods

Bile canalicular (BCM) and sinusoidal membrane (SM) fractions were isolated by Mg^{2+} precipitation and sucrose density centrifugation, respectively (Rossario et al., 1988). Relative enrichment of marked enzymes is shown in Fig. 2. Importantly, neither NaK-ATPase nor glucagon-stimulated adenylate cyclase activity were present in BCM fractions. However, minimal enrichment of BCM enzymes was found in the SM fraction indicating small amount of contamination of SM with canalicular components.

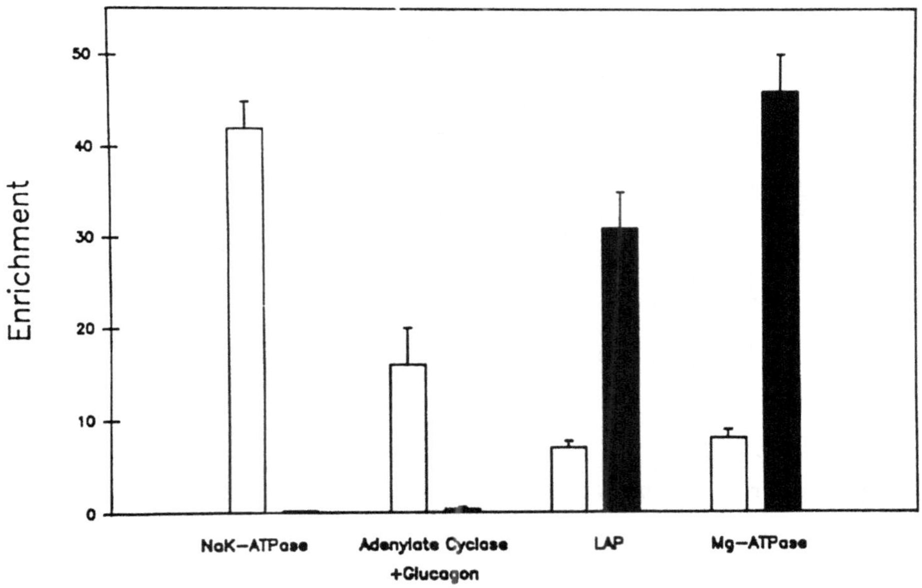

Figure 2: Relative enrichment of marker enzymes in liver plasma membrane subfractions. □ = SM, ■ = BCM

Lipids were measured by gas chromatography and fluidity determined by fluorescence polarization, as previously described (Rossario et al., 1988). Enzyme activities were measured by standard assays (Molitoris and Simon, 1985).

Lipid Composition and Fluidity of Liver Surface Membrane Subfractions

A comparison of cholesterol, phospholipid and fatty acid composition is presented in Fig. 3. Consistent with studies from renal proximal tubule (Molitoris and Simon, 1985) and intestinal epithelial cells (Brasitus and Schachter, 1980) the cholesterol to phospholipid and sphingomyelin to phosphatidylcholine molar ratios are significantly greater in the apical than the basolateral membrane fractions. However, surprisingly the percentage of fatty acid saturation is similar in the two subfractions.

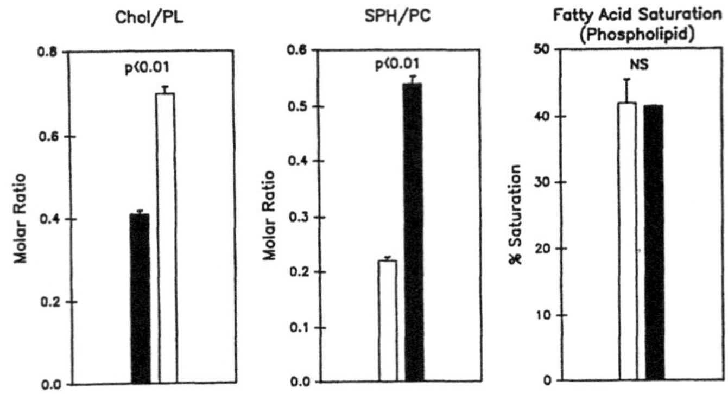

Figure 3: Comparison of the lipid composition in liver surface membrane subfractions. □ = SM, ■ = BCM

The effect of these differences in lipid composition on membrane lipid fluidity was examined using dynamic fluorescence polarization. Membrane fluidity consists of two major types of motion (Schachter, 1984; Harris and Simon, 1987). One component is order parameter, which is the structural component and relates to the packing of fatty acyl chains; the second is microviscosity or the rotational rate and relates to the dynamic/kinetic component. These components may change independently and, thus, selectively alter function.

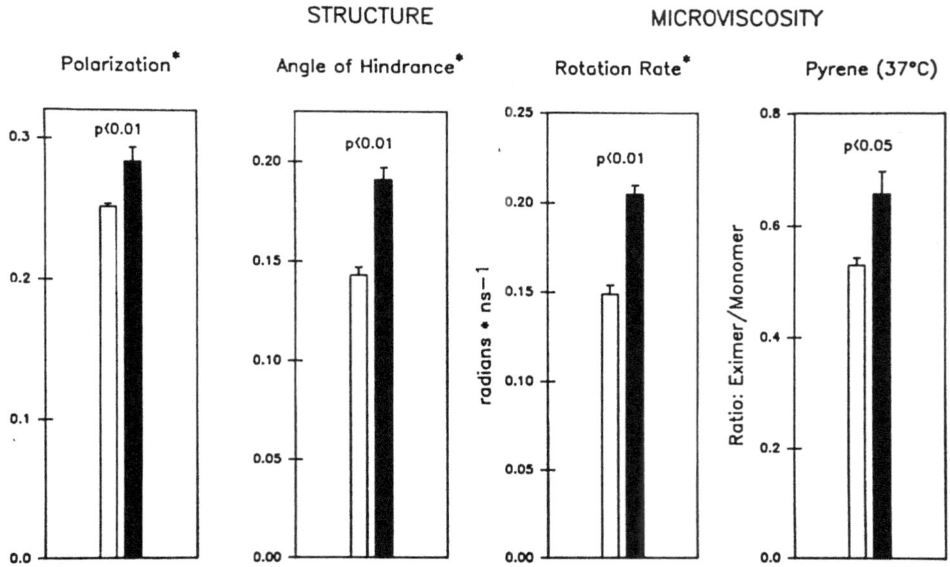

Figure 4: Comparison of fluidity measurements in liver plasma membrane subfractions. Bulk lipid fluidity was measured with the probe diphenylhexatriene at 37°C and is shown as polarization. Components of fluidity were determined by dynamic depolarization. □ = SM, ■ = BC

Fluidity, as previously shown (Storch et al., 1984; Whetton et al., 1983; Lowe and Coleman, 1982), was lower in the BCM than in the SM fractions. As expected, BCM fractions showed significantly greater order than the SM. However, in contrast to renal proximal tubule subfractions (Molitoris and Holien, 1988), the dynamic component of fluidity was increased (faster rotation rate). The latter component relates more to lateral diffusion. We, therefore, quantitated lateral mobility in liver subfractions using the pyrene probe. The ratio of pyrene excimer (490) to monomer (395) is a measure of lateral mobility. Since formation of the intramolecular excimer is dependent upon lateral movement of its two components, the high ratio measured in BCM fractions indicates that this fraction has increased lateral mobility of the probe, compared to SM fractions. Although for the most part hepatocytes' lipid composition and fluidity in the apical and basolateral domains are similar to that previously reported for other transporting epithelia, important differences in fatty acid composition and microviscosity are present. These differences in the components of membrane lipid composition and fluidity may provide important restraints on conformational changes and/or coupling of plasma membrane carriers and receptors.

Regulation of NaK-ATPase Activity

Surface membrane proteins of most epithelial are generally localized either to the apical or basolateral poles (Simons and Fuller, 1985). In contrast, although NaK-ATPase is generally believed to be localized at the basolateral surface of kidney cells (Kyte, 1976) and intestinal absorptive cells (Sterling, 1972), it is also present on the apical surface of the choroid plexus (Ernst et al., 1986). In the liver, using histochemical, biochemical, and cell fractionation techniques, NaK-ATPase activity is generally found on the basolateral surface, while specific antibody immunocytochemistry localized NaK-pump sites to the apical surface as well as the sinusoidal surface (Blitzer and Boyer, 1978; Takemur et al., 1984).

Since it is well known that the activity of many membrane enzymes is regulated by their lipid environment (Sanderman, 1978; Sinensky et al., 1979), we have examined whether decreasing BCM lipid fluidity might unmask latent NaK-ATPase catalytic activity. In BCM fractions which are apparently devoid of NaK-ATPase activity (Fig. 2), addition of A2C, benzyl alcohol, and Triton WR-1339, agents known to fluidize membranes; latent NaK-ATPase activity was now measured (Sutherland et al., 1987). In contrast, glucagon stimulated adenylate cyclase activity, another SM enzyme was not detected in BCM fractions. Similarly, BCM enzyme activities, Mg^{2+}-ATPase, leucine aminopeptidase, and alkaline phosphatase were unaltered. Thus, NaK-ATPase is apparently uniquely located on both poles of hepatocytes, but its activity is inhibited because of the high lipid order of this organelle. This dual surface membrane location of NaK-ATPase is unique to hepatocytes, since neither renal nor jejunal brush border membrane fractions contained significant activity after fluidization with A2C. We next explored whether differences in structural properties of the NaK-ATPase molecules accounted for their dual location in the liver, kidney, and jejunum. Liver plasma membrane subfractions were also isolated, the polypeptides separated by SDS-PAGE and immunoblotted with different polyclonal antibodies to NaK-ATPase. The catalytic subunit was demonstrated in each tissue, and in particular both the BCM as well as the SM had similar bands at 100 kDa. In contrast, although ß-subunits were recognized in the kidney and jejunum, none were seen in the liver. Failure to identify a ß-subunit in hepatocytes has previously been reported (Hubert et al., 1986; Mercer et al., 1987), suggesting the possibility that a functional role for the ß-subunit may be to direct or retain the α-subunit in the basolateral membrane.

Cholestasis

Previous studies using various models of cholestasis have focused attention on the liver surface membrane, and, in particular, the canalicular domain as the primary site of abnormalities in bile secretory failure (Simon and Arias, 1983). However, these studies used a mixed plasma membrane fraction. We have, therefore, begun to re-examine the relative roles played by sinusoidal and canalicular domains in the pathogenesis of cholestasis.

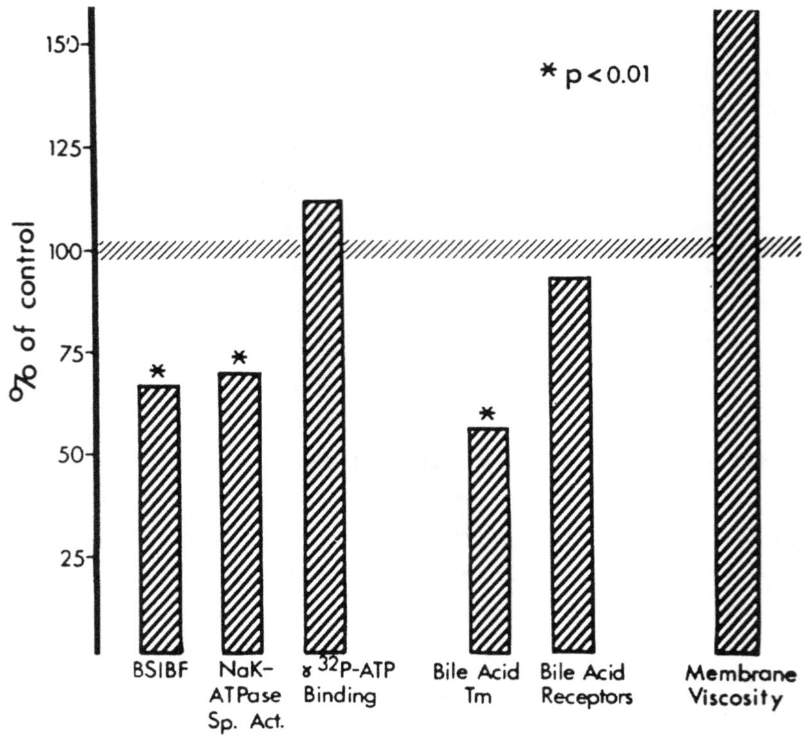

Figure 5: Effects of ethinyl estradiol administration on biliary secretory function, the number of carriers, and plasma membrane viscosity.

A summary of our previous studies (Davis et al., 1978; Simon et al., 1980) is depicted in Fig. 5. Ethinyl estradiol (EE) (5 mg/kg BW) was administered for 5 days. Bile flow and organic anion excretion was reduced, but neither NaK-ATPase units in the number of putative bile acid carriers were altered. Since membrane fluidity was decreased (i.e. an increase in the rotational correlation time) and both transport function and fluidity were corrected with Triton WR 1339, we proposed that the functional changes were secondary to alterations in bile canalicular fluidity (Davis et al., 1978). However, when SM, BCM, and microsomal membrane fractions were isolated and fluidity measured by DPH fluorescence polarization, we demonstrated decreased fluidity only in the SM fraction. Decreased SM fluidity was not associated with changes in neutral lipid content, but rather in alterations in fatty acid composition (Rossario et al., 1988).

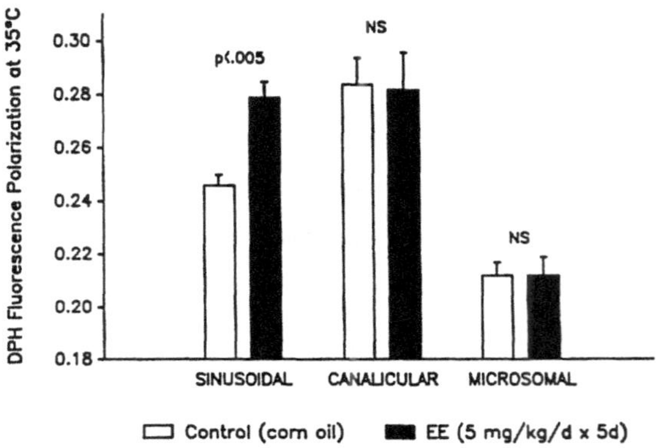

Figure 6: Lipid membrane fluidity of liver fractions. Steady state DPH fluorescence polarization of SM, BCM, and microsomal fractions after EE treatment (5 mg/kg x 5 d).

The relation of these changes to the pathogenesis of cholestasis was explored by examining the time course for changes in bile flow, NaK-ATPase, and order parameter. The results are shown in Fig. 7 and indicate that as early as one day after starting EE administration, order parameter is significantly increased. This change in fluidity is coordinately changed with a decrease in bile flow and NaK-ATPase activity. Furthermore, bile duct ligation, in contrast, results in increased BCM fluidity and no change in SM fluidity (Nibel et al., 1988). Taken together, these results strongly suggest that the change in SM fluidity is the primary abnormality in EE-induced cholestasis.

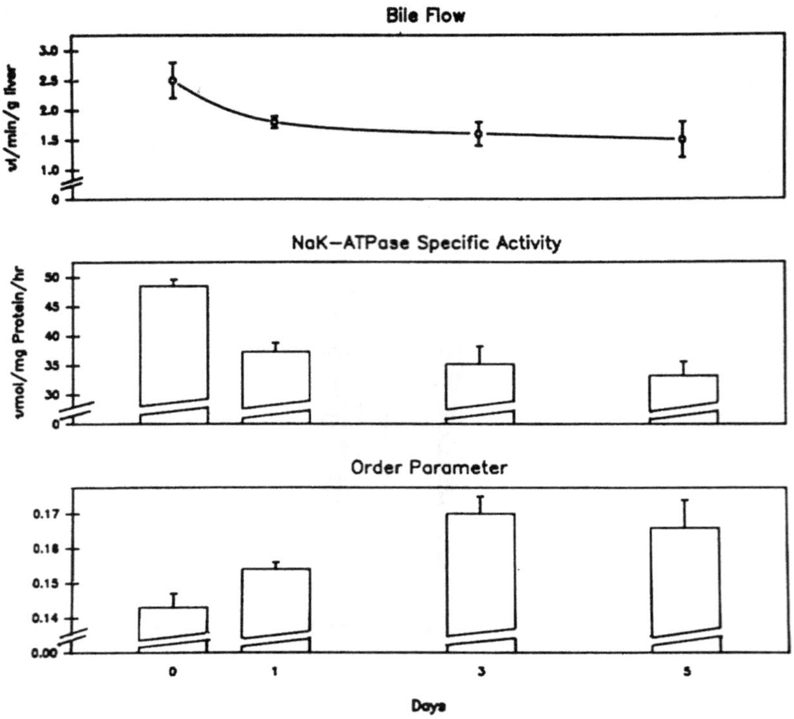

Figure 7: Change in bile flow, NaK-ATPase specific activity, and order parameter following ethinyl estradiol administration (5 mg/kg BW[1]).

Additional evidence linking altered sinusoidal membrane fluidity and cholestasis is the dramatic decrease in taurocholate transport and the increased activation energy required for its uptake into isolated hepatocytes from EE-treated rats (Table I) (Berr et al., 1984). Thus, intrahepatic cholestasis following EE administration results from sinusoidal membrane abnormalities in lipid composition and structure. This alteration impairs bile acid and cation transport, leading to decreased biliary secretion of biliary water and organic anions. However, at present it is unclear whether other forms of intrahepatic cholestasis such as taurolithocholate and chlorpromazine are also associated with selective defects at the sinusoidal pole of the hepatocyte.

TABLE I: EFFECT OF ETHINYL ESTRADIOL TREATMENT ON BILE ACID TRANSPORT AND ACTIVATION ENERGY

	Taurocholate initial uptake (V_{max}) (n mol/min/10^6cells)	Activation energy (kcal/mol/10^6cells)
Control	3.7 + 0.2	17 + 4
Ethinyl estradiol	1.0 + 0.1	23 + 4
p value	0.001	0.05

Summary

Hepatocytes, similar to other transporting epithelial cells, have protein and lipid components which are localized at either the apical or basolateral surfaces. This asymmetric location of components is important in hepatocyte vectorial functions such as bile secretion. Although little is understood about the establishment and maintenance of this surface membrane asymmetry, it is now clear that lipid plays an important role in regulating the apparent functional distribution of selective proteins. In addition, disorders of bile secretion secondary to estrogen administration are associated with a selective alteration of sinusoidal membrane fluidity. Thus, at least one form of bile secretory failure results from altered membrane lipid composition and structure. Further studies utilizing biochemical, biophysical, and physiologic approaches should greatly advance our understanding of the special role lipids play in regulating hepatic transport of organic solutes.

Acknowledgements

Studies were supported by VAH Merit Funds and NIH Grants AM-15851 and AM-34914. I am grateful to the technical assistance of E. Sutherland and Larry Zocarro, and the fruitful discussions with H. Leffert, B. Molitoris, R. Davis, and A. Harris.

References

Berr F, Simon FR, Reichen J (1984) Ethynylestradiol impairs bile salt uptake and NaK pump function of rat hepatocytes. Am J Physiol 247:G437-G443

Blitzer BL, Boyer JL (1978) Cytochemical localization of Na,K^+-ATPase in the rat hepatocyte. J Clin Invest 62:1104-1108

Brasitus TA, Schachter D (1980) Lipid dynamics and lipid protein interactions in rat enterocyte basolateral and microvillus membranes. Biochemistry 19:2763-2769

Davis RA, Kern F, Showalter R, Sutherland E, Sinensky M, Simon F (1978) Alterations of hepatic Na^+,K^+-ATPase and bile flow by estrogen: Effects on liver surface membrane lipid structure and function. Proc Natl Acad Sci USA 75:4130-4134

Ernst SA, Palacios JR, Siegel GJ (1986) Immunocytochemical localization of Na^+,K^+-ATPase catalytic polypeptide in mouse choroid plexus. J Histochem Cytochem 34:189-195

Evans WH (1980) A biochemical dissection of the functional polarity of the plasma membrane of the hepatocyte. Biochim Biophys Acta 604:27-64

Harris RA, Simon FR (1987) What should hepatologists know about membrane fluidity? Hepatology 7:177-180

Hubert JJ, Schenk DB, Skelly H, Leffert HL (1986) Rat hepatic (Na^+,K^+)-ATPase: α-subunit isolation by immunoaffinity chromatography and structural analysis by peptide mapping. Biochemistry 25:4156-4163

Kyte J (1976) Immunoferritin determination of the distribution of (Na^+,K^+)-ATPase over the plasma membranes of renal convoluted tubules. II. Proximal segment. J Cell Biol 68:304-318

Leffert HL, Schenk DB, Hubert JJ, Skelly H, Schumacher M, Ariyasu R, Ellisman M, Koch KS, Keller GA (1985) Hepatic (Na^+,K^+)-ATPase: A current view of its structure, function and localization in rat liver as revealed by studies with monoclonal antibodies. Hepatology 5:501-507

Lowe PJ, Coleman R (1982) Fluorescence anisotropy from diphenylhexatriene in rat liver plasma membranes. Biochim Biophys Acta 689:403-409

Meier PJ, Sztul S, Reuben A, Boyer JL (1984) Structural and functional polarity of canalicular and basolateral plasma membrane vesicles isolated in high yield from rat liver. J Cell Biol 98:991-1000

Mercer RW, Schneider JW, Savitz A, Emanuel J, Benz EJ, Levenson R (1986) Rat brain Na,K-ATPase chain gene: Primary structure, tissue-specific expression, and amplification in ouabain-resistant HeLa C^+ cells. Mol Cell Biol 6:3884-3890

Molitoris BA, Simon FA (1985) Renal cortical brush border and basolateral membranes: Cholesterol and phospholipid composition and relative turnover. J Membr Biol. 83:207-216

Molitoris BA, Hoilien C (1987) Static and dynamic components of renal cortical brush border and basolateral membrane fluidity: Role of cholesterol. J Membr Biol 99:165-172

Nibel D, Sutherland E, Dixion B, Zaccarro L, Burke C, Simon FR (1988) Alterations in distribution of sinusoidal membrane proteins: glucagon-stimulated adenylate cyclase and Na,K-ATPase. Clin Res 36:184A

Rossario J, Sutherland E, Zaccaro L, Simon FR (1988) Ethinyl estradiol administration selectively alters liver sinusoidal membrane lipid fluidity and protein. Biochemistry, in press

Sandermann H (1978) Regulation of membrane enzymes by lipids. Biochim Biophys Acta 515:209-237

Schachter D (1984) Fluidity and function of hepatocyte plasma membranes. Hepatology 4:140-151

Simons K, Fuller SD (1985) Cell surface polarity in epithelia. Ann Rev Cell Biol I:243-288

Simon FR, Arias IM (1973) Alteration of bile canalicular enzymes in cholestasis. J Clin Invest 52:765-775

Simon FR, Gonzalez M, Sutherland E, Accatino L, Davis RA (1980) Reversal of ethinyl estradiol-induced bile secretory failure with triton WR1339. J Clin Invest 65:851-860

Sinensky M, Pinkerton F, Sutherland E, Simon FR (1979) Rate limitation of $(Na^+ + K^+)$-stimulated adenosinetriphosphatase by membrane acyl chain ordering. Proc Natl Acad Sci USA 76:4893-4897

Stirling CE (1972) Radioautographic localization of sodium pump sites in rabbit intestine. J Cell Biol 53:704-714

Storch J, Schachter D, Inoue M, Wolkoff AW (1983) Lipid fluidity of hepatocyte plasma membrane subfractions and their differential regulation by calcium. Biochim Biophys Acta 727:209-212

Sutherland E, Dixon B, Leffert HL, Skally H, Zaccaro L, Simon FR (1987) Biochemical localization of hepatic plasma membrane NaK-ATPase activity is dependent on membrane lipid fluidity. Gastroenterology 92:1783

Takemura S, Omori K, Tanaka K, Omori K, Matsuura S, Tashiro Y (1984) Quantitative immunoferritin localization of (Na^+, K^+)-ATPase on canine hepatocyte cell surface. J Cell Biol 99:1502-1510

van Meer G, Simons K (1987) Lipid polarity and sorting in epithelial cells. J Cell Biochem 36:51-58

Whetton AD, Houslay MD, Dodd NJF, Evans WH (1983) The lipid fluidity of rat liver membrane subfractions. Biochem J 214:851-854

ATP in and Around the Bile Canaliculus

Irwin M. Arias, M.D.
Departments of Physiology and Medicine
Tufts University School of Medicine
Boston, MA 02111
U.S.A.

ATP is a major high energy phosphate-containing compound which is critical in metabolism and energy-dependent transport. In recent years, ATP has been recognized as a putative neurotransmitter and there has been a surge of interest in the presence and function of extracellular ATP. ATP has traditionally been considered solely as an intracellular constituent. However, the development of sensitive methods for measurement of ATP, detection of ATP "receptors" on the plasma membrane of many cell types, and description of cellular effects of ATP have renewed interest in extracellular ATP (Cornwell et al., 1986b; Daly, 1983).

At least 30 years ago, enzyme histochemists, such as the late Alex Novikoff, demonstrated ATPase activity in the bile canaliculus of many species. More precise localization was made by combining lead phosphate precipitation, enzyme activity, and transmission electron microscopy. The specific enzyme(s) responsible for the "apparent" ATPase activity in the bile canaliculus were unknown. Biliary physiologists at this time postulated that the transport of organic anions and other substituents into bile probably involved active transport mechanisms associated with canalicular ATPase activity.

Hepatic Transport in Organic Substances
Ed. by E. Petzinger, R. K.-H. Kinne, H. Sies

The development of techniques for relative separation of the sinusoidal (basal-lateral) and canalicular domains of the plasma membrane of hepatocytes has advanced knowledge regarding the nature and function of canalicular ATPase activity.

This report will consider:
(i) Specific ATPases associated with the bile canaliculus; (ii) the function of canalicular ectoenzymes which have ATPase activity; (iii) Gp 170, a newly described canalicular membrane transport protein which has ATPase activity; (iv) effects of extracellular ATP on hepatocytes, and (v) a hypothesis regarding the possible role of canalicular ATP in liver injury, particularly cholestasis.

Studies using hepatocyte plasma membranes, which are highly enriched in sinusoidal or canalicular domains, have partially localized the major transport ATPases and are reviewed in other chapters. Consistent with the results of other investigators, we also demonstrated that the $Na^{+}K^{+}$ATPase activity is predominantly associated with the sinusoidal plasma membrane domain; lesser activity is often associated with bile canalicular preparations. Enzymatic and immunologic studies also confirm these observations. Recent studies indicate that the catalytic subunit of the sodium pump is present in the canalicular domain in liver, whereas it is absent from the brush border of proximal tubular cells of the kidney.

There is increasing evidence that Ca^{++}ATPase may also be in the sinusoidal domain. Despite the presence of a well-characterized H^{+}ATPase in clathrin-coated endocytic vesicles, which are derived from the plasma membrane, there is no direct evidence for a H^{+}ATPase in the sinusoidal domain of the plasma membrane of hepatocytes. Current methods of analysis, which are based upon transport characteristics in membrane vesicles, may not be sufficiently sensitive to detect a relatively small number of proton pump molecules. Recent purification of the clathrin-coated vesicle proton pump and development of monoclonal antibodies may shortly resolve its plasma membrane localization (Forgac and Berne, 1986; Arai et al., 1988). Numerous studies of bile canalicular membrane vesicles fail to reveal any active transport system, particularly a primary transporter ATPase of the type which forms a phosphorylated intermediate. This group includes the Na^+K^+ATPase, Ca^{++}ATPase, and H^+ATPase. Most studies with canalicular vesicles indicate that the primary driving force for transport of bile acids and their conjugates, bilirubin conjugates, glutathione, and other organic anions is probably the electrochemical gradient (Akerboom et al., 1984; Arias and Forgac, 1984; Inoue et al., 1982a; 1982b; 1983; 1984a; 1984b; 1984c). The gradient is largely generated by Na^+K^+ATPase which is predominantly localized on the sinusoidal domain of the plasma membrane of hepatocytes.

The bile canaliculus contains an abundant ATPase which requires Ca^{++} and Mg^{++} for catalytic activity. This enzyme activity is responsible for the major portion of "apparent" canalicular ATPase activity and histochemical studies reveal that it is primarily located in the canalicular membrane where it is an integral protein. Recently, $Ca^{++}Mg^{++}$ATPase was purified from rat liver and the long-standing argument as to whether or not it is a transport protein appears to have been settled. Sue Lin at Harvard demonstrated that canalicular $Ca^{++}Mg^{++}$ATPase is an ectoenzyme and not a transporter or primary pump (Lin, 1985; Lin and Fain, 1984; Lin and Russell, 1987). In right side out bile canalicular membrane vesicles, papain digestion eliminated $Ca^{++}Mg^{++}$ATPase activity without altering other transport activity, vesicular integrity, or immunoreactivity of the enzyme protein. This effect was not seen with inside out canalicular membrane vesicles indicating that the catalytic site of $Ca^{++}Mg^{++}$ATPase essentially resides within the bile canalicular space. These characteristics should, in retrospect, not seem remarkable because $Ca^{++}Mg^{++}$ATPase is an ubiquitous enzyme which is intrinsic in the plasma membrane of many cells, particularly endothelial and secretory cells. The latter include pancreatic acinar, adrenal medullary, mast and nerve cells.

These secretory cells as well as the bile canaliculus also host a 5'-nucleotidase (AMPase), which was demonstrated enzymatically many years ago. AMPase is also an ectoenzyme and is attached to the membrane by a glycosyl phosphatidyl inositol anchor (Low et al., 1986). Papain digestion of right side out vesicles also reduces or removes catalytic activity. A monoclonal antibody has been raised which has specificity for the hydrophyllic domain of the protein which extends into the bile canaliculus. Inhibition of catalytic activity was observed using this antibody with right side out vesicles. The major substrate for the enzyme is AMP which is the product of ATP hydrolysis by $Ca^{++}Mg^{++}$ATPase. Both ecto-enzymes are found in the same plasma membrane domain of all cells in which they exist. In endothelial cells the two ectoenzymes face the circulation and are believed to be important in removal of circulating extracellular ATP (Chaudry, 1982). For example, both enzymes are abundant in pulmonary endothelial cells. After injection of ATP into the pulmonary circulation, there is prompt removal within a single circulation probably reflecting the abundance and activity of $Ca^{++}Mg^{++}$ATPase and AMPase in pulmonary endothelial cells (Chaudry, 1982; Cornwell et al., 1986b). Liver endothelial cells have relatively low activities for both enzymes (Arias, I.M., unpublished data). ATP is not as rapidly removed from the portal blood during a single pass through the liver (Chaudry, 1982). The two enzymes are abundant in the plasma membrane of secretory cells, particularly if the secretory granule contains a high concentration of ATP. For example, the secretory granules of the adrenal medulla or enterochromaffin system contain up to 10 mM ATP. This "guilt-by-association" suggests that the two ectoenzymes act in synchrony to degrade high concentrations of ATP which occur after fusion of a secretory vacuole with the plasma membrane.

If we apply this analogy to the liver, it seems plausible that the two ectoenzymes in the bile canaliculus hydrolyze extra-cellular ATP. We examined this hypothesis more fully using bile canalicular membrane vesicles of both sidedness. Various purines were used as substrates for measurement of canalicular membrane transport. Adenosine transport was time, temperature, and concentration dependent and was substantially greater in canalicular membrane vesicles than in sinusoidal membrane vesicles. In canalicular membrane vesicles, transport in a 100 mmol/Na^+ gradient was significantly greater than that observed in a 100 mmol K^+ gradient. Preloading of vesicles with sodium also inhibited transport. Incubation with hypertonic raffinose reduced vesicular volume as well as adenosine radioactivity associated with vesicles. The results were analyzed to differentiate non-specific binding from transport. Electrogenic transport is suggested by the effect of various anions on the initial transport overshoot peak. Transport overshoot was greatest when thiocyanate was the anion, and was progressively reduced in nitrate, chloride, and sulphate. The more permeant anion (thiocyanate) generated more rapid initial uptake. In a sodium thiocyanate gradient, the apparent V_{max} for adenosine transport was 1600 pmol per mg per 15 sec; the apparent K_m was 40 μmol. The transport process was purine-specific; pyrimidines were not transported. Of interest is that nitrobenzylthioinosine, a photoaffinity reagent which inhibits adenosine transport in erythrocytes, labeled a 45 kDa protein but did not inhibit purine transport in canalicular membranes. The 45 kDa protein, which has been postulated to be a purine transporter, is apparently not involved in unidirectional purine transport in canalicular membrane vesicles. The transport system was not inhibited by dipyramidol and appears to be unique. Many other membranes, including the sinusoidal domain of the plasma membrane of rat hepatocytes, transfer purines bidirectionally by a low affinity system which is inhibited by dipyramidol and has carrier-mediated kinetics. The bile canalicular mechanism is unique (Li and Hochstadt, 1976a; 1976b; Miras-Portugal et al., 1986; Paterson et al., 1981; Tong et al., 1986; Wohlhueter and Plagemann, 1982).

The sequential action of the two ectoenzymes and the purine transporter was studied by incubating canalicular membrane vesicles with Mg^{++}ATP and sequentially quantitating specific

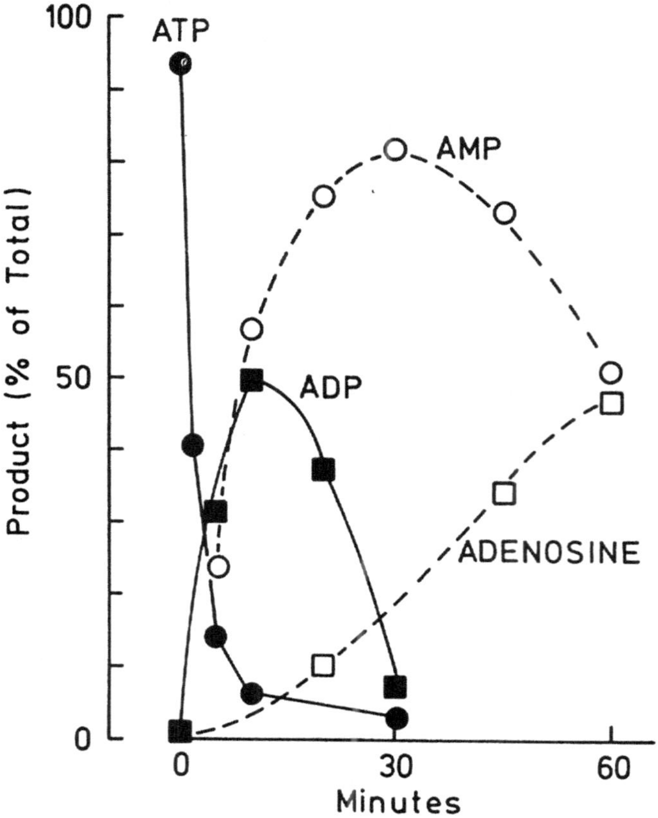

Figure 1: Sequential degradation of ATP to adenosine and its subsequent transport by rat liver canalicular membrane vesicles. Vesicles were at least 90% right side-out and were incubated with ATP for various time intervals. Aliquots were subjected to HPLC for separation and quantitation of ATP, ADP, AMP, and adenosine. Adenosine transport was also measured as a function of time in canalicular membrane vesicles.

nucleotide and nucleoside production using HPLC (Fig. 1). Incubation of right side out bile canalicular membrane vesicles with ATP resulted in rapid and sequential conversion to ADP and AMP (reflecting the action of $Ca^{++}Mg^{++}$ATPase), to adenosine (reflecting the action of AMPase), and transport of adenosine into the vesicle. The system appears to be coupled and may function in a similar manner in vivo. According to this hypothesis, the system can rapidly inactivate extracellular ATP and conserve purines. However, there is no evidence that ATP is present in the bile canaliculus. The inaccessibility of the bile canaliculus to quantitative micropuncture, inability to inhibit $Ca^{++}Mg^{++}$ATPase and AMPase specifically, and other methodologic limitations create difficult problems in testing this hypothesis.

Extracellular ATP has profound effects on tissues and cells including isolated hepatocytes (Charest et al., 1985; Exton, 1988). At 10^{-6} mmol/l, ATP rapidly mobilizes calcium and increases the concentration of free cytosolic calcium within 1-2 s. The response is dose-dependent and is maximal within 10 s. It is potentiated by glucagon and phorbol esters. Both extracellular ATP and ADP rapidly increase the cellular level of isositol 1,4,5 triphosphate (IP3), the messenger for calcium mobilization from the endoplasmic reticulum. P2-purinergic receptors have high affinity for the binding of ATP and appear to be involved in the response to extracellular ATP. At higher concentrations, extracellular ATP causes loss of K^+ and cytoplasmic blebbing takes place. We do not know whether any component of the canaliculus is particularly vulnerable to increased concentrations of extracellular ATP.

If there is intracanalicular ATP, the most likely source is the hepatocyte. There are no nerve endings, secretory cells, or other special sources in proximity to the bile canaliculus. The ATP concentration in hepatocytes is influenced by metabolic activity; however, concentrations of 8 mM are frequent. Loss of ATP through vesicular flow, exocytosis, and transport may occur normally and be controlled by ecto-ATPase and AMPase in the bile canaliculus.

These observations suggest that ATP may be closely involved in the function of the bile canaliculus and in its major expression of disease, namely cholestasis. We postulate that hepatocellular anoxia results in rapid breakdown of ATP, largely to adenosine, which is transferred into the circulation and increases flow and oxygen. Hypoxia may render the canaliculus more susceptible to loss of intracellular ATP, through permeability change or vesicular exocytosis, resulting in increased intracanalicular ATP concentrations which exceed the capacity of the canalicular ectoenzymes to degrade ATP. Increased intracanalicular ATP may facilitate injury to components of the bile secretory mechanism resulting in cholestasis. We have observed that intraportal infusion of low concentrations of ATP promptly reduces or blocks bile flow in a reversible manner (Harvey, K., unpublished data).

With one exception, all intrinsic enzymes which are associated with the bile canaliculus are ectoenzymes. This list includes gamma glutamyl transpeptidase and leucine aminopeptidase, which are widely used as markers of canalicular membrane enzymatic activity. The single exception is Gp 170, which is the product of the multi drug resistance (MDR) gene and will be considered separately. Kobata (1986) has performed structural analysis including sequencing of the family of canalicular ectoenzymes. An unique carbohydrate sequence was described which suggests that this determinant is acquired in the Golgi apparatus and targets the enzyme to the canalicular domain of the plasma membrane. Further analysis of the primary carbohydrate sequence of canalicular intrinsic proteins and its biologic effects is likely to be rewarding.

Gp 170 is a newly described inhabitant of the bile canalicular membrane. It is the product of the MDR gene. Bile canalicular Gp 170 appears to be similar, if not identical, to the MDR plasma membrane protein associated with resistance of various cell lines to anti-cancer drugs. Over 10 years ago, several investigators (Biedler and Riehm, 1970; Ling and Thompson, 1973) discovered the multi drug resistance phenomenon using epithelial or CHO cell lines which were cultured in colchicine or an anti-cancer drug, such as adriamycin. Cells which became resistant to either drug were resistant to a large number of other anti-cancer drugs which seemingly have little structural relationship. Thus, cells which became resistant to adriamycin were partially or totally resistant to drugs such as actinomycin, vinca alkaloids, cycloheximide, taxol, colchicine, and others. Using photoaffinity probes of the drugs, several investigators identified a 170 kD plasma membrane protein which was associated with multi drug resistant phenotype (Cornwell et al., 1986a; 1986b; 1986c). Molecular biologists used in-gel DNA renaturation methods to clone a full length sequence of the MDR (from rodents) and MDR (from primates) genes (Gros et al., 1986a; 1986b; Roninson et al., 1986). DNA sequence analysis suggested that Gp 170 spans the plasma membrane twelve times and has two internal sites which are consistent with nucleotide binding domains. Various resistant cell lines and human neoplasms were probed using a full length cDNA and expressed elevated mRNA levels for the MDR gene (Fojo et al., 1986; 1987; Roninson et al., 1984).

These observations resulted in the hypothesis that the MDR gene product is an efflux pump in the plasma membrane and transfers the anti-cancer drugs out of the cell at a rate which exceeds their ability to bind to DNA and other essential macromolecules (Dano, 1973; Fojo et al., 1985; Roninson et al., 1984). Drug levels are usually lower in resistant than in sensitive cells. Drug uptake and metabolism by resistant cells appeared to be unimpaired. In addition, cells grown in various metabolic inhibitors lost their apparent resistance and rapid drug efflux. Quinidine, verapamil, reserpine, Tween 80, and other drugs were shown to reduce anti-cancer drug efflux and restore drug sensitivity in resistant cells in culture (Cornwell et al., 1987). Kinetic studies in whole cells suggest that numerous drugs competitively inhibit the efflux of anti-cancer drugs from resistant cells thereby rendering them sensitive. This observation has resulted in clinical trials on the effect of quinidine as an adjunct in chemotherapy of several malignancies.

With development of molecular probes, it became possible to determine that the gene is amplified in some, but not all, drug-resistant cell lines (Moscow and Cowan, 1987). In

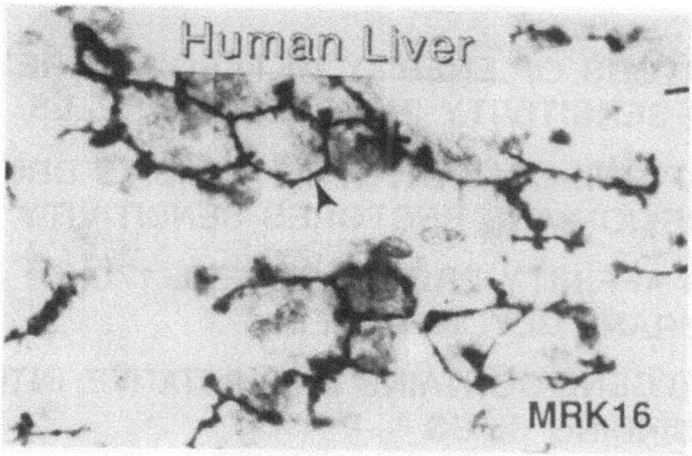

Figure 2: Immunohistochemical peroxidase localization of Gp 170 in normal human liver using MRK 16 monoclonal antibody (from Thiebaut et al., 1987).

addition, the MDR gene was localized to chromosome 7 (Fojo et al., 1986). Normal tissues were also screened for expression of MDR mRNA. Relatively high mRNA levels were detected in human adrenal cortex and medulla, liver, small and large intestine, and kidney. Ontogenetic maturation was observed in rat liver, which also showed increased mRNA expression during regeneration following partial hepatectomy.

Immunomorphologic studies of normal human tissue were performed using a monoclonal antibody coupled with horse radish peroxidase (Thiebaut et al., 1987). The reaction product was restricted to the bile canaliculus (Fig. 2), brush border of the proximal tubules and small intestine, luminal domain of the plasma membrane of colonic epithelial cells, and the secretory surface of adrenal cells. No other major intracellular localization was detected in the liver or other tissues. Gp 170 is not secreted by hepatocytes and the protein in the bile canaliculus appears to have a relatively long half life suggesting that it does not circulate intracellularly as a transporter or receptor.

1. **INHIBITORS OF ENERGY METABOLISM RESTORE DRUG SENSITIVITY IN RESISTANT CELLS**
2. **VERAPAMIL COMPETITIVELY INHIBITS DRUG SECRETION AND RESTORES SENSITIVITY**
3. **DRUG AFFINITY PROBES REVEAL 170KD MEMBRANE GLYCOPROTEIN**
4. **Gp170 GENE CONTAINS TWO PUTATIVE INTERNAL ATP BINDING SITES**
5. **Gp170 GENE TRANSFECTION CONVERTS SENSITIVE TO RESISTANT CELLS**
6. **Gp170 mRNA IN NORMAL TISSUES**

Figure 3: Demonstrated characteristics of Gp 170 and the MDR system.

Strong evidence that Gp 170 alone can confer drug resistance was obtained when the MDR gene was transfected into sensitive NIH 3T3 or CHO cells. Surviving cells manifested Gp 170 as determined by immunoblotting and acquired resistance to anti-cancer drugs in culture (Deuchars et al., 1987; Shen et al., 1986; Ueda et al., 1987) (Fig. 3).

Following demonstration of the canalicular localization of Gp 170 in rats, we studied its role as a transporter. An abstract describing some of this work has been presented (Hsu et al., 1987) and details will shortly be published. Our studies demonstrated ATP-dependent transport of daunomycin and other anti-cancer drugs in inside-out canalicular membrane vesicles. In canalicular vesicles, there was no transport in the absence of ATP or presence of non-hydrolyzable ATPs. Sinusoidal membrane vesicles did not transport daunomycin in the presence of ATP. The transport system is not affected by inhibitors of plasma membrane or mitochondrial active-site phosphorylation

ATPases, secondary transporters (such as Na^+-coupled symports), or ionic changes. Verapamil, chloroquine, vinblastin, and other drugs competitively inhibit the transport system which is temperature-dependent and saturated by substrate. Western blots using anti-Gp 170 monoclonal antibody reveal Gp 170 in bile canalicular but not sinusoidal membrane vesicles. Affinity probes of ATP bind to Gp 170 in the canalicular membrane.

Additional studies have shown that various fluorescent dyes, such as Rhodamine 1,2,3 and others, are retained in sensitive cells but are rapidly secreted by resistant cells (Mazzanti, R. and Arias, I.M., unpublished). Addition of verapamil to resistant cells results in retention of the fluorescent dyes. Transfection of a human hepatocellular carcinoma cell line with MDR resulted in enhanced anti-cancer drug resistance, expression of Gp 170 in bile canaliculi, and rapid exit of fluorescent test substances.

The mechanism whereby ATP is coupled to drug efflux has not been directly demonstrated. Active site phosphorylation of Gp 170 does not take place. The MDR gene has up to 50% homology with several bacterial ATP-dependent membrane systems which transport sugars into bacteria (Ferro-Luzzi Ames, 1986). Phosphoenol pyruvate is required as an energy source and the transported sugar is phosphorylated, which permits it to be retained inside bacteria. Despite the structural homology between MDR and these bacterial transporters, it is evident that the latter transport into the cell, whereas Gp 170 transports ligands out of the cell. In the bacterial systems, the high energy phosphate is believed to produce a tertiary configurational change in the transporter. The molecular mechanism of function of Gp 170 is not known.

In view of the expression of Gp 170 in normal tissues, it was appropriate to search for a putative natural substrate(s) for this efflux system. Structural studies of the anti-cancer drugs and competing agents suggest that the restrictive structure has broad limits. The limiting characteristics are an amphipathic molecule of approximately 300 - 800 molecular weight which has at least one weakly charged hydrophyllic terminal. We are currently testing various candidates; however, it should be recalled that most detoxification systems consist of families of enzymes with overlapping substrate specificities. The MDR gene product(s) may be similar and form a series of efflux pumps which have overlapping substrate specificity and facilitate detoxification and removal of the diverse structures to which we are exposed. The high homology to bacterial membrane transport proteins suggests that Gp 170(s) may serve common disposal functions in an evolutionary sense.

Gp 170 is the only bile canalicular protein which has been shown to have transport function which depends on ATP as an energy source. ATPase activity has been claimed for a partially purified Gp 170 from a resistant cell line (Hamada and Tsuruo, 1988).

We have much to learn about ATP in and around the bile canaliculus.

References

Akerboom T, Inoue M, Sies H, Kinne R, Arias IM (1984) Biliary transport of glutathione disulfide studied with isolated rat liver canalicular membrane vesicles. Eur J Biochem 141:211-215

Arai H, Terres G, Pink S, Forgac M (1988) Topography and subunit stoichiometry of the coated vesicle proton pump. J Biol Chem, in press

Arias IM, Forgac M (1984) The sinusoidal domain of the plasma membrane of rat hepatocytes contains an amiloride-sensitive Na^+/H^+ antiport. J Biol Chem 259:5406-5408

Biedler JL, Riehm H (1970) Cellular resistance to actinomycin D in Chinese hamster cells in vitro: cross-resistance, radioautographic, and cytogenetic studies. Cancer Res 30:1174-1184

Charest R, Blackmore PF, Exton JH (1985) Characterization of responses of isolated rat hepatocytes to ATP and ADP. J Biol Chem 260:15789-15794

Chaudry IH (1982) Does ATP cross the cell plasma membrane? Yale J Biol Med 55:1-10

Cornwell MM, Gottesman MM, Pastan IH (1986a) Increased vinblastine binding to membrane vesicles from multidrug-resistant KB cells. J Biol Chem 261:7921-7928

Cornwell MM, Pastan I, Gottesman MM (1987) Certain calcium channel blockers bind specifically to multidrug-resistant human KB carcinoma membrane vesicles and inhibit drug binding to P-glycoprotein. J Biol Chem 262:2166-2170

Cornwell MM, Safa AR, Felsted RL, Gottesman MM, Pastan I (1986b) Membrane vesicles from multidrug-resistant human cancer cells contain a specific 150- to 170-kDa protein detected by photoaffinity labeling. Proc Natl Acad Sci 83:3847-3850

Daly JW (1983) Role of ATP and adenosine receptors in physiologic processes. In: Daly JW et al. (eds) Physiology and pharmacology of adenosine derivative. Raven Press, New York, p 275-291

Dano K (1973) Active outward transport of daunomycin in resistant Ehrlich ascites tumor cells. Elsevier, Amsterdam

Deuchars KL, Du RP, Naik M, Evernden-Porelle D, Kartner N, van der Bliek AM, Ling V (1987) Expression of hamster P-glycoprotein and multidrug resistance in DNA-mediated transformants of mouse LTA cells. Mol Cell Biol 7:718-724

Exton JH (1988) Role of inositol triphosphate and diacylglycerol in the regulation of liver function. In: Arias IM, Jakoby WB, Popper H, Schachter D, Shafritz DA (eds) The liver: Biology and pathobiology, 2nd Edition. Raven Press, New York, p 785-791

Ferro-Luzzi Ames G (1986) The basis of multidrug resistance in mammalian cells: Homology with bacterial transport. Cell 47:323-324

Fojo AT, Akiyama SI, Gottesman MM, Pastan I (1985) Reduced drug accumulation in multiply drug-resistant human KB carcinoma cell lines. Cancer Res 45:3002-3007

Fojo AT, Lebo R, Shimizu N, Chin JE, Roninson IB, Merlino GT, Gottesman MM, Pastan I (1986) Localization of multidrug resistance-associated DNA sequences to human chromosome 7. Somat Cell Mol Genet 12:415-420

Fojo AT, Ueda K, Slamon DJ, Poplack DG, Gottesman MM, Pastan I (1987) Expression of a multidrug-resistance gene in human tumors and tissues. Proc Natl Acad Sci 84:265-269

Forgac M, Berne M (1986) Structural characterization of the ATP-hydrolyzing portion of the coated vesicle proton pump. Biochemistry 25:4275-4280

Gros P, Croop J, Housman D (1986a) Mammalian multidrug resistance gene: complete cDNA sequence indicates strong homology to bacterial transport proteins. Cell 47:371-380

Gros P, Croop J, Roninson I, Varshavsky A, Housman DE (1986b) Isolation and characterization of DNA sequences amplified in multidrug-resistant hamster cells. Proc Natl Acad Sci 83:337-341

Gros P, Neriah YB, Croop JM, Housman DE (1986c) Isolation and expression of a complementary DNA that confers multidrug resistance. Nature 323:728-731

Hamada H, Tsuruo T (1988) Purification of the 170- to 180-kilodalton membrane glycoprotein associated with multidrug resistance. J Biol Chem 263:1454-1458

Hsu J, Gatmaitan Z, Willingham M, Gottesman MM, Pastan I, Cornwell M, Arias IM (1987) The multidrug resistance system and anti-cancer drug transport by rat liver canalicular membrane vesicles. Hepatology 7:1104

Inoue M, Akerboom TP, Sies H, Kinne R, Tran T, Arias IM (1984a) Biliary transport of glutathione-S-conjugate by rat liver canalicular membrane vesicles. J Biol Chem 254:4998-5002

Inoue M, Kinne R, Tran T, Arias IM (1982a) Taurocholate transport by liver sinusoidal membrane vesicles: evidence of sodium cotransport. Hepatology 2:572-579

Inoue M, Kinne R, Tran T, Arias IM (1983) The mechanisms of biliary secretion of reduced glutathione: analysis of transport process in isolated rat liver canalicular membrane vesicles. Eur J Biochem 134:467-471

Inoue M, Kinne R, Tran T, Arias IM (1984b) Glutathione transport across hepatocyte plasma membranes - analysis using isolated rat-liver sinusoidal-membrane vesicles. Eur J Biochem 138:491-495

Inoue M, Kinne R, Tran T, Arias IM (1984c) Taurocholate transport by rat liver canalicular membrane vesicles: evidence for the presence of a Na^+-independent transport system. J Clin Invest 73:659-663

Inoue M, Kinne R, Tran T, Biempica L, Arias IM (1982b) Rat liver canalicular membrane vesicles: isolation and topological characterization. J Biol Chem 258:5183-5188

Kobata A (1986) Current status of carbohydrate constituents and prospects. In: Seigers WH, Walz DA (eds) Prothrombin and other vitamin K dependent proteins, Vol. 1. CRC Press, p 81-94

Li CC, Hochstadt J (1976a) Transport mechanisms in isolated plasma membranes. J Biol Chem 251:1175-1180

Li CC, Hochstadt J (1976b) Membrane-associated enzymes involved in nucleoside processing by plasma membrane vesicles isolated from L_{929} cells grown in defined medium. J Biol Chem 251:1181-1187

Lin SH (1985) Novel ATP-dependent calcium transport component from rat liver plasma membranes. J Biol Chem 260:7850-7856

Lin SH, Fain JM (1984) Purification of (Ca^{2+}-Mg^{2+})-ATPase from rat liver plasma membranes. J Biol Chem 259:3016-3020

Lin SH, Russell WE (1987) The high affinity Ca^{2+}(Mg^{2+})-ATPase (Ca^{2+}(Mg^{2+})-nucleotidase) of rat liver plasma membrane is an ecto-enzyme. J Biol Chem, in press

Ling V, Thompson LH (1973) Reduced permeability in CHO cells as a mechanism of resistance to colchicine. J Cell Physiol 83:103-116

Low MG, Ferguson MAJ, Futerman AH, Silman I (1986) Covalently attached phosphatidylinositol as a hydrophobic anchor for membrane proteins. TIBS 11:212-215

Miras-Portugal MT, Torres M, Rotlan P, Aunis D (1986) Adenosine transport in bovine chromaffin cells in culture. J Biol Chem 261:1712-1719

Moscow JA, Cowan KH (1987) Multidrug resistance. J Natl Canc 80:14-20

Paterson ARP, Kolassa M, Cass CE (1981) Transport of nucleoside drugs in animal cells. Pharmacol Ther 12:515-536

Roninson IB, Abelson HT, Housman DE, Howell N, Varshavsky A (1984) Amplification of specific DNA sequences correlates with multi-drug resistance in Chinese hamster cells. Nature 309:626-628

Roninson IB, Chin JE, Choi K, Gros P, Housman DE, Fojo A, Shen DW, Gottesman MM, Pastan I (1986) Isolation of human mdr DNA sequences amplified in multidrug-resistant KB carcinoma cells. Proc Natl Acad Sci 83:4538-4542

Shen DW, Fojo A, Roninson IB, Chin JE, Soffir R, Pastan I, Gottesman MM (1986) Multidrug resistance of DNA-mediated transformants is linked to transfer of the human mdri gene. Mol Cell Biol 4039-4045

Thiebaut F, Tsuruo T, Hamada H, Gottesman MM, Pastan I, Willingham MC (1987) Cellular localization of the multi-drug-resistance gene product P-glycoprotein in normal human tissues. Proc Natl Acad Sci 84:7735-7738

Tong A, Gatmaitan Z, Arias IM (1986) On the function of Ca^{++} Mg^{++} ATPase and 5'nucleotidase in bile canaliculi. Hepatology 5: 1223

Ueda K, Cardarelli C, Gottesman MM, Pastan I (1987) Expression of a full-length cDNA for the human "MDRI" gene confers resistance of colchicine, doxorubicin and vinblastine. Proc Natl Acad Sci 84:3004-3008

Wohlhueter RM, Plagemann PGW (1982) On the functional symmetry of nucleoside transport in mammalian cells. Biochim Biophys Acta 689:249-260

Part 2
Physiology, Pharmacology, and Biochemistry of Hepatic Transport Systems

Hepatic Transport of Glutathione-Conjugates

Glutathione Conjugate Transport in Hepatic Elimination of Electrophiles

H. Sies
Institut für Physiologische Chemie I
Universität Düsseldorf
Moorenstrasse 5
D-4000 Düsseldorf 1
West Germany

Introduction

Glutathione has several functions in liver and in the organism. This tripeptide is of interest regarding the hepatic elimination of drugs, subsequent to the formation of glutathione S-conjugates. The reaction is catalysed by glutathione S-transferases in the first step of mercapturic acid biosynthesis (Boyland and Chasseaud, 1969; Jacoby, 1978; Levine, 1978). These thioethers in general are less toxic and more water-soluble than the original foreign compound and are eliminated via the bile, since the S-conjugates are anions with a molecular weight exceeding 300 (Chasseaud, 1974). The glutathione S-conjugates generated from the glutathione S-transferase reactions are also eliminated from many other types of cell in the organism. A sufficient capacity for transport of glutathione conjugates is required not only for the inter-organ relationships regarding mercapturate formation, but also for the maintenance of a low intracellular steady state concentration of these compounds, as glutathione S-conjugates have potentially harmful effects. The present article focuses on the competition of glutathione conjugates with other transport systems at the cell surface (Akerboom et al, 1982) as well as with site in the interior of the cell such as the inhibition of glutathione reductase (Bilzer et al, 1984). Some of these topics have been reviewed (Inoue, 1985; Ishikawa and Sies, 1988; Kaplowitz et al, 1985; Sies, 1983); the material presented here is based on a recent review (Sies, 1988).

Hepatic Transport in Organic Substances
Ed. by E. Petzinger, R. K.-H. Kinne, H. Sies

Biliary S-Conjugate Excretion in the Perfused Rat Liver

Based on information obtained by Klaassen and Fitzgerald (1974) using diethylmaleate and by Barnhart and Combes (1978) using ethacrynate, we investigated the properties of conjugate transport in the intact organ (Wahlländer and Sies, 1979). An S-conjugate that is generated intracellularly at high rates in order to facilitate the maintenance of steady state conditions and that is readily distinguishable spectrophotometrically from the unconjugated parent compound was studied; as a suitable model compound, the conjugate formed from the widely-used substrate, 1-chloro-2,4-dinitrobenzene (CDNB), was examined. This conjugate, S-(2,4-dinitrophenyl)glutathione, proved quite useful in obtaining basic information. The infusion of 13 µM CDNB led to a steady state rate of S-conjugate release after about 10 min; corresponding to a rate of infusion of 52 nmol CDNB/min per g of liver. When the concentration of CDNB in the infusion was increased to 66 µM, a plateau value of S-conjugate release was not attained due to a limitation on intracellular glutathione which is depleted during this process.

Separate sampling of the bile and the caval perfusate revealed a large concentration difference for the S-conjugate. The biliary concentration was about 3,000-fold higher than that in the caval perfusate, when the rate of substrate infusion was about 100 nmol CDNB/min per g of liver wet weight. The partitioning of S-conjugate release between bile and caval perfusate was such that the S-conjugate was transferred quantitatively to the bile at low rates of S-conjugate formation, and a gradually increasing percentage was released into the caval perfusate as the rates of intracellular formation of the conjugate increased above about 30 nmol/min per g liver wet weight.

The S-conjugate concentration in bile rose to 36 mM in the saturation curve of biliary S-conjugate concentration. The infusion of the S-conjugate itself led to very low biliary S-conjugate excretion, there being apparently almost no uptake of the conjugate by the hepatocytes.

A similar study was carried out with perfused rat heart (Ishikawa and Sies, 1984). When CDNB was infused into the influent perfusate, the corresponding S-conjugate was observed in the effluent. Steady state rates of the release were reached about 3 min after onset of the infusion. The rate remained stable for at least 5 min. There was saturation of S-conjugate release, and the maximal rate of transport was 40 nmol/min per g of heart. The conjugation reaction is not considered as a rate-limiting step, because glutathione S-transferase activity in the heart is 300 times higher than the maximal release rate and the apparent K_m value of the enzyme in heart homogenate is 150 µM for CDNB. It is noteworthy that when this substrate was infused at 2 µM for 5 min, more than 90 % of the compound was released from heart as glutathione-S-conjugate.

S-Conjugate Transport in Rat Hepatocyte Membrane Vesicles

Isolated canalicular membrane vesicles were shown to transport S-(2,4-dinitrophenyl)glutathione (Inoue et al, 1984). The initial rate of transport followed saturation kinetics with respect to conjugate concentration. The apparent K_m was 1 mM, and the V_{max} was 5.1 nmol/mg protein per min. These results show that the transport of the glutathione S-conjugate across the canalicular membranes is a carrier-mediated process. Sodium chloride in the transport medium could be replaced by KCl; LiCl or choline chloride were without any changes in transport activity. The rate of conjugate transport was enhanced by a valinomycin-induced K^+ diffusion potential (the inside of the vesicle being positive). The rate of conjugate uptake was enhanced by replacing KCl in the transport medium with potassium gluconate, providing a less permeant anion, and was decreased by replacing KCl with KSCN, providing a more permeant anion. These data indicated that conjugate transport is electrogenic and involves the transfer of negative charge (Inoue et al, 1984). Further, the transport of S-(2,4-dinitrophenyl)glutathione was inhibited by S-benzylglutathione, GSSG, or GSH (Inoue et al, 1984).

Mutual Competition between Glutathione Conjugates and Glutathione Disulfide or Taurocholate for Canalicular Transport

In 1978, we observed that glutathione disulfide (GSSG) efflux from liver occurs via excretion into bile, and that, as mentioned above, the glutathione conjugate obtained from 1-chloro-2,4-dinitrobenzene also occurred predominantly into bile (Sies et al, 1978 a,b). We suggested that the transport of glutathione disulfide and of glutathione S-conjugates may have common features, viewing the disulfide as a glutathione "homoconjugate" (Sies et al, 1980). In experiments to study the interaction of conjugate transport and disulfide transport, it was observed that the transport of GSSG into bile was restricted during biliary S-conjugate transport; vice versa, S-conjugate transport into bile was restricted during stimulated GSSG transport resulting from the infusion of t-butyl hydroperoxide, as described by Akerboom et al (1982). This established a mutual competition between transport of GSSG and of S-conjugates from perfused rat liver into bile. Similar observations were made with isolated canalicular membrane vesicles (Inoue et al, 1984). Likewise, there was a mutual competition for transport in the isolated perfused heart (Ishikawa et al, 1986). The export of GSSG was diminished when the glutathione S-conjugate was formed from 4-hydroxynonenal by the cardiac cells. Thus, GSH conjugates may have an unfavorable effect on the GSH/GSSG redox status in the cell, firstly by the consumption of GSH required for conjugate formation, and secondly by the inhibition of GSSG transport from the cell by the conjugate.

There also is an inhibition of bile acid efflux from canalicular membrane vesicles by S-(2,4-dinitrophenyl)glutathione, as was studied examining taurocholate export (Griffiths et al, 1987). Therefore, the potential exists that increased level of S-conjugates may have unfavorable effects on a major liver function, the elimination of bile acids. Whether this has any long-term effects, such as cholostasis, is currently unknown.

It is noteworthy that incubation of a rat hepatocyte plasma fractions with µM concentrations of either GSSG or various glutathione S-conjugates resulted in a several-fold increase in the rate of ATP hydrolysis (Nicotera et al, 1985). It was proposed that this newly discovered ATPase may function in the cellular extrusion of both glutathione disulfide and glutathione S-conjugates. The glutathione conjugates employed were those of 1-chloro-2,4-dinitrobenzene, hexachlorobutadiene, N-(4-ethoxy-phenyl)-p-benzoquinoneimine. The K_m for these S-conjugates for stimulation of ATP hydrolysis was in the range of 6 µM; the V_{max} was in the range of 3 nmol phosphate liberated/mg protein per min (Nicotera et al, 1985).

Glutathione S-Conjugates as Enzyme Inhibitors and Binding to GSH Transferase

In addition to the influence of S-conjugates on the glutathione/ disulfide status of hepatocytes by inhibition of GSSG transport into bile, there is the potential of effects on other GSSG binding sites, e.g. the active site of GSSG reductase (Bilzer et al, 1984). Human glutathione reductase is competitively inhibited by the S-conjugate, and the inhibition is fully reversible by dialysis against buffer. The S-conjugate combines to more than one site on the enzyme. Fifty percent inhibition was observed at an S-conjugate concentration of 30 µM with NADPH present at saturating concentration and GSSG concentration approximating the K_m. There was no indication that the S-conjugate binds to the NADPH site of glutathione reductase, because non-competitive inhibition was observed when NADPH was the variable substrate. Studying the S-(2,4-dinitrophenyl)glutathione-enzyme complex by x-ray crystallography, the binding of the S-conjugate to GSSG reductase was demonstrated in the form of a difference electron density map, representing the difference between ligand-enzyme complex and enzyme. The major binding site of the glutathione S-conjugate overlaps the binding site of the substrate, GSSG. However, the glutathione moiety of the bound S-conjugate does not coincide with either of the glutathione moieties of GSSG. It was concluded that the overall orientation of the glutathione portion of the conjugate depends on the non-glutathione moiety. In the case of the S-(2,4-dinitrophenyl)conjugate, the conformation is determined by its bulky dinitrophenyl residue (Bilzer et al, 1984).

The binding of an S-conjugate, S-(2-chloro-4-nitrophenyl) glutathione, to GSH transferase has been reported by Jacobson et al (1979). The dissociation constant K_d was 0.5 µM, whereas that of GSH was 7 µM, and there was one mole of S-conjugate bound per subunit. The binding was competitive with the substrate, GSH. Thus, it can be concluded that enzyme activity can be modified by the binding of the S-conjugate. However, since the physiological levels of GSH range between 5 and 10 mM, an inhibition by the S-conjugate may be overcome.

Another enzyme that was found to respond to glutathione S-conjugates by changing its properties is the gamma-glutamylcysteine synthetase. It was found by Kondo et al (1984) that the feedback inhibition of this enzyme by

GSH was released by the addition of the glutathione S-conjugate, S-(2,4-dinitrophenyl)-glutathione. A half-maximal effect of glutathione S-conjugate on gamma-glutamylcysteine synthetase activity was obtained at approximately 1 µM. Fifty µM conjugate in the presence of 1 mM GSH actually increased the enzyme activity two-fold above uninhibited levels. The conjugate had no effect on the enzyme activity in the absence of GSH. Since glutathione is consumed by the process catalyzed by glutathione S-transferases, this release of the feedback inhibition and the loss of GSH both favor the resynthesis of GSH in the situation of increased requirement.

In attempts to study the significance of GSH for cellular metabolism, several GSH depleting agents have been used (see Plummer et al, 1981, for review). Such compounds were diethylmaleate (Boyland and Chasseaud, 1970) or phorone, diisopropylidene acetone (Van Doorn et al, 1978). Both these compounds were shown to lead to an induction of heme oxygenase (Burk and Correia, 1979; Yoshida et al, 1987). While it is possible that both these alpha-, beta- unsaturated carbonyl compounds directly cause the induction of the heme oxygenase, it is an interesting question whether the products, i.e. the respective S-conjugates, might by themselves be responsible for the effect. It should be noted that phorone was able to further deplete hepatocytes of GSH over the effect obtained with buthionine sulfoximine (Romero and Sies, 1984).

Toxicological Implications of Various S-Conjugates

The glutathione conjugation of menadione leads to the formation of thiodione. A first product is the reduced compound, thiodiol, as a product of the nucleophilic Michael addition reaction resulting in the formation of the hydroquinone conjugate. As discussed in detail by Nickerson et al (1963) and Gant et al (1986), the conjugate may then be oxidised either by the unreacted parent quinone or by oxygen or other oxidants to the corresponding quinone. In the case of menadione, the resulting S-conjugate would be thiodione. This compound is the major form of excretion of menadione from the intact liver (Akerboom et al, 1988). It is clear that the 1,4-positions are left unchanged upon the formation of thioether at carbon 3 of the molecule. Indeed, Wefers and Sies (1983) have demonstrated that like menadione the compound thiodione is also capable of redox cycling. Therefore, there is no immediate "detoxication" by making

the thioether. This has also been studied by Ross et al (1985).
Biologically active glutathione S-conjugates are formed in different classes of compounds, as has been emphazised by Igwe (1986) and Kauffman (1987). One such class is, of course, the formation of the glutathione conjugate in the leukotriene pathway, leading to the highly active glutathione conjugate, leukotriene C_4 (Corey et al, 1980). At this point it may suffice to indicate that glutathione S-conjugates in special cases may have critical influence on the biological response to xenobiotics, as biologically more active GSH derivatives may be formed from such compounds. Therefore, a system for control of the concentrations of glutathione conjugates in cells may indeed serve protective functions. On the other hand, as is indicated for the case of menadione, biologically active compounds might also be transported out of cells to spread toxicological potential away from the site of generation.

Concluding Remarks

The role of the transport systems for glutathione S-conjugates has been emphazised here. In addition to supplying substrate for the mercapturate pathway outside of the initial cell that generated the thioether, the transport systems probably are crucial in maintaining a low level of the conjugates. Some evidence of interaction between such conjugates and specific enzymes has been given. It appears likely that further physiological and toxicological functions of thioethers will become known in the future. For example, the different subcellular localisation of different glutathione transferases and the almost ubiquitous presence of these enzymes point to fundamental roles of these reactions.

Acknowledgements

Work from the author's laboratory was supported by Deutsche Forschungsgemeinschaft and by National Foundation for Cancer Research.

References

Akerboom TPM, Bilzer M, Sies, H (1982) Competition between transport of glutathione disulfide (GSSG) and glutathione S-conjugates from perfused rat liver into bile. FEBS Lett 140:73-76

Akerboom TPM, Bultmann, T, Sies H (1988) Inhibition of biliary taurocholate excretion during menadione metabolism in perfused rat liver. Arch Biochem Biophys 263:10-18

Bannai S, Tateishi, N (1986) Role of membrane transport in metabolism and function of glutathione in mammals. J Membrane Biol 89:1-8

Barnhart JL, Combes B (1978) Choleresis associated with metabolism and biliary excretion of diethyl maleate in rat and dog. J Pharmacol Exp Ther 206:614-623

Bilzer M, Krauth-Siegel RL, Schirmer RH, Akerboom TPM, Sies H (1984) Interaction of a glutathione S-conjugate with glutathione reductase. Kinetic and X-ray crystallographic studies. Eur J Biochem 138:373-378

Boyland E, Chasseaud LF (1969) The role of glutathione and glutathione S-transferases in mercapturic acid biosynthesis. Adv Enzymol 32:172-219

Boyland E, Chasseaud LF (1970) The effect of some carbonyl compounds on rat liver glutathione levels. Biochem Pharmacol 19:1526-1528

Burk RF, Correia MA (1979) Stimulation of rat hepatic microsomal heme oxygenase by diethyl maleate. Res Commun Chem Pathol Pharmacol 24: 205-207

Chasseaud LF (1974) Glutathione S-transferase. In: Flohé L, Benöhr HC, Sies H, Waller HD, Wendel A (eds) Glutathione. G. Thieme, Stuttgart, p 90-108

Chen WJ, DeSmidt PC, Armstrong RN (1986) Stereoselective product inhibition of glutathione S-transferase. Biochem Biophys Res Commun 141:892-897

Corey EJ, Clark DA, Goto G, Marfat C, Mioskowski B, Samuelsson B, Hammarström B (1980) Stereospecific total synthesis of a "slow reacting substance" of anaphylaxis, leukotriene C-1. J Am Chem Soc 102:1436-1439

Finley KT (1974) The addition and substitution chemistry of quinones. In: Patai S (ed) The chemistry of the quinonoid compounds, Part II, J. Wiley, London, p 878-1144

Gant TW, d'Arcy Doherty M, Odowole D, Sales KD, Cohen GM (1986) Semiquinone anion radicals formed by the reaction of quinones with glutathione or amino acids. FEBS Lett 201:296-300

Griffiths JC, Sies H, Meier PJ, Akerboom TPM (1987) Inhibition of taurocholate efflux from rat hepatic canalicular membrane vesicles by glutathione disulfide. FEBS Lett 213:34-38

Igwe OJ (1986) Biologically active intermediates generated by the reduced glutathione conjugation pathway. Toxicological implications. Biochem Pharmacol 35:2987-2994

Inoue M (1985) Interorgan metabolism and membrane transport of glutathione and related compounds. In: Kinne RKH (ed) Elsevier, Amsterdam, New York, Oxford, p 225-269

Inoue M, Akerboom TPM, Sies H, Kinne R, Thao T, Arias IM (1984) Biliary transport of glutathione S-conjugate by rat liver canalicular membrane vesicles. J Biol Chem 259:4998-5002

Ishikawa T, Esterbauer H, Sies H (1986) Role of cardiac glutathione transferase and of glutathione S-conjugate export system in biotransformation of 4-hydroxynonenal in the heart. J Biol Chem 261:1576-1581

Ishikawa T, Sies H (1984) Cardiac transport of glutathione disulfide and S-conjugate. Studies with isolated perfused rat heart during hydroperoxide metabolism. J Biol Chem 259:3838-3843

Ishikawa T, Sies H (1988) Glutathione as an antioxidant: toxicological aspects. In: Dolphin D, Poulson R, Avramovic D (eds) J. Wiley, New York, in press

Jacobson I, Warholm M, Mannervik B (1979) The binding of substrates and a product of the enzymatic reaction to glutathione S-transferase A. J Biol Chem 254:7085-7089
Jakoby WB (1978) The glutathione S-transferases: A group of multifunctional detoxification proteins. Adv Enzymol 46:383-414
Jongen WMF, Alink GM, Koeman JH (1978) Mutagenic effect of dichloromethane on Salmonella typhimurium. Mutation Res 56:245-248
Kaplowitz N, Aw TY, Ookhtens M (1985) The regulation of hepatic glutathione. Ann Rev Pharmacol Toxicol 25:715-744
Kauffman FC (1987) Conjugation-deconjugation reactions in drug metabolism and toxicity. Fed Proc 46:2434-2445
Klaassen CD, Fitzgerald TJ (1974) Metabolism and biliary excretion of ethacrynic acid. J Pharmacol Exp Ther 191:548-556
Kondo T, Taniguchi N, Kawakami Y (1984) Significance of glutathione S-conjugate for glutathione metabolism in human erythrocytes. Eur J Biochem 145:131-136
Levine WG (1978) Biliary excretion of drugs and other xenobiotics. Ann Rev Pharmacol Toxicol 18:81-96
Meister A, Anderson ME (1983) Glutathione. Ann Rev Biochem 52:711-760
Nickerson WJ, Falcone G, Strauss G (1963) Studies on quinone-thioethers. 1. Mechanism of formation and properties of thiodione. Biochemistry 2: 537-546
Nicotera P, Baldi C, Svensson S-Å, Larsson R, Bellomo G, Orrenius S (1985) Glutathione S-conjugates stimulate ATP hydrolysis in the plasma membrane fraction of rat hepatocytes. FEBS Lett 187:121-125
Plummer JL, Smith BR, Sies H, Bend JR (1981) Chemical depletion of glutathione in vivo. Meth Enzymol 77:50-59
Romero F, Sies H (1984) Subcellular glutathione contents in isolated hepatocytes treated with L-buthionine sulfoximine. Biochem Biophys Res Commun 123:1116-1121
Ross D, Thor H, Orrenius S, Moldeus P (1985) Interaction of menadione (2-methyl-1,4-naphthoquinone) with glutathione. Chem Biol Interact 55: 177-184
Sies H (1983) Reduced and oxidized glutathione efflux from liver. In: Sakamato Y, Higashi T, Tateishi N (eds) Glutathione: storage, transport and turnover in mammals. Japan Scientific Soc Press, Tokyo, p 63-88
Sies H (1988) Intracellular effects of glutathione conjugates and their transport from the cell. In: Sies H, Ketterer B (eds) Glutathione conjugation: its mechanism and biological significance. Academic Press, London, p 175-192
Sies H, Wahlländer A, Linke I, Marklstorfer A (1978a) Glutathione disulfide (GSSG) efflux from liver occurs via excretion into bile. Hoppe-Seyler's Z Physiol Chem 359:1151
Sies H, Wahlländer A, Waydhas C (1978b) Properties of glutathione disulfide (GSSG) and glutathione-S-conjugate release from perfused rat liver. In: Sies H, Wendel A (eds) Functions of glutathione in liver and kidney. Springer, Heidelberg, New York, p 120-126
Sies H, Wahlländer A, Waydhas C, Soboll S, Häberle D (1980) Functions of intracellular glutathione in hepatic hydroperoxide and drug metabolism and the role of extracellular glutathione. Adv Enzyme Regul 18:303-320
van Bladeren PJ, Breimer DD, Rotteveel-Smijs GTM, Mohn GR (1980) Mutagenic activation of dibromomethane and diiodomethane by mammalian microsomes and glutathione S-transferase. Mutation Res 74:341-346
van Doorn R, Leijdekkers ChM, Henderson PTh (1978) Synergistic effects of phorone on the hepatotoxicity of bromobenzene and paracetamol in mice. Toxicology 11:225-233

Wahlländer A, Sies H (1979) Glutathione S-conjugate formation from 1-chloro-2,4-dinitrobenzene and biliary excretion in the perfused rat liver. Eur J Biochem 96:441-446

Wefers H, Sies H (1983) Hepatic low-level chemiluminescence during redox cycling of menadione and the menadione-glutathione conjugate relation to glutathione and NAD(P)H: Quinone reductase (DT-diaphorase) activity. Arch Biochem Biophys 224:568-578

Yoshida T, Oguro T, Numazawa S, Kuroiwa Y (1987) Effects of phorone (diisopropylidene acetone), a glutathione (GSH) depletor, on hepatic enzymes involved in drug and heme metabolism in rats: evidence that phorone is a potent inducer of heme oxygenase. Biochem Biophys Res Commun 145: 502-508

Hepatobiliary Cysteinyl Leukotriene Elimination

M. Huber, T. Baumert, A. Guhlmann, D. Mayer*, and D. Keppler
Division of Tumor Biochemistry and *Division of Cytopathology
Deutsches Krebsforschungszentrum, Im Neuenheimer Feld 280
D-6900 Heidelberg 1, Federal Republic of Germany

Introduction

The cysteinyl leukotrienes, LTC_4, LTD_4, LTE_4, and *N*-acetyl-LTE_4, are highly potent lipid mediators which are derived from arachidonate (Lewis and Austen, 1984; Hammarström et al., 1985; Piper and Samhoun, 1987; Samuelsson et al., 1987). 5-Lipoxygenase/LTA_4 synthase (Dixon et al., 1988; Matsumoto et al., 1988) converts arachidonate to the epoxide LTA_4 which is subsequently conjugated with glutathione by a particulate LTC_4 synthase (Söderström et al., 1988; Yoshimoto et al., 1988). LTC_4 is metabolized to LTD_4, LTE_4, and *N*-acetyl-LTE_4 (Fig. 1). Predominant cellular sources of LTC_4 are eosinophils, mast cells, macrophages, and monocytes (Weller et al., 1983; Lewis and Austen, 1984; Williams et al., 1984). The cysteinyl leukotrienes are potent smooth muscle constrictors involved in inflammatory and anaphylactic reactions, in the release of luteinizing hormone, and in cell proliferation (Lewis and Austen, 1984; Baud et al., 1985; Hammarström et al., 1985; Kragballe et al., 1985; Piper and Samhoun, 1987; Samuelsson et al., 1987). Their effects are mediated through well-defined receptors on the cell surface (Mong et al., 1988; Winkler et al., 1988). In human isolated bronchi and lung parenchyma LTE_4 is less potent than either LTC_4 or LTD_4 (Piper and Samhoun, 1987). Although 25-fold less active than LTD_4 in the guinea-pig pulmonary parenchymal strip (Lewis et al., 1981), *N*-acetyl-LTE_4 is still a potent constrictor of rat mesenteric vessels (Siren et al., 1988). Enhanced production *in vivo* of endogenous cysteinyl leukotrienes has been demonstrated under various pathophysiological conditions including tissue trauma (Denzlinger et al., 1985), acute hepatitis (Hagmann et al., 1987), hepatorenal syndrome (Huber et al., 1987b; Keppler et al., 1988), anaphylactic shock (A. Keppler et al., 1987), endotoxin shock (Hagmann et al., 1984,1985), infusion of tumor necrosis factor-α (Huber et al., 1988), administration of staphylococcal enterotoxin B (Denzlinger et al., 1986), and injection of platelet-activating factor (Huber and Keppler, 1987).

Hepatic Transport in Organic Substances
Ed. by E. Petzinger, R. K.-H. Kinne, H. Sies

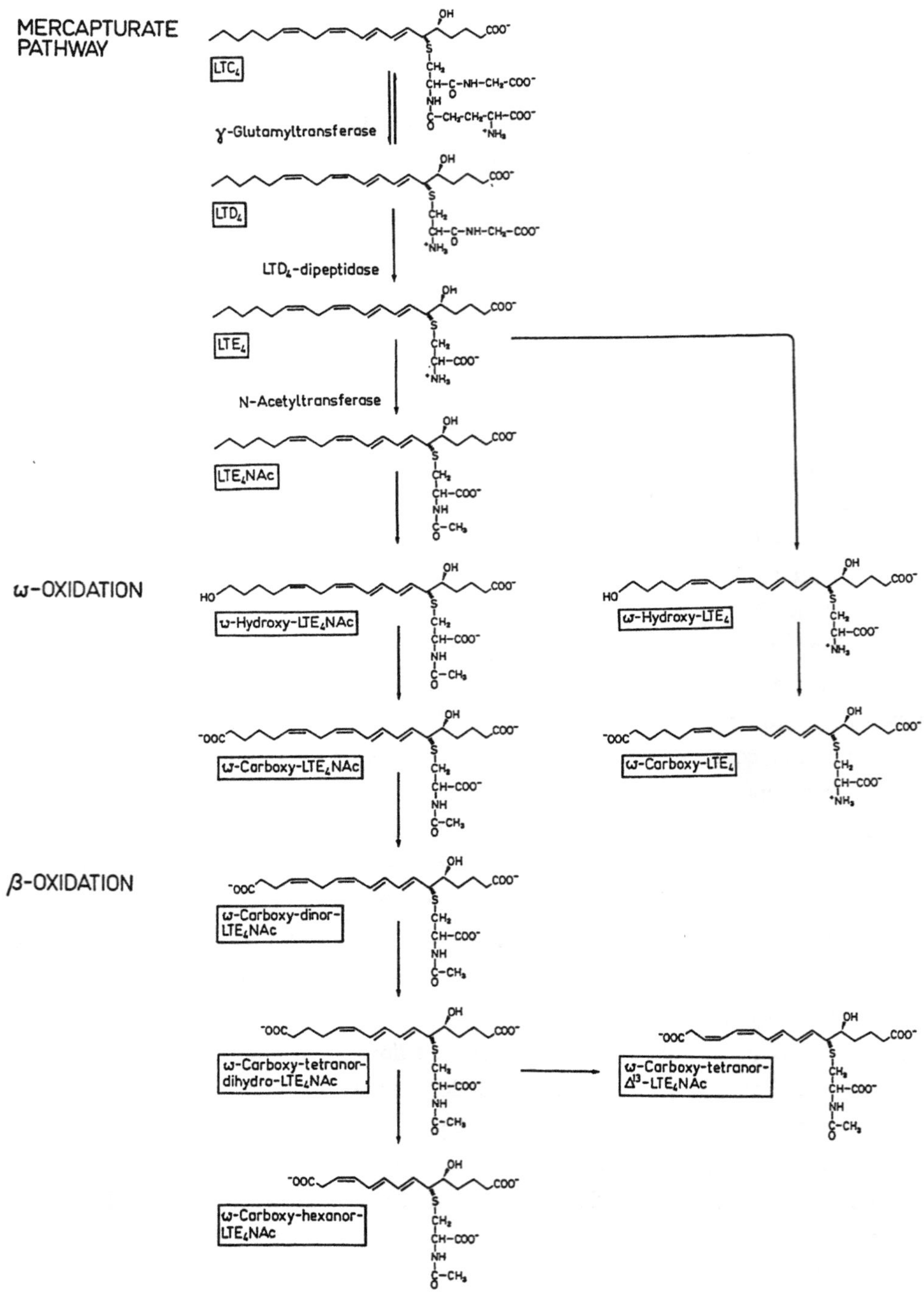

Fig. 1. Metabolism of LTC_4. The glutathionyl conjugate LTC_4 is degraded to the cysteinylglycine derivative LTD_4 and to the cysteinyl derivative LTE_4 by the

Deactivation of Cysteinyl Leukotrienes

Leukotriene deactivation is determined by the following factors: (i) ectoenzyme-catalyzed extracellular degradation, (ii) uptake by hepatocytes and kidney cells, (iii) intracellular metabolism, and (iv) excretion into bile and urine. In the blood circulation LTC_4 is rapidly metabolized to LTD_4 and LTE_4 by the ectoenzymes γ-glutamyltransferase (EC 2.3.2.2) and dipeptidase (EC 3.4.13.11) (Fig. 1). LTC_4, LTD_4, and LTE_4 are eliminated from the vascular space with initial half-lives of 30 to 40 seconds (Fig. 2) and taken up by the liver and by the kidneys (Huber and Keppler, 1988).

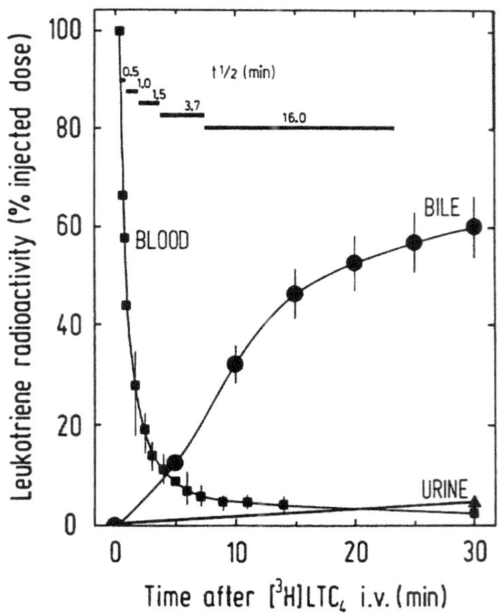

Fig. 2. Elimination of intravenously injected $[^3H]LTC_4$ in the rat. Half-life times of leukotriene radioactivity circulating in blood are indicated by *horizontal bars* and *superscript numbers*. Hepatic and renal uptake contribute to leukotriene removal from the circulation; leukotriene metabolites are mainly eliminated into bile, only small amounts are recovered from urine (Denzlinger et al., 1985; Keppler et al., 1985; Huber et al., 1987a).

- *Fig. 1 - continued -* ectoenzymes γ-glutamyltransferase and LTD_4 dipeptidase, respectively. Intracellular catabolism of LTE_4 results in the formation of ω-oxidized metabolites (Ball and Keppler, 1987; Foster et al., 1987; Örning 1987a,b) and products of β-oxidation from the ω-end (Stene and Murphy, 1988). *N*-Acetylation of the cysteinyl residue is an additional important step in cysteinyl leukotriene catabolism of the rat (Hagmann et al., 1986; Örning et al., 1986). LTE_4NAc, *N*-acetyl-LTE_4.

The intracellular LTE_4 metabolism includes *N*-acetylation (Hagmann et al., 1986; Örning et al., 1986) and ω-oxidation (Örning, 1987a;b) followed by β-oxidation from the ω-end (Stene and Murphy, 1988). Incubation of isolated rat hepatocytes with LTE_4 results in the formation of the mercapturate *N*-acetyl-LTE_4 and of the oxidation products ω-carboxy-*N*-acetyl-LTE_4, ω-carboxy-dinor-*N*-acetyl-LTE_4, ω-carboxy-tetranor-dihydro-*N*-acetyl-LTE_4, ω-carboxy-tetranor-Δ^{13}-*N*-acetyl-LTE_4, and ω-carboxy-hexanor-*N*-acetyl-LTE_4 (Fig. 1) (Stene and Murphy, 1988). After intracellular metabolism leukotriene metabolites are excreted into bile and urine. In all species investigated so far hepatobiliary elimination predominates over renal excretion. One hour after intravenous administration of $[^3H]LTC_4$ about 80 % of the radioactivity are recovered from rat bile and 1.5 % from urine (Fig. 2). In rat bile LTC_4 metabolites of defined structure comprise LTD_4, *N*-acetyl-LTE_4 (Hagmann et al., 1986), ω-hydroxy-*N*-acetyl-LTE_4, and ω-carboxy-*N*-acetyl-LTE_4 (Foster et al., 1987). In the monkey about 40 % of the radioactivity from intravenously injected $[^3H]LTC_4$ are recovered from bile and about 20 % from urine within 5 hours (Denzlinger et al., 1986). In this species biliary leukotriene metabolites include LTD_4, LTE_4 (Denzlinger et al., 1986), ω-hydroxy-LTE_4, and ω-carboxy-LTE_4 (Ball and Keppler, 1987). Cysteinyl leukotriene metabolites eliminated via bile into the intestine partially undergo enterohepatic circulation (Denzlinger et al., 1986; Guhlmann et al., 1987).

Hepatic Transport of Cysteinyl Leukotrienes

Freshly isolated rat hepatocytes possess a temperature- and ATP-dependent uptake system for cysteinyl leukotrienes. This transport system at the sinusoidal membrane is involved in the uptake of LTC_4, LTD_4, and LTE_4 (Uehara et al., 1983). LTB_4, glutathione disulfide, and the glutathione-S-conjugate of acetaminophen do not interfere with the hepatocellular uptake of cysteinyl leukotrienes. We have recently measured the uptake of LTE_4, the predominant leukotriene metabolite in the circulating blood, by hepatocytes in primary culture. During culture hepatocytes lose their capacity for LTE_4 uptake with time; 72 hours after cell isolation and culture, LTE_4 uptake was less than 5 % as compared to the uptake by rat hepatocytes that were in primary culture for 6 hours (Fig. 3).

The hepatocellular uptake of cysteinyl leukotrienes is followed by intracellular catabolism and by an efficient excretion of leukotriene metabolites into bile. After intravenous injection of ^3H-labeled cysteinyl leukotrienes into the rat, 50 % of the

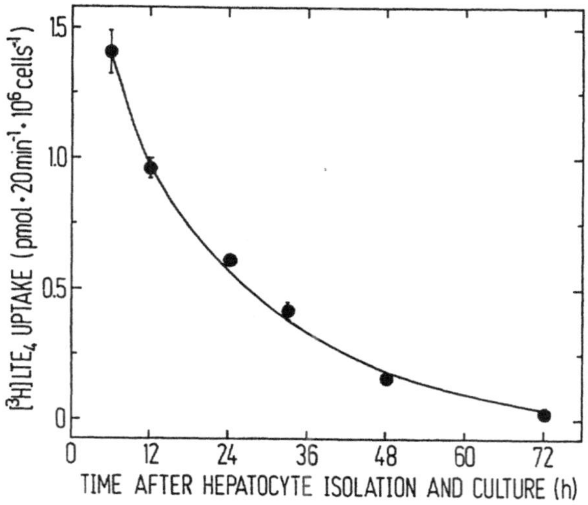

Fig. 3. Time-dependent loss of leukotriene uptake by cultured hepatocytes. Rat hepatocytes were cultured and uptake experiments were performed at different times after cell isolation. After incubation with 4 nmol/l LTE_4 (1.48 x 10^{15} Bq/mol) for 20 minutes, hepatocytes were washed twice and cell-associated radioactivity was counted. Control experiments at 0°C were performed to determine non-specific binding of radioactivity and the measurements were corrected for this.

radioactivity are recovered from bile within the first 20 minutes. The transport system at the canalicular membrane seems to be identical with the system excreting amphiphilic organic anions such as dibromosulfophthalein and conjugated bilirubin (Huber et al., 1987a).

Impaired Leukotriene Elimination

A few examples and experimental conditions have been observed with impaired hepatocellular leukotriene uptake or impaired biliary excretion of leukotriene metabolites. Two epithelial liver cell lines (c_1I and c_2I) which were cloned from a primary culture of hepatocytes (Mayer and Schäfer, 1982) have lost the uptake of LTE_4. This is in agreement with the time-dependent decrease of leukotriene uptake in cultured hepatocytes (Fig. 3). Moreover, suspensions of freshly collected AS-30D rat ascites hepatoma cells are completely deficient in the uptake of cysteinyl leukotrienes (Weckbecker and Keppler, 1986; Keppler et al., 1987). As shown in control experiments these hepatocyte-derived hepatoma cells take up glutamine and adenosine and convert extracellular LTC_4 to LTD_4 and LTE_4.

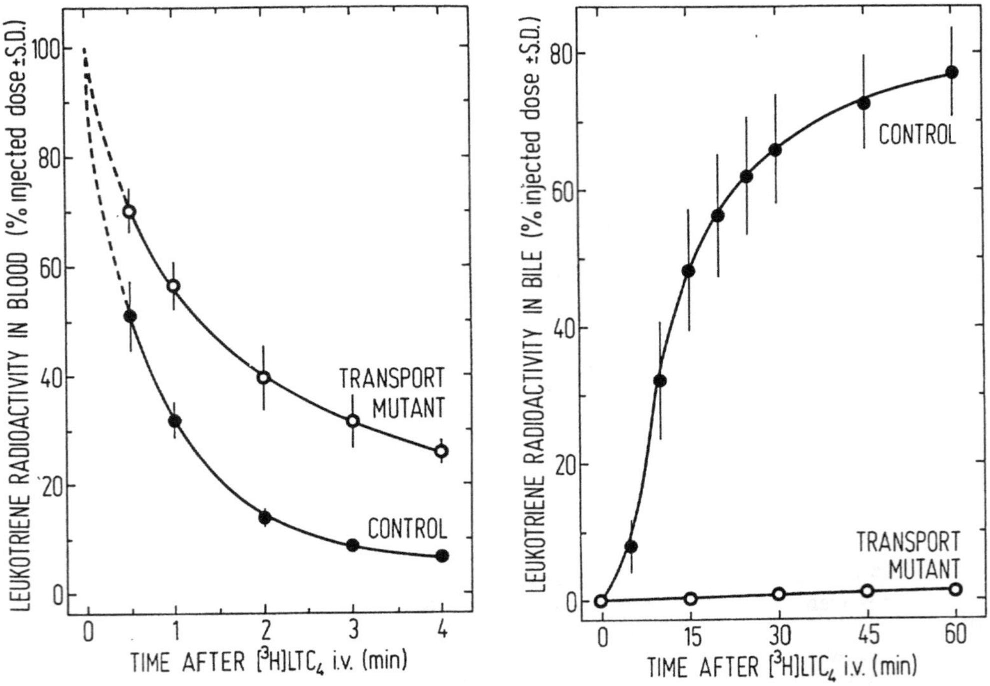

Fig. 4. Leukotriene elimination in transport mutant rats as compared to normal control animals. *Left panel:* Elimination of leukotriene radioactivity from circulating blood after intravenous injection of [^{3}H]LTC$_4$. *Right panel:* Hepatobiliary excretion of leukotriene radioactivity after intravenous injection of [^{3}H]LTC$_4$ (Huber et al., 1987a).

A loss of hepatobiliary leukotriene excretion was found in a mutant rat strain with a hereditary impairment of the hepatobiliary elimination of conjugated bilirubin, dibromosulfophthalein, and ouabain (Huber et al., 1987a). After intravenous injection of [14,15-^{3}H$_2$]LTC$_4$, the initial half-life of radioactivity in the circulating blood increases from 31 seconds in normal Wistar rats to 79 seconds in transport mutant rats whereas the leukotriene radioactivity recovered from bile within 1 hour is reduced from 77 % in control animals to 1.4 % in transport mutant rats (Fig. 4). The recovery in bile of dibromosulfophthalein and ouabain is reduced to 28 % and 47 %, respectively, as compared to normal Wistar rats (Jansen et al., 1987); the biliary excretion of bilirubin is 40 % of normal (Jansen et al., 1985). The hepatobiliary elimination of taurocholate is not affected in these transport mutant rats (Jansen et al., 1985,1987). The reduced biliary leukotriene elimination is accompanied by an enhanced leukotriene excretion into urine and by an enhanced content of radioactive leukotriene metabolites in various organs

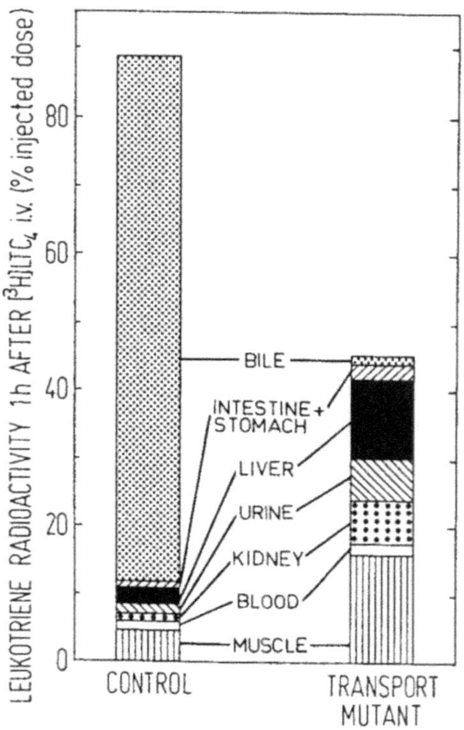

Fig. 5. Recovery of tritium radioactivity one hour after administration of [^{3}H]LTC$_4$ in control Wistar rats and transport mutants. After injection of [14,15-$^3H_2$]LTC$_4$ bile and urine were sampled continuously. At the end of the experiment, blood was collected, and the organs were excised, homogenized, and leukotriene metabolites were extracted (Huber et al., 1987a).

(Fig. 5). The incomplete recovery of radioactivity from [14,15-$^3H_2$]LTC$_4$ in the mutant strain may result from a more rapid ω-oxidation (Fig. 1) and elimination of tritium from the leukotriene molecule.

Endotoxin is not only a stimulus for endogenous cysteinyl leukotriene production but also a potent inhibitor of hepatobiliary cysteinyl leukotriene elimination (Hagmann et al., 1984,1985). Leukotriene elimination is much more inhibited by endotoxin than bile flow or biliary taurocholate excretion. In rats endotoxin treatment reduces bile flow and biliary excretion of [^{14}C]taurocholate by 11 % but biliary [^{3}H]leukotriene excretion by 44 % as compared to controls (Hagmann et al., 1984). The impaired secretion of leukotrienes into bile is associated with an enhanced hepatic leukotriene content.

Another interesting observation comes from leukotriene measurements in the urine of patients with liver cirrhosis or hepatorenal syndrome. Urinary leukotriene concentrations were enhanced 3-fold in patients with cirrhosis and 30-fold in patients with hepatorenal syndrome (Huber et al., 1987b; Keppler et al., 1988). This may indicate that impaired hepatocellular uptake and inactivation of cysteinyl leukotrienes, possibly in association with enhanced leukotriene production, results in a diversion of hepatobiliary elimination to renal excretion with subsequent renal failure and severe illness.

Concluding Remarks

The biologically highly potent cysteinyl leukotrienes are eliminated from the body mainly by trans-hepatocellular excretion. This deactivation pathway depends on an efficient uptake system at the sinusoidal membrane and on a transport system for excretion at the canalicular membrane of the hepatocyte. Inhibition of hepatobiliary excretion has been observed early after endotoxin injection and in mutant rats with defective hepatic anion excretion. This inhibition of excretion into bile leads to increased concentrations of leukotriene metabolites in the liver and in the blood circulation. However, in view of the pathophysiological actions of the cysteinyl leukotrienes, the hepatocellular uptake and the intracellular metabolic inactivation seem to play the most important role, particularly under conditions of impaired hepatobiliary elimination. It is of importance, therefore, whether increased amounts of leukotriene metabolites in the blood circulation consist of biologically active compounds or of metabolites inactivated intracellularly and released subsequently into the blood circulation.

References

Ball HA, Keppler D (1987) ω-Oxidation products of leukotriene E_4 in bile and urine of the monkey. *Biochem Biophys Res Commun* **148**:664-670

Baud L, Sraer J, Perez J, Nivez M-P, Ardaillou R (1985) Leukotriene C_4 binds to human glomerular epithelial cells and promotes their proliferation in vitro. *J Clin Invest* **76**:374-377

Denzlinger C, Rapp S, Hagmann W, Keppler D (1985) Leukotrienes as mediators in tissue trauma. *Science* **230**:330-332

Denzlinger C, Guhlmann A, Scheuber PH, Wilker D, Hammer DK, Keppler D (1986) Metabolism and analysis of cysteinyl leukotrienes in the monkey. *J Biol Chem* **261**:15601-15606

Dixon RAF, Jones RE, Diehl RE, Bennett CD, Kargman S, Rouzer CA (1988) Cloning of the cDNA for human 5-lipoxygenase. *Proc Natl Acad Sci USA* **85**:416-420

Foster A, Fitzsimmons B, Rokach J, Letts G (1987) Evidence of in-vivo ω-oxidation of peptide leukotrienes in the rat: biliary excretion of 20-CO_2H N-acetyl LTE_4. Biochem Biophys Res Commun 148:1237-1245

Guhlmann A, Hagmann W, Keppler D (1987) Enterohepatic circulation of N-acetyl-leukotriene E_4. Prostaglandins 34:63-70

Hagmann W, Denzlinger C, Keppler D (1984) Role of peptide leukotrienes and their hepatobiliary elimination in endotoxin action. Circ Shock 14:223-235

Hagmann W, Denzlinger C, Keppler D (1985) Production of peptide leukotrienes in endotoxin shock. FEBS Lett 180:309-313

Hagmann W, Denzlinger C, Rapp S, Weckbecker G, Keppler D (1986) Identification of the major endogenous leukotriene metabolite in the bile of rats as N-acetyl leukotriene E_4. Prostaglandins 31:239-251

Hagmann W, Steffan A-M, Kirn A, Keppler D (1987) Leukotrienes as mediators in frog virus 3-induced hepatitis in rats. Hepatology 7:732-736

Hammarström S, Örning L, Bernström K (1985) Metabolism of leukotrienes. Mol Cell Biochem 69:7-16

Huber M, Guhlmann A, Jansen PLM, Keppler D (1987a) Hereditary defect of hepatobiliary cysteinyl leukotriene elimination in mutant rats with defective hepatic anion excretion. Hepatology 7:224-228

Huber M, Kästner S, Schölmerich J, Keppler D (1987b) Enhanced urinary excretion of cysteinyl leukotrienes in patients with hepatorenal syndrome. J. Hepatol 5 Suppl 1:S34

Huber M, Keppler D (1987) Inhibition of leukotriene D_4 catabolism by D-penicillamine. Eur J Biochem 167:73-79

Huber M, Beutler B, Keppler D (1988) Tumor necrosis factor-α stimulates leukotriene production in vivo. Submitted 1988

Huber M, Keppler D (1988) Leukotrienes and the mercapturate pathway. In: Sies H, Ketterer B (eds) Glutathione conjugation: its mechanism and biological significance. Academic Press, London, pp 449-470

Jansen PLM, Peters WH, Lamers WH (1985) Hereditary chronic conjugated hyperbilirubinemia in mutant rats caused by defective hepatic anion transport. Hepatology 5:573-579

Jansen PLM, Groothuis GMM, Peters WHM, Meijer DFM (1987) Selective hepatobiliary transport defect for organic anions and neutral steroids in mutant rats with hereditary-conjugated hyperbilirubinemia. Hepatology 7:71-76

Keppler A, Örning L, Bernström K, Hammarström S (1987) Endogenous leukotriene D_4 formation during anaphylactic shock in the guinea pig. Proc Natl Acad Sci USA 84:5903-5907

Keppler D, Hagmann W, Rapp S, Denzlinger C, Koch HK (1985) The relation of leukotrienes to liver injury. Hepatology 5:883-891

Keppler D, Huber M, Weckbecker G, Hagmann W, Denzlinger C, Guhlmann A (1987) Leukotriene C_4 metabolism by hepatoma cells and liver. Adv Enzyme Regul 26:211-224

Keppler D, Huber M, Hagmann W, Ball HA, Guhlmann A, Kästner S (1988) Metabolism and analysis of endogenous cysteinyl leukotrienes. Ann NY Acad Sci 524:68-74

Kragballe K, Desjarlais L, Voorhees JJ (1985) Leukotrienes B_4, C_4 and D_4 stimulate DNA synthesis in cultured human epidermal keratinocytes. Br J Derm 113:43-52

Lewis RA, Drazen JM, Austen KF, Toda M, Brion F, Marfat A, Corey EJ (1981) Contractile activities of structural analogs of leukotrienes C and D: role of the polar substituents. Proc Natl Acad Sci USA 78:4579-4583

Lewis RA, Austen KF (1984) The biologically active leukotrienes. Biosynthesis, metabolism, receptors, functions, and pharmacology. J Clin Invest 73:889-897

Matsumoto T, Funk CD, Rådmark O, Höög J-O, Jörnvall H, Samuelsson B (1988) Molecular cloning and amino acid sequence of human 5-lipoxygenase. Proc Natl Acad Sci USA 85:26-30

Mayer D, Schäfer B (1982) Biochemical and morphological characterization of glycogen-storing epithelial liver cell lines. Exp Cell Res 138:1-14

Mong S, Miller J, Wu H-L, Crooke ST (1988) Leukotriene D_4 receptor-mediated hydrolysis of phosphoinositide and mobilization of calcium in sheep tracheal smooth muscle cells. J Pharmacol Exp Ther 244:508-515

Örning L (1987a) ω-Hydroxylation of N-acetylleukotriene E_4 by rat liver microsomes. Biochem Biophys Res Commun 143:337-344

Örning L (1987b) ω-Oxidation of cysteine-containing leukotrienes by rat-liver microsomes. Isolation and charcterization of -hydroxy and ω-carboxy metabolites of leukotriene E_4 and N-acetylleukotriene E_4. Eur J Biochem 170:77-85

Örning L, Norin E, Gustafsson B, Hammarström S (1986) In vivo metabolism of leukotriene C_4 in germ-free and conventional rats. Fecal excretion of N-acetylleukotriene E_4. J Biol Chem 261:766-771

Piper PJ, Samhoun MN (1987) Leukotrienes. Br Med Bull 43:297-311

Samuelsson B, Dahlén S-E, Lindgren J-Å, Rouzer CA, Serhan CN (1987) Leukotrienes and lipoxins: structures, biosynthesis, and biological effects. Science 237:1171-1176

Siren A-L, Letts G, Feuerstein G (1988) N-Acetyl-leukotriene E_4 is a potent constrictor of rat mesenteric vessels. Eur J Pharmacol 146:331-335

Söderström M, Hammarström S, Mannervik B (1988) Leukotriene C synthase in mouse mastocytoma cells. An enzyme distinct from cytosolic and microsomal glutathione transferases. Biochem J 250:713-718

Stene DO, Murphy RC (1988) Metabolism of leukotriene E_4 in isolated rat hepatocytes. Identification of β-oxidation products of sulfidopeptide leukotrienes. J Biol Chem 263:2773-2778

Uehara N, Ormstad K, Örning L, Hammarström S (1983) Characteristics of the uptake of cysteine-containing leukotrienes by isolated hepatocytes. Biochim Biophys Acta 732:69-74

Weckbecker G, Keppler D (1986) Leukotriene C_4 metabolism by hepatoma cells deficient in the uptake of cystein leukotrienes. Eur J Biochem 154:559-562

Weller PF, Lee CW, Foster DW, Corey EJ, Austen KF, Lewis RA (1983) Generation and metabolism of 5-lipoxygenase pathway leukotrienes by human eosinophils: predominant production of leukotriene C_4. Proc Natl Acad Sci USA 80:7626-7630

Williams JD, Czop JK, Austen KF (1984) Release of leukotrienes by human monocytes on stimulation of their phagocytic receptor for particulate activators. J Immunol 132:3034-3040

Winkler JD, Mong S, Crooke ST (1988) Leukotriene D_4-induced homologous desensitization of calcium mobilization in rat basophilic leukemia cells. J Pharmacol Exp Ther 244:449-455

Yoshimoto T, Soberman RJ, Spur B, Austen KF (1988) Properties of highly purified leukotriene C_4 synthase of guinea pig lung. J Clin Invest 81:866-871

Biliary Excretion of Thiols and Their Role in Elimination of Methylmercury

Curtis D. Klaassen and Zoltan Gregus
Department of Pharmacology, Toxicology and Therapeutics
University of Kansas Medical Center
Kansas City, Kansas 66103 U.S.A.

Introduction

Liver is the main site of glutathione (GS) synthesis and it transports this tripeptide both into blood and bile (Kaplowitz et al., 1985). Hepatobiliary transport of GS is also thought to play an important role in the biliary excretion of some toxic metals such as methylmercury (MM; Ballatori and Clarkson, 1985) which is mainly excreted from the body via the bile and feces (Norseth and Clarkson, 1971; Gregus and Klaassen, 1986). MM is found in rat bile complexed with GS and cysteine (Ohsawa and Magos, 1974; Refsvik and Norseth, 1975). Complexation of MM with these non-protein thiols (SH) in bile may explain the direct relationship between the biliary excretion rates of SH and MM in rats. For example, increased biliary excretion of SH during the postnatal period (Ballatori and Clarkson, 1982) or diminished SH excretion following depletion of hepatic GS (Refsvik, 1978) are associated with respective changes in the biliary excretion of MM.

Whereas a correlation between the biliary excretion of MM and SH exists, it is not clear whether potential enzymatic reactions involved in the formation and degradation of MM thiol complexes are important in determining the rate of excretion of MM. Glutathione S-transferases (GST) are abundant soluble

Hepatic Transport in Organic Substances
Ed. by E. Petzinger, R. K.-H. Kinne, H. Sies

enzymes in the liver that catalyze conjugation of various electrophilic chemicals with the nucleophilic GS. A possible role of GSTs in the biliary excretion of MM-GS complex has been suggested (Magos et al., 1979; Refsvik, 1983). In addition, liver cells contain gamma-glutamyltransferase (GGT) in the bile canalicular membrane (Szewczuk et al., 1980; Inoue et al., 1983) that can initiate hydrolysis of GS resulting in formation of GS-derived thiols such as cysteinylglycine (CG) and cysteine (Cys). In fact, more recent data demonstrate that GS is hydrolyzed in the biliary tree (Abbot and Meister, 1986; Stein et al., 1986). GGT-initiated hydrolysis of GS in kidney is a prerequisite for preventing urinary loss of SH (Rankin et al., 1983). However, it is not certain what roles hepatic GGT plays in the biliary excretion of SH and what implications the possible GGT-mediated hydrolysis of biliary GS may have on the biliary elimination of MM.

The present report summarizes recently conducted studies (Gregus and Varga, 1985; Gregus et al., 1987a; 1987b; Stein et al., 1988) on the role of hepatic GST and GGT in the biliary excretion of thiols and MM. Chemicals that induce GSTs, such as phenobarbital, benzo(a)pyrene, pregnenolone-16 alpha-carbonitrile (PCN) and <u>trans</u>-stilbene oxide (TSO) were used as experimental tools to determine the importance of GSTs in the biliary excretion of MM. The role of GGT in the composition and excretion of biliary SH and in the biliary excretion of MM was assessed by examining the relationship between biliary excretion of SH and hepatic GGT during postnatal development, in various animal species, and in response to inhibition of GGT by acivicin as well as by analyzing the relationship between biliary MM and SH excretion under these conditions.

Methods

Biliary excretion of SH and MM were measured in anesthetized, bile duct cannulated animals. In the studies with inducers, total biliary SH was determined using the Ellman's reagent. In the other investigations, reduced and oxidized GS, CG and Cys were separated by HPLC on a reversed-phase column and detected electrochemically (Stein et al., 1986). Hepatic GGT activity was determined as described by Meister et al. (1981). Activity of hepatic GST toward 1-chloro-3,4-dinitrobenzene was determined according to Habig et al. (1974). MM labelled with ^{203}Hg was quantitated by measuring radioactivity with a gamma spectrometer. Other details have been described elsewhere (Gregus and Varga, 1985; Gregus et al., 1987a, 1987b; Stein et al., 1988).

Results and Discussion

1. Role of GST in the biliary excretion of SH

The effects of treatments with inducers on hepatic GST activity and the biliary excretion of SH in rats were compared (Fig. 1). Benzo(a)pyrene slightly increased, phenobarbital and PCN doubled, and TSO tripled hepatic GST activity. The biliary excretion of SH was more than doubled after the rats were treated with PB and PCN, whereas the other inducers had minimal effects. Thus, there was a poor correlation between GST activity and SH excretion, suggesting that biliary SH excretion is not dependent on GST activity.

2. Role of GGT in determining the composition of biliary SH

(a) Age-dependent composition of biliary SH in rats. In addition to GS, rat bile contains significant amounts of CG and Cys. Because these latter thiols occur in liver at a very low concentration compared to GS (Gregus et al., 1987b), it is most likely that they are formed from GS during or after its hepatobiliary translocation. Both the excretion rate and the relative proportion of these thiols are dependent on age (Fig. 2). The total SH excretion increased 9-fold between 2 and 10 weeks of age (data not shown). GS was the predominant thiol in bile in 2-3- and 7-10-week-old rats, however, at 4-5 weeks of age the excretion rates of CG and Cys exceeded that of GS. The increased biliary excretion of CG and Cys in 4-5-week-old rats is associated with a 3-4-fold elevation of hepatic GGT activity during this age (Fig. 3). Correlation of hepatic GGT activity with the biliary excretion rate of CG plus Cys suggests

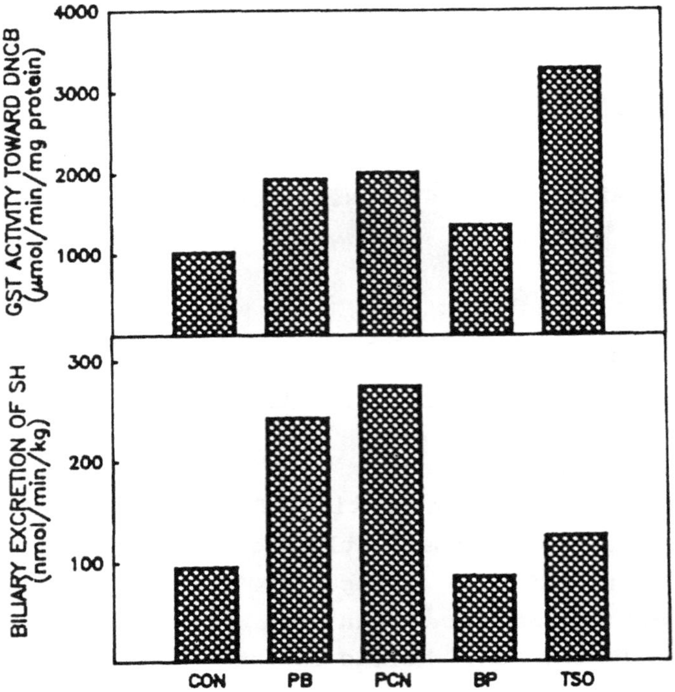

Figure 1. Effect of microsomal enzyme inducers on hepatic GST activity and biliary excretion of SH. Rats were pretreated intraperitoneally with phenobarbital (PB; 75 mg/kg/day), pregnenolone-16 alpha-carbonitrile (PCN; 75 mg/kg/day), benzo(a)-pyrene (BP; 20 mg/kg/day) and _trans_-stilbene oxide (400 mg/kg/day) for 5 days. (Modified after Gregus and Varga, 1985)

that the latter thiols originate from GGT-mediated hydrolysis of GS. In addition, the observation that the biliary excretion rate of GS hydrolysis products does not increase with the higher SH excretion in the rats at 7-10 weeks of age may indicate that at higher hepatobiliary transport rates of GS, the GGT-mediated GS hydrolytic process becomes capacity limited. Thus, at high rates of transport more GS can escape hydrolysis. This may explain the predominance of GS among biliary SH in adult rats.

(b) _Species-dependent composition of biliary SH_. There is a marked species difference in the biliary excretion of total GS-related SH and disulfides (ΣS) noted in the top panel of Fig. 4. The biliary excretion rate of ΣS was highest in mice,

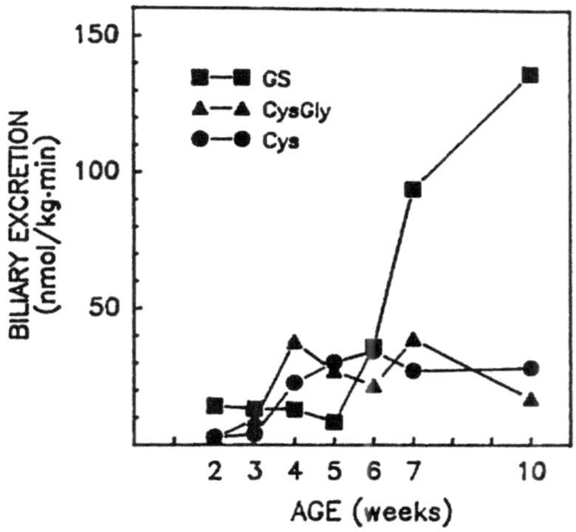

Figure 2. Age-dependent biliary excretion of reduced and oxidized glutathione (GS), cysteinylglycine (CysGly) and cysteine (Cys) in rats. (Modified after Gregus et al., 1987a)

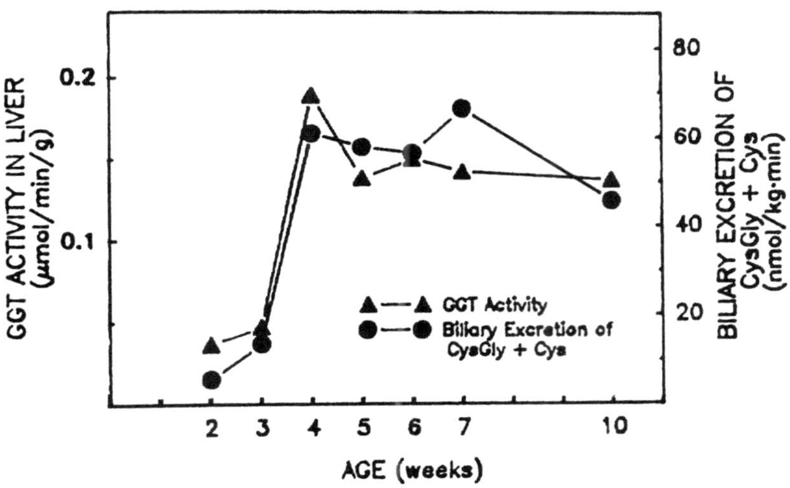

Figure 3. Hepatic activity of gamma-glutamyltranspeptidase (GGT) and biliary excretion of cysteinylglycine (CysGly) and cysteine (Cys) in rats of various ages. (Modified after Gregus et al., 1987a)

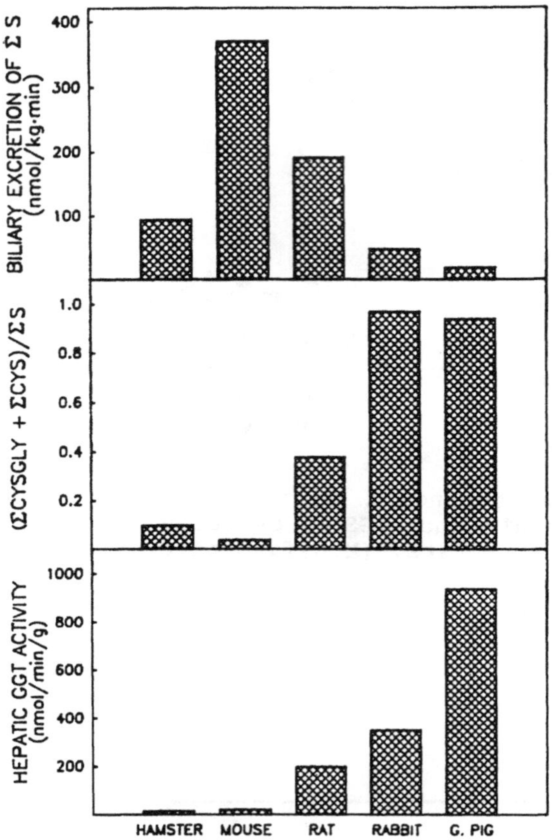

Figure 4. Species variations in biliary excretion of total glutathione-related thiols and disulfides (ΣS), the ratio of glutathione hydrolysis products (ΣCysGly and ΣCys) and ΣS, and hepatic gamma-glutamyltranspeptidase (GGT) activity. (Modified after Stein et al., 1988)

followed by rats, hamsters, rabbits and guinea pigs with the excretion rate decreasing by approximately 50% from one species to the next. The composition of biliary ΣS is also markedly different among these species. GS hydrolysis products (i.e., reduced and oxidized CG and Cys) make up only 4 and 10 percent of ΣS, respectively, in mouse and hamster bile (Fig. 4, middle panel); i.e., these species excrete ΣS in bile predominantly as reduced and oxidized GS. The low degree of hydrolysis of biliary GS in mice and hamsters is associated with the very low hepatic GGT activities in these species (Fig. 4, bottom panel). In contrast, almost all biliary ΣS in rabbit and guinea pig bile is in the form of GS hydrolysis products which parallels with the high GGT activities in the liver of rabbits and guinea pigs. In regard to both the proportion of biliary GS hydrolysis products and hepatic GGT activities, rats were intermediate of the species examined. Thus, a correlation exists between the hepatic GGT activity and the concentration of hydrolysis products in the bile which suggests a causative relationship. A similar correlation has been observed among these species between hepatic GGT activity and the extent of hydrolysis of acetaminophen glutathione conjugate excreted in bile (Gregus et al., 1988), indicating that both endogenous GS and GS conjugates of xenobiotics undergo GGT-dependent hydrolysis in the bile.

3. Role of GGT in determining the biliary excretion rate of SH

Acivicin (AT-125) is an irreversible inhibitor of GGT (Allen et al., 1980). Administration of 0.1 mmol/kg acivicin i.v. resulted in 88 and 99% decrease in hepatic and biliary GGT activity in rats (Gregus et al., 1987b). The GGT inhibition markedly altered the biliary excretion rates of the individual components of ΣS (Fig. 5). Following acivicin treatment of 4-week-old rats the excretion of GS hydrolysis products (i.e. reduced and oxidized CG and Cys) declined which was associated with a significant increase in the biliary excretion of reduced and oxidized GS. Because these opposing changes compensated each other, the biliary excretion rate of ΣS remained unaltered following acivicin treatment. Thus, while GGT located at the luminal surface of the biliary tree can markedly modify the composition of biliary SH, it does not significantly influence the rate of excretion of total SH. This observation also suggests that hepatic GGT does not mediate significant reabsorption of SH from the biliary tree. Thus, the role of GGT in liver contrasts with that in kidney where this enzyme is instrumental in the reabsorption of GS-derived thiols.

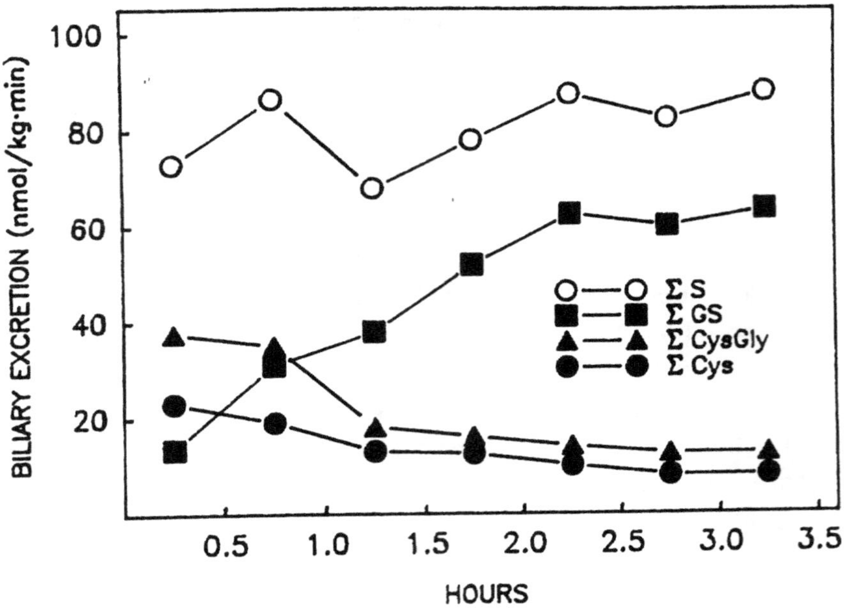

Figure 5. Effect of acivicin (0.1 mmol/kg iv, injected at 0.5 hr) on biliary excretion of glutathione-related thiols and disulfides (ΣS), reduced and oxidized glutathione (ΣGS), cysteinylglycine (ΣCysGly) and cysteine (ΣCys). (Modified after Gregus et al., 1987b)

4. Role of GST in the biliary excretion of MM

Treatments with the inducers phenobarbital and PCN significantly increased the biliary excretion of both SH and MM (Fig. 6). In contrast, benzo(a)pyrene and TSO, the latter being the most effective inducer of GST, failed to enhance the biliary excretion of SH or MM. Thus, biliary excretion rate of MM was correlated with the excretion of SH in bile rather than hepatic GST activity. Therefore, this observation further supports the role of biliary excretion of SH but not the role of hepatic GSTs in the hepatobiliary transport of MM. Similar conclusions have been made regarding the biliary excretion of zinc and cadmium (Gregus and Varga, 1985) that are also transported into bile in a GS-dependent manner. Since MM (Simpson, 1961), cadmium and zinc (Perrin and Watt, 1971) have high affinities for GS, it is conceivable that complexation of these metals with GS does not require enzymatic (GST) catalysis.

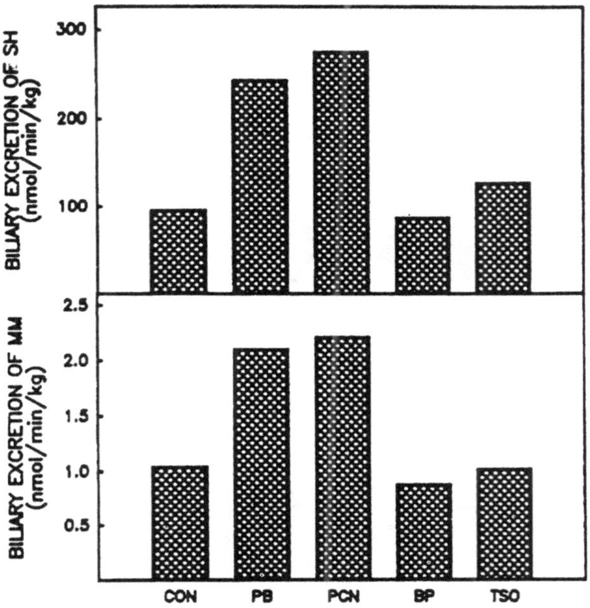

Figure 6. Effect of microsomal enzyme inducers on biliary excretion of nonprotein thiols (SH) and methylmercury (MM) in rats. Abbreviations for the inducers and doses are given under Figure 1. (Modified after Gregus and Varga, 1985)

5. Role of GGT and biliary SH in the biliary excretion of MM

In order to investigate the role of hepatic GGT in biliary excretion of MM, the excretion of MM was measured in rats pretreated with the GGT inhibitor acivicin. As noted earlier, inhibition of GGT by acivicin markedly increased the biliary excretion of GS but did not alter the excretion of total GS-related ΣS (Fig. 5). Acivicin also did not affect the biliary excretion of MM significantly (Fig. 7). This observation indicates that biliary excretion of MM in rats is not influenced by hepatic GGT-catalyzed hydrolysis of biliary GS and also supports the view that the ΣS excretion (which probably represents the amount of ΣGS originally transported from liver cells to bile canaliculi), and not solely the biliary excretion of GS, may be important in influencing elimination of MM through the bile. In contrast, inhibition of renal GGT by acivicin resulted in a marked glutathionuria together with a 100-fold

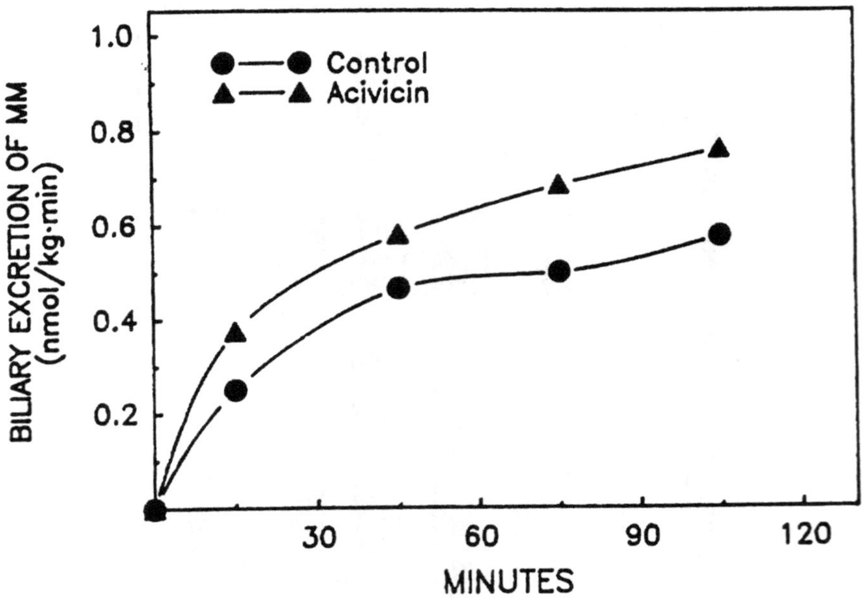

Figure 7. Effect of acivicin (0.1 mmol/kg iv), on the biliary excretion of methylmercury (MM; 10 µmol/kg iv 1 hr after acivicin). (Modified after Gregus et al., 1987b)

increase in urinary excretion of SH as well as a 30-fold enhancement of MM excretion into urine (Gregus et al., 1987b), indicating the different influences hepatic and renal GGT has on the excretion of thiols and MM into bile and urine, respectively.

Species differed from each other not only in terms of excretion rate and composition of biliary SH (Fig. 4) but also in their capacity to excrete MM into bile (Fig. 8). The three species with the highest biliary excretion rate of SH (i.e., mouse, rat, hamster) eliminated more MM into bile than rabbits and guinea pigs, which exhibited the lowest biliary excretion rates of SH. However, within these groups there is no correlation between biliary SH excretion and MM excretion. For example, both the composition and excretion rate of biliary thiols in rats were significantly different from these parameters in hamsters and mice (Fig. 4), yet the excretion rates of MM in bile in these three species were very similar

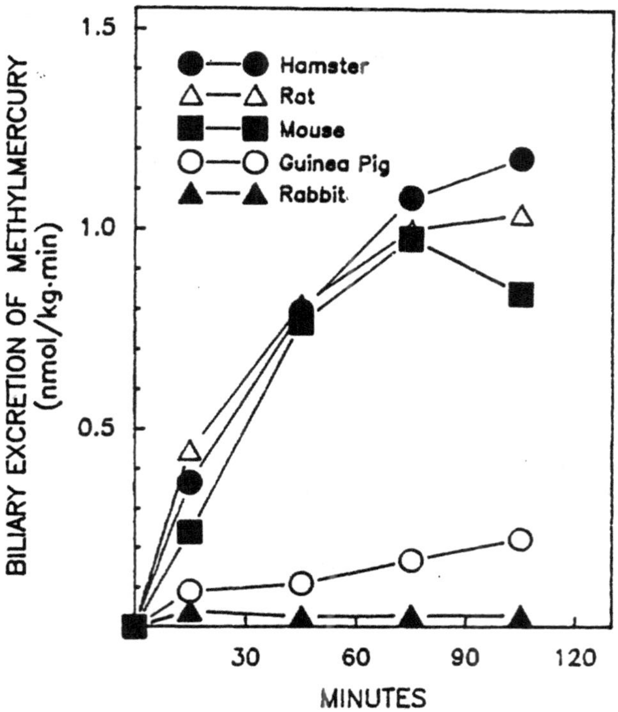

Figure 8. Species differences in the biliary excretion of methylmercury (10 μmol/kg iv). (Modified after Stein et al., 1988)

(Fig. 8). This observation suggests that the biliary excretion rate of SH may be an important determinant in the biliary elimination of MM, but the species difference in hepatobiliary transport of MM is not determined solely by the ability to excrete GS or GS-related thiols in bile.

In summary, the present results suggest that GST-mediated formation and GGT-mediated hydrolysis of GS conjugates have limited importance in the biliary excretion of MM. Increased MM excretion into bile following treatment with inducers is probably due to enhanced hepatobiliary transport of thiols rather than enhanced activity of GSTs in liver. Significant portions of GS transported from liver to bile may undergo hydrolysis initiated by hepatic GGT. The extent of GS hydrolysis is determined largely by the activity of hepatic GGT that is age- and species-dependent. However, GGT-initiated hydrolysis of GS apparently does not mediate significant thiol reabsorption from the biliary tree nor does it influence the biliary excretion of MM in rats. Biliary excretion of SH may be an important determinant of biliary elimination of MM in rats, yet it is not the sole determinant of species variations in hepatobiliary transport of MM.

References

Abbot WW, Meister A (1986) Intrahepatic transport and utilization of biliary glutathione and metabolites. Proc Natl Acad Sci USA 83:1246-1250

Allen L, Meck R, Yunis A (1980) The inhibition of gamma-glutamyl transpeptidase from human pancreatitic carcinoma cells by (S,5S)-alpha-amino-3-chloro-4,5-dihydro-5-isoxasoleacetic acid (AT-125; NSC-163501). Res Commun Chem Pathol Pharmacol 27:175-182

Ballatori N, Clarkson TW (1982) Developmental changes in the biliary excretion of methylmercury and glutathione. Science 216:61-62

Ballatori N, Clarkson TW (1985) Biliary secretion of glutathione and glutathione-metal complexes. Fund Appl Toxicol 5:816-832

Gregus, Z, Klaassen, CD (1986) Disposition of metals in rats: comparative study of fecal, urinary and biliary excretion and tissue distribution of eighteen metals. Toxicol Appl Pharmacol 85:24-38

Gregus Z, Madhu C, Klaassen CD (1988) Species variation in toxication and detoxication of acetaminophen _in vivo_: A comparative study of biliary and urinary excretion of acetaminophen metabolites. J Pharmacol Exp Therap 244:91-99

Gregus Z, Stein AF, Klaassen CD (1987a) Age-dependent biliary excretion of glutathione-related thiols in rats: role of gamma-glutamyl transferase. Am J Physiol 253:G86-G92

Gregus Z, Stein AF, Klaassen CD (1987b) Effect of inhibition of gamma-glutamyltranspeptidase on biliary and urinary excretion of glutathione-derived thiols and methylmercury. J Pharmacol Exp Therap 242:27-32

Gregus Z, Varga F (1985) Role of glutathione and hepatic glutathione S-transferase in the biliary excretion of methylmercury, cadmium and zinc: a study with enzyme inducers and glutathione depletors. Acta Pharmacol Toxicol 56:398-403

Habig WH, Pabst MJ, Jakoby WB (1974) Glutathione transferases. The first enzymatic step in mercapturic acid formation. J Biol Chem 249:7130-7839

Inoue M, Kinne R, Iran T, Biempica L, Arias IM (1983) Rat liver canalicular membrane vesicles. Isolation and topological characterization. J Biol Chem 258:5183-5188

Kaplowitz N, Aw TJ, Ookhtens M (1985) The regulation of hepatic glutathione. Ann Rev Pharmacol Toxicol 25:715-744

Magos L, Clarkson TW, Allen J, Snowden R (1979) The effects of bromosulfophthalein, indocyanine green and bilirubin on the biliary excretion of methylmercury. Chem Biol Interact 26:317-320

Meister A, Tate SS, Griffith OW (1981) Gamma-glutamyl-transpeptidase. Methods Enzymol 77:237-253

Norseth T, Clarkson TW (1971) Intestinal transport of ^{203}Hg-labeled methylmercuric chloride. Role of biotransformation in rats. Arch Environ Health 22:568-577

Ohsawa M, Magos L (1974) The chemical form of methylmercury complex in the bile of the rat. Biochem Pharmacol 23:1903-1905

Perrin DD, Watt AE (1971) Complex formation of zinc and cadmium with glutathione. Biochem Biophys Acta 230:96-104

Rankin BB, McIntyre TM, Curthoys NP (1983) Role of the kidney in the interorgan metabolism of glutathione. In: Larsson A, Holmgren A, Orrenius S, Mannervik B (eds) Functions of glutathione: Biochemical, Physiological, Toxicological and Clinical Aspects. Raven Press, New York, p 31-38

Refsvik T, Norseth T (1975) Methylmercuric compounds in rat bile. Acta Pharmacol Toxicol 36:67-78

Refsvik T (1978) Excretion of methylmercury in rat bile: the effect of diethylmaleate, cyclohexene oxide and acrylamide. Acta Pharmacol Toxicol 42:135-141

Refsvik T (1983) The mechanism of biliary excretion of methyl mercury: studies with methylthiols. Acta Pharmacol Toxicol 53:153-158

Simpson RB (1961) Association constants of methylmercury with sulfhydryl and other bases. J Amer Chem Soc 83:4711-4717

Stein AF, Dills RL, Klaassen CD (1986) High-performance liquid chromatographic analysis of glutathione and its thiol and disulfide degradation products. J Chromatog 381:259-270

Stein AF, Gregus Z, Klaassen CD (1988) Species variations in biliary excretion of glutathione-related thiols and methylmercury. Toxicol Appl Pharmacol 93:351-359

Szewczuk A, Milnerowitz H, Polosatov MV, Sobiech KA (1980) Immunofluorescent localization of gamma-glutamyl transferase in rat and bovine tissues. Acta Histochem 66:152-159

Hepatic Transport of Amino Acids

Glutamine and Glutamate Transport in Perfused Liver

Dieter Häussinger
Department of Internal Medicine
University of Freiburg
D-7800 Freiburg
W.-Germany

Introduction

Studies on glutamine and glutamate transport in the intact liver must take into account hepatocyte heterogeneity in nitrogen metabolism (for reviews see Sies & Häussinger 1984; Häussinger et al. 1988). Accordingly, also amino acid transport systems may exhibit a zonal distribution inside the liver acinus. Even when the transporters would be equally distributed along the acinus, under physiological conditions they may bring about net movements of amino acids in opposite directions in different subacinar hepatocyte populations, as a consequence of hepatocyte heterogeneity with respect to amino acid metabolizing enzymes. Studies on amino acid transport across the plasma membrane in the intact organ, albeit reflecting a more physiological situation compared to membrane vesicle or hepatocyte studies on the initial rate of uptake, however, are complicated by the rapid metabolism of the compounds under investigation. These problems can be overcome, when simultaneously measurements of flux through the corresponding metabolic pathways and of the concentration gradients across the plasma membrane are performed and the effect of metabolism in various compartments on the transmembrane concentration gradients and vice versa are determined. This article reviews some of our studies on hepatic glutamine and glutamate transport in the intact perfused rat liver in relation to glutamine and glutamate metabolism.

Hepatic Transport in Organic Substances
Ed. by E. Petzinger, R. K.-H. Kinne, H. Sies

Hepatocyte Heterogeneity in Glutamine Transport and its Characterization in Isolated Perfused Rat Liver

Glutamine metabolizing enzymes exhibit a remarkable zonal distribution along the liver acinus: periportal hepatocytes contain urea cycle enzymes and glutaminase, but no glutamine synthetase, whereas glutamine synthetase is restricted to a small perivenous hepatocyte population (comprising about 7% of all hepatocytes in an acinus) near the outflow of the sinusoidal bed (Häussinger 1983; Gebhardt & Mecke 1983). This structural organization of hepatic nitrogen metabolism at the acinar level and its functional implications are of crucial importance for the maintenance of whole body nitrogen and bicarbonate homeostasis (Häussinger 1983; for reviews see Sies & Häussinger 1984; Häussinger et al. 1988). In the intact liver, periportal glutaminase and perivenous glutamine synthetase are simultaneously active (Häussinger & Sies 1979). Thus, the reciprocal distribution of glutaminase and glutamine synthetase implies a net import of glutamine into periportal glutamine-consuming hepatocytes and a simultaneous net export of glutamine from perivenous glutamine-synthesizing hepatocytes. These opposing net glutamine fluxes across the plasma membrane of periportal and perivenous hepatocytes, respectively, have been studied in detail during metabolic steady states including the influence of various hormones, substrates and the extracellular pH (Häussinger et al 1983; Sies & Häussinger 1984; table 1). They finally determine whether the glutamine balance across the whole liver is positive or negative (table 1). In isolated perfused rat liver, glutamine uptake is inhibited by about 80% after replacement of Na^+ by choline (Häussinger et al. 1985). This is in line with a predominant transport of glutamine across the plasma membrane by the Na^+-dependent system N (Joseph et al. 1978; Kilberg et al 1980). In addition, a Na^+-independent system was suggested to be involved in facilitated diffusion of glutamine from hepatocytes (Fafournoux et al. 1983). However, studies in isolated perfused rat liver showed that periportal glutamine uptake as well as the perivenous release of newly synthesized glutamine, were largely (i.e. 80%) dependent on Na^+ in the perfusion fluid (Häussinger et al. 1985). In addition, histidine inhibited periportal glutamine uptake as well as perivenous glutamine release in the intact liver, with K_i values near 0.2 mM, i.e. in the physiological concentration range.

Glutamine is also transported across the mitochondrial membrane into the mitochondria, where glutaminase is localized (for review see Kovačević and McGivan 1984). The glutamine transport systems in the plasma and mitochondrial membrane generate glutamine concentration gradients between the respective compartments. In rat liver in vivo and isolated rat liver being perfused with a physiological extracellular glutamine concentration of 0.6 mM, the cytosolic and mitochondrial glutamine concentrations are 6.6±0.8 and 19.1±2 mM, respectively (Häussinger et al. 1985). These values largely reflect the much larger periportal compartment of the liver acinus. Thus, under physiological conditions glutaminase is operating with substrate concentrations near its K_m of 22-28 mM (McGivan et al. 1984). Although it is not possible at present to determine the subcellular glutamine concentrations in the small perivenous hepatocyte population containing glutamine synthetase, their total glutamine content can be estimated by measuring whole tissue glutamine levels in presence and absence of methionine sulfoximine, the inhibitor of glutamine synthetase, and after washout of preexisting glutamine from the periportal hepatocytes (which are unable to synthesize their glutamine because they lack glutamine synthetase). Under these conditions glutamine can only be produced and accumulate in perivenous cells at the outflow of the sinusoidal bed from endogenous ammonia (Häussinger 1983) and the glutamine content of this perivenous hepatocyte pool was calculated to be 3.5 μmol/g perivenous liver tissue and was increased about 3-fold when NH_4^+ was added to augment glutamine synthesis. These values are above the glutamine content of whole livers being perfused with glutamine (0.6 mM), suggesting a higher intra/extracellular glutamine concentration gradient in perivenous than in periportal hepatocytes (Häussinger et al. 1985). This could explain the different directions of net glutamine transport in both subacinar compartments.

When livers are perfused with a physiological portal glutamine concentration (0.6 mM) and glutamine synthetase is inhibited by methionine sulfoximine, glutamine tissue levels largely reflect the much larger periportal compartment. Under these conditions, glutamine tissue levels are strongly dependent on the extracellular pH (fig. 1). This pH dependence is largely abolished by amiloride (fig. 1, Lenzen et al. 1987; Soboll et al. 1988). In addition, at each pH value tested, the glutamine tissue levels were decreased by amiloride (fig. 1); at

ADDITIONS	PERIPORTAL NET GLUTAMINE INFLUX	PERIVENOUS NET GLUTAMINE EFFLUX	TOTAL LIVER GLUTAMINE BALANCE
None	67±3	148±10	81±11
NH_4Cl (0.6 mM)	168±9	248±12	80±8
Glucagon (0.1 μM)	231±32	84±20	-147±30
Phenylephrine (5 μM)	88±4	113±9	25±8
Glucagon plus NH_4Cl	379±36	122±27	-254±33
Glucagon plus Phenylephrine	152±15	64±3	-88±22
Phenylephrine plus NH_4Cl	299±7	83±5	-216±15

Table 1. Simultaneous net glutamine uptake by periportal hepatocytes and net glutamine release from perivenous hepatocytes during metabolic steady states in isolated perfused rat liver.
The influent glutamine concentration was 0.6 mM. Data (nmol/g/min) are given as means±SEM (n=4-15). Total liver glutamine balance refers to hepatic glutamine uptake (-) or glutamine release (+).

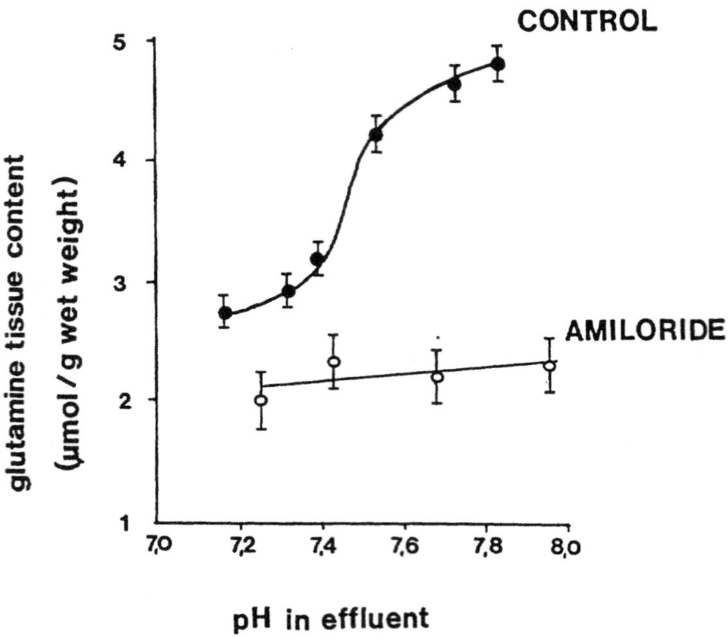

Fig. 1: Effect of extracellular pH and amiloride (1 mM) on glutamine tissue levels in perfused rat liver.
Livers were perfused with a physiological glutamine concentration (0.6 mM) and glutamine synthetase was inhibited by methionine sulfoximine. (From Soboll et al. 1988)

pH 7.4 the respective values were 3.56±0.15 (n=5) and 2.33±0.33 (n=5) μmol/g liver wet weight in the absence and presence of amiloride (1 mM), respectively. These findings may suggest an involvement of the Na^+/H^+ exchanger in glutamine transport. In line with this was the finding that phorbol 12-myristate 13-acetate (PMA 10^{-7}M), which is known to activate the Na^+/H^+ exchanger increased the glutamine tissue level from 3.65±0.28 (n=3) to 6.33±0.31 (n=5) μmol/g liver wet weight.

Glutamine Transport is an Important Factor Controlling its Metabolism

Studies on the initial rate of glutamine transport across the plasma and mitochondrial membranes in isolated membrane vesicles, mitochondria or hepatocytes showed that glutamine transport was always much faster than its known rates of metabolism (Kovačević & Bajin 1982; Fafournoux et al.1983; Kilberg 1980), favouring the view that transport of glutamine is not controlling its metabolism. Recent studies in the intact perfused liver, however, gave evidence for such a control of glutamine breakdown by its transport even under physiological conditions (Häussinger et al.1985; Lenzen et al.1987). (i) In presence of a physiological portal glutamine concentration (0.6 mM) a 6-fold stimulation of glutaminase flux by glucagon and NH_4^+ resulted in a 40% and 50% decrease of the cytosolic and mitochondrial glutamine concentrations, respectively. (ii) Conditions accelerating glutamine breakdown via cytosolic glutamine transamination or mitochondrial glutaminase resulted in a marked decrease of the intra/extracellular glutamine concentration ratio. These concentration ratios were further decreased by histidine in physiological concentrations whereby simultaneously the flux through the glutamine-degrading pathways decreased. Flux through mitochondrial glutaminase was inhibited by about 80% after stimulation of cytosolic glutamine transamination with ketomethionine. This was accompanied by a considerable fall of glutamine tissue levels, suggesting that glutamine entering the cell was already consumed in the cytosol, before it reached the mitochondrial space and that glutamine import into the cell could not keep up with its metabolism. (iii) In presence of a constant portal glutamine concentration (0.6 mM), an increase of the extracellular pH from 7.3 to 7.7 increased the mitochondrial glutamine concentration from 15 mM to about 50 mM, being paralleled by a simultaneous increase of flux through glutaminase. Such a setting of the mitochondrial glutamine concentration by the transport systems in the plasma and the mitochondrial membrane critically affects flux through glutaminase with its high K_m(glutamine) = 22-28 mM. These studies (Lenzen et al. 1987) suggested that the pH control of glutaminase flux in liver is largely mediated by the activities of the glutamine transport systems, which determine the steady state concentrations of glutamine in the cytosolic and mitochondrial compartment.

Glutamate Transport in Perfused Rat Liver

From the high intra/extracellular glutamate concentration gradient, the plasma membrane of hepatocytes was thought to be impermeable for glutamate (Hems et al. 1968). In line with this was the finding that in glutamate-free liver perfusions an effluent glutamate concentration of only about 10 μM is observed although the glutamate tissue content remains fairly high (about 2.5 μmol/g liver). Even when glutamate accumulated inside the liver up to tissue levels of 20 μmol/g wet weight during the metabolism of added glutamine (5 mM), glutamate release was less than 100 nmol/g/min, corresponding to an about 1000 fold intra/extracellular glutamate concentration gradient. Neither addition of glucagon or NH_4Cl nor hypoxia affected the low basal rates of glutamate release. On the other hand, addition of 1-^{14}C-glutamate in physiological concentrations (0.1 mM) to the perfusate resulted in the production of $^{14}CO_2$, indicating the uptake of extracellular glutamate. The maximal rate of $^{14}CO_2$ production of about 0.2 μmol/g/min was observed with glutamate concentrations above 0.5 mM (Häussinger & Gerok 1983) and was low compared to the rates of glutamate oxidation being observed, when the latter compound was generated inside the cell during glutamine breakdown. These studies also demonstrated that the metabolism of vascular glutamate is controlled by the glutamate transport across the liver plasma membrane. $^{14}CO_2$ production from labeled glutamate, however, exceeded net glutamate uptake by the liver, indicating also a simultaneous release of glutamate. Isotope studies with the isolated perfused liver showed that about 70% of total hepatic uptake of vascular glutamate occurred by the small perivenous hepatocyte population containing glutamine synthetase (Häussinger & Gerok 1983). Thus, the transport system in the plasma membrane for glutamate uptake may be heterogeneously distributed in the liver acinus and an about 20-fold higher glutamate uptake activity was calculated for perivenous glutamine synthetase containing hepatocytes compared to periportal ones. Even when glutamate accumulated inside the liver at a rate of about 0.4 μmol/g/min during the metabolism of added glutamine (5 mM), $^{14}CO_2$ production from simultaneously added ^{14}C-glutamate (0.1 mM) remained practically unaffected, indicating that extracellular glutamine and glutamate were metabolized in different subacinar compartments. The perivenous localization of the glutamate uptake system in the plasma membrane was also confirmed in experiments with zonal cell damage. Induction of a perivenous liver cell necrosis by CCl_4 completely abolished hepatic glutamine synthesis and decreased $^{14}CO_2$ production from labeled glutamate by about 80%, whereas periportal urea synthesis was not affected. In contrast, the uptake of proline predominantly occurs in periportal hepatocytes capable of urea synthesis, and was not inhibited after CCl_4 treatment. The high capacity of perivenous hepatocytes to take up vascular glutamate and to synthesize glutamine may be seen as another example for a common subacinar localization of functionally linked processes.

Surprisingly, a variety of physiologically occurring ketoacids was able to stimulate rapidly and reversibly glutamate release from the liver. As expected, this effect was accompanied by a decrease of the tissue glutamate levels. Most effective were the ketoanalogues of leucine, isoleucine, methionine, phenylalanine, tyrosine,and 2-ketobutyrate, whereas the corresponding amino acids, the ketoanalogue of valine, pyruvate, oxoglutarate and oxaloacetate were ineffective (Häussinger & Gerok 1984). Recent studies (Häussinger & Colombo, unpublished) showed that benzoate was even slightly more potent than ketoisocaproate (KIC) (fig. 2). The stimulation of glutamate efflux from the liver by the mentioned compounds is very sensitive; already at concentrations of 50 μM, KIC and benzoate led to a doubling of hepatic glutamate output. Glutamate efflux under the influence of the ketoacids seems not to be related to the metabolism of the ketoacids (Häussinger & Gerok 1984), but this does not exclude a glutamate-/carboxylate exchange. The amount of glutamate released from the liver under the influence of the ketoacids or benzoate depends on the hepatic capacity to regenerate glutamate inside the cell. Also, when glutamine (5 mM) is present in order to regenerate intracellular glutamate, ketoacid- and benzoate-induced glutamate efflux is a saturable process; double reciprocal analysis of the data gave maximal rates of glutamate efflux of 0.90, 0.92, 1.08 and 0.74 μmol/g/min for benzoate, KIC, ketomethionine and phenylpyruvate, respectively, with corresponding K_m values of about 0.8, 1.2, 3 and 3.8 mM, respectively. The effect of the ketoacids on glutamate efflux is also observed after replacement of Na^+ by choline, indicating the Na^+ independence of the process. Thus it is unlikely that the Na^+-dependent glutamate transport system found in membrane vesicles (Sips et al. 1982) or in cultured hepatocytes (Gebhardt & Mecke 1983) is involved, whereby the latter authors also suggested the existence of a Na^+-independent transport system. Recent studies on glutamate efflux under the influence of benzoate (Häussinger & Colombo, unpublished) showed an accompanying K^+ efflux and an inhibition of benzoate-induced glutamate efflux by quinidine. This could point to the involvement of a glutamate-transporting anion channel.

Whereas glutamate is predominanly taken up by a small perivenous hepatocyte population, glutamate efflux from the liver under the influence of ketoacids also involves the much larger periportal compartment. Whereas KIC stimulates glutamate efflux from the liver, it probably only exerts, if at all, little effect on glutamate uptake, as suggested by experiments on $^{14}CO_2$ production from added 1-^{14}C-glutamate. In addition, there is some evidence from these isotope data for a simultaneous periportal glutamate release and perivenous reuptake (Häussinger & Gerok 1984). Addition of KIC to perfused rat liver inhibits urea synthesis due to an increased glutamate withdrawal by export from the liver. When, however, the intracellular glutamate supply was high enough to meet the glutamate requirements for urea synthesis as well as for KIC-induced glutamate efflux, urea synthesis was no longer inhibited. Branched chain ketoacids and benzoate have been successfully used for treatment of hyperammonemia, however, their mechanism of action is not fully understood. It is remarkable that just these compounds proved to be most effective in stimulating glutamate export from the perfused liver and one could speculate about a link between their beneficial effects in hyperammonemic states and their potency to stimulate glutamate efflux: these compounds could improve hepatic ammonia detoxication by activating an alternative pathway, namely glutamate formation and export from the liver, when the capacities for glutamine and urea synthesis are exhausted in a diseased liver.

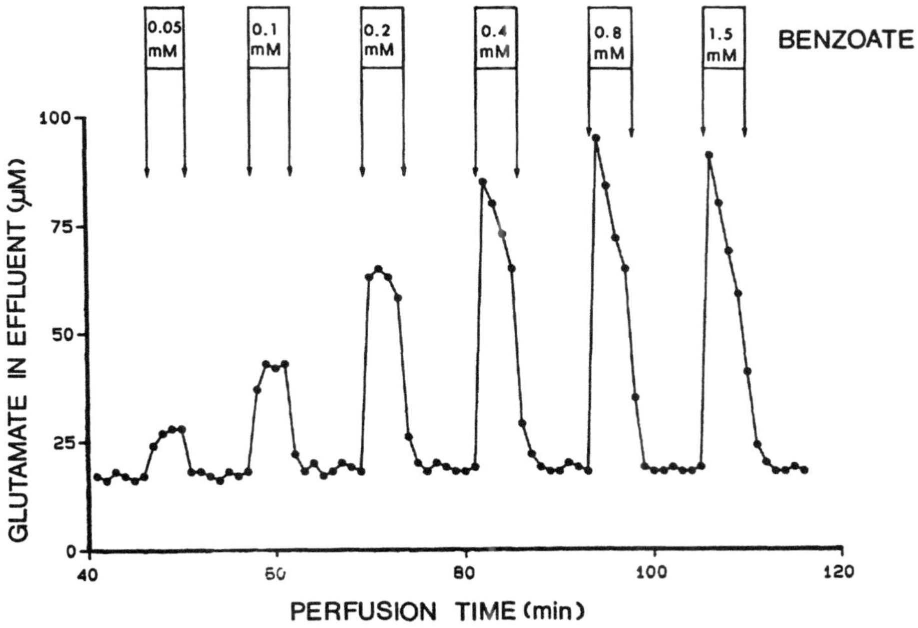

Fig. 2: Effect of benzoate on glutamate release from perfused rat liver

References

Fafournoux P, Demigné C, Rémésy C, Le Cam ALE (1983) Bidirectional transport of glutamine across the cell membrane in rat liver. Biochem J 216: 401-408

Gebhardt R, Mecke D (1983) Heterogeneous distribution of glutamine synthetase among rat liver parenchymal cells in situ and in primary culture. EMBO J 2: 567-570

Gebhardt R, Mecke D (1983) Glutamate uptake by cultured rat hepatocytes is mediated by hormonally inducible, sodium-dependent transport systems. FEBS Lett 161: 275-278

Häussinger D, Sies H (1979) Hepatic glutamine metabolism under the influence of the portal ammonia concentration in the perfused rat liver. Eur J Biochem 101: 179-184

Häussinger D, Gerok W, Sies H (1983) Regulation of flux through glutaminase and glutamine synthetase in isolated perfused rat liver. Biochim Biophys Acta 755: 272-278

Häussinger D (1983) Hepatocyte heterogeneity in glutamine and ammonia metabolism and the role of an intercellular glutamine cycle during ureogenesis in perfused rat liver. Eur J Biochem 133: 269-275

Häussinger D & Gerok W (1983) Hepatocyte heterogeneity in glutamate uptake by isolated perfused rat liver. Eur J Biochem 136: 421-425

Häussinger D, Gerok W (1984) Regulation of hepatic glutamate metabolism. Role of oxoacids in glutamate release from isolated perfused rat liver. Eur J Biochem 143: 491-497

Häussinger D, Soboll S, Meijer AJ, Gerok W, Tager JM, Sies H (1985) Role of plasma membrane transport in hepatic glutamine metabolism. Eur J Biochem 152: 597-603

Häussinger D, Meijer AJ, Gerok W, Sies H (1988) Hepatic nitrogen metabolism and pH homeostasis. In: Häussinger D (ed) pH Homeostasis. p 337-377 Academic Press London

Hems R, Stubbs M, Krebs HA (1968) Restricted permeability of rat liver for glutamate and succinate. Biochem J 107: 807-815

Joseph SK, Bradford NM, McGivan JD (1978) Characteristics of the transport of alanine, serine and glutamine across the plasma membrane of isolated rat liver cells. Biochem J 176: 827-836

Kilberg, MS, Handlogten ME, Christensen HN (1980) Characteristics of an amino acid transport system in rat liver for glutamine, asparagine, histidine and closely related analogs. J Biol Chem 255: 4011-4019

Kovačević Z, Bajin K (1982) Kinetics of glutamine efflux from liver mitochondria loaded with the ^{14}C-labeled substrate. Biochim Biophys Acta 687 : 291-295.

Kovačević Z, McGivan JD (1984) Glutamine transport across biological membranes. In: Häussinger D, Sies H (eds) Glutamine metabolism in mammalian tissues. p 49-59, Springer Verlag Heidelberg

Lenzen C, Soboll S, Sies H, Häussinger D (1987) pH control of hepatic glutamine degradation: role of transport. Eur J Biochem 166: 483-488

McGivan JD, Bradford NM, Verhoeven AJ, Meijer AJ (1984) Liver glutaminase. In: Häussinger D, Sies H (eds) Glutamine metabolism in mammalian tissues, p 122-137 Springer Verlag Heidelberg

Sies H, Häussinger D (1984) Hepatic glutamine and ammonia metabolism. In: Häussinger D, Sies H (eds) Glutamine metabolism in mammalian tissues. p 78-97, Springer Verlag Heidelberg

Sips HJ, DeGraaf PA, Van Dam K (1982) Transport of L-aspartate and L-glutamate in plasma mambrane vesiclesfrom rat liver. Eur J Biochem 122:259-264.

Soboll S, Lenzen C, Häussinger D (1988) Regulation of glutamine transport in isolated perfused rat liver. In: Soeters P, Meijer AJ, Holm E (eds) Advances in hepatic encephalopathy and urea cycle diseases. Elsevier Amsterdam in press.

Regulation of Hepatic Amino Acid Transport and Partial Purification of the System A Carrier

M.S. Kilberg
Department of Biochemistry and Molecular Biology
University of Florida
School of Medicine
Box J-245, JHMHC
Gainesville, Florida 32610, USA

Amino acid transport across the plasma membrane is an important site for metabolic regulation. Indeed, in many instances transport actually represents the first step of catabolism. Evidence from several laboratories suggests that transport into the hepatocyte is the rate-limiting step for alanine metabolism when tested at physiological levels of the substrate (Christensen, 1983). The general importance of amino acid flows between the major tissues of the body has been reviewed by Christensen (1982). The concept of metabolic control via translocation of nutrients across the plasma membrane is not new; Exton et al. (1970) proposed that substrate supply was a key regulator of amino acid-dependent gluconeogenesis nearly two decades ago. More recently, Christensen (1983) has postulated that concurrent changes in hepatic amino acid metabolism and accumulation ensure that transport across the plasma membrane remains rate-limiting over a wide range of extracellular substrate concentrations. Data from our laboratory indicate that the hepatic utilization of several amino acids for *de novo* glucose synthesis may be limited by availability. The rate of incorporation into

Hepatic Transport in Organic Substances
Ed. by E. Petzinger, R. K.-H. Kinne, H. Sies

glucose, as one measure of the rate of amino acid catabolism, appears to be limited by transport for several amino acids including glycine, cysteine, alanine, serine, threonine, proline, and asparagine (Kilberg et al., 1985a).

One property that this group of amino acids has in common is that they are among the best substrates for the hepatic System A transporter. System A activity, first described by Oxender and Christensen (1963) for the Ehrlich ascites tumor cell, is a Na^+-dependent carrier which mediates the translocation of all neutral amino acids but exhibits the greatest affinity for those with small, unbranched side-chains. Carrier activity can be monitored selectively by measuring the Na^+-dependent uptake of the non-metabolizable amino acid analog, 2-aminoisobutyric acid (AIB). In addition to its alkalization dependence, a number of other features have served to distinguish hepatic System A from other neutral amino acid transporters (Kilberg et al., 1985a). In liver tissue System A is subject to complex regulation by a wide variety of hormones as well as by substrate availability, the latter process is commonly referred to as adaptive regulation (Kilberg, 1986). Although a few reports have demonstrated a rapid membrane potential-related induction of hepatic System A in response to hormones (Edmondson and Lumeng, 1980), most of the increased activity following hormone treatment or substrate starvation requires several hours and has been shown to result from <u>de novo</u> synthesis of mRNA and protein (Kilberg, 1986). The dependence on macromolecular synthesis for adaptive induction of System A activity in H4 hepatoma cells is illustrated in Fig. 1 (taken from Kilberg et al., 1985b).

The addition of the protein synthesis inhibitor cycloheximide, at the time of initiating amino acid starvation, completely blocks the adaptive induction of System A activity (Fig. 1A). If one waits for 90-120 min following initiation of amino acid starvation and then adds cycloheximide, the increase in transport activity remains equal to that of the starved cells at the time that the inhibitor was added. The RNA synthesis

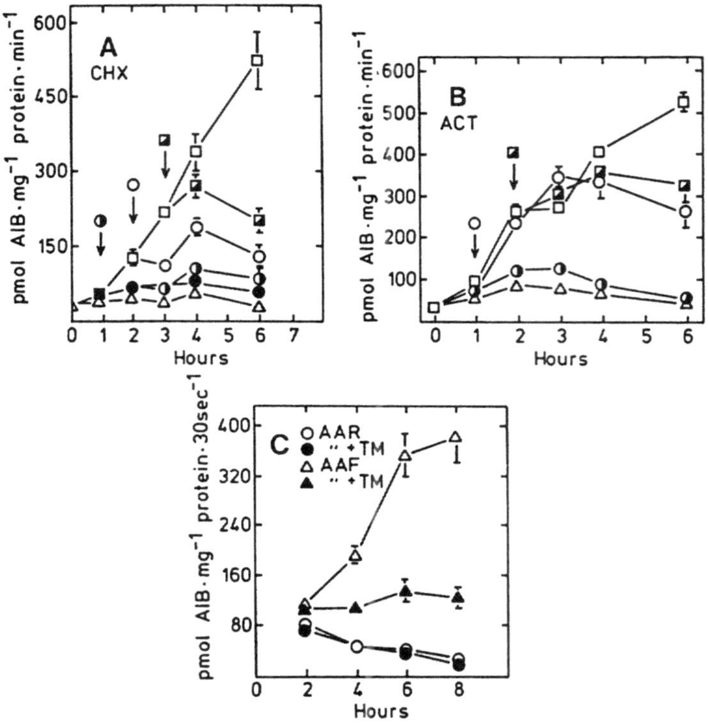

Figure 1: (Parts A and B) Effect of cycloheximide (CHX) or actinomycin (ACT) on the adaptive regulation of hepatic System A. Rat H4 hepatoma cells were transferred to amino acid-containing medium (△) or amino acid-free medium (□). In some cases 0.1 mM cycloheximide (Part A) was added after 0 (●), 1 (◑), 2 (○), or 3 h (◪), or 4 μM actinomycin (Part B) was added after 0 (◑), 1 (○), or 2 h (◪) of incubation in the amino acid-free medium. The Na^+-dependent uptake of 0.05 mM ^{3}H-AIB was measured for 1 min at 37°C. (Part C) H4 hepatoma cells were incubated in either amino acid-containing medium (○, ●) or amino acid-free medium (△, ▲) in the presence or absence of 1 μg/ml tunicamycin (●, ▲). The Na^+-dependent uptake of 0.05 mM ^{3}H-AIB was measured for 30 seconds at 37°C by measuring the uptake in the presence or absence of sodium (Kilberg et al., 1985b). Where not shown, the standard deviation bars are contained within the symbol.

inhibitor actinomycin also prevents the adaptive induction of System A activity (Fig. 1B). In this case, starving the cells for 60 min or more prior to the addition of the inhibitor allows for continued synthesis at the control rate (starved cells) for approximately 1-2 h, then the increase in activity slows to a value below that of the starved control (Fig. 1B). These data are in agreement with those of Gazzola et al. (1981) and indicate that within 60 min following amino acid starvation of cultured cells, a large portion of the System A mRNA required to initiate and maintain adaptive induction has already been synthesized. It is also evident, that both the carrier protein and the corresponding mRNA have relatively short half-lives. Previous work has shown that the starvation-induced carrier activity, when repressed by the addition of amino acid substrates, decays with a half-life of about 1.5 h (Bracy et al., 1986).

Our laboratory has also demonstrated that the asparagine-linked glycoprotein biosynthesis inhibitor tunicamycin completely prevents the induction of System A in response to amino acid starvation (Fig. 1C). Collectively, our results as well as those from a number of other laboratories indicate that the stimulation of System A activity following substrate deprivation of cells results from the derepression of a gene coding for a plasma membrane-bound glycoprotein. As a working hypothesis we assume this glycoprotein to be the carrier itself. This interpretation is supported by the fact that from the amino acid-starved cells one can isolate plasma membrane vesicles which retain the elevated levels of System A activity (Fig. 2A).

Similar results to those shown for amino acid-starved cells can be obtained when one monitors the System A activity in liver plasma membrane vesicles prepared from glucagon-treated rats (Fig. 2B). Plasma membrane vesicles prepared 4-6 h following glucagon injection of donor animals retain the elevated System A activity seen in intact hepatocytes (Schenerman and Kilberg, 1986). The induction of hepatic System A activity by glucagon either in vivo or in vitro is analogous to that following amino acid starvation of cells in vitro in that it requires de novo synthesis of a mRNA and a plasma membrane-bound glycoprotein (Kilberg et al., 1985a). The ability to induce System A activity via hormone treatment in vivo prior to preparation of plasma membranes has allowed the isolation of vesicles with sufficient activity to develop a reconstitution assay for the System A carrier.

As mentioned above, the transport activity in freshly isolated plasma membrane vesicles from hormone-treated rats is proportional to that of intact hepatocytes from the same animal (Schenerman and Kilberg, 1986). The induction of hepatic

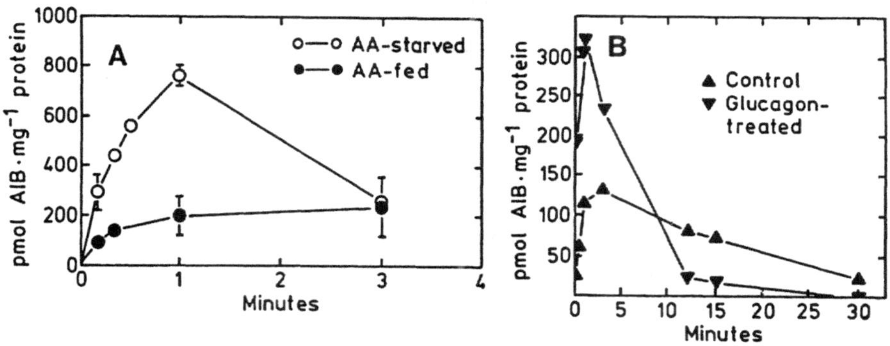

Figure 2: (Part A) Adaptive regulation in plasma membrane vesicles from substrate-starved rat hepatoma cells. Fao cells in monolayer culture were incubated for 6 h in amino acid-free medium (Krebs-Ringer bicarbonate buffer) or amino acid-containing medium (Krebs-Ringer bicarbonate buffer + 20 mM asparagine). Following this incubation, plasma membranes were isolated by the method of Prpic et al. (1984), and assayed for System A activity by measuring the Na^+-dependent uptake of 0.05 mM ^{3}JAIB at 37°C (Bracy et al., 1987). The Na^+-dependent transport was determined by measuring the uptake in the presence or absence of a Na^+ gradient using a nitrocellulose filter assay (0.12 M sucrose, 10 mM HEPES-KOH, pH 7.5, and either 100 mM NaSCN or KSCN). Where not shown, the standard deviations are contained within the symbol. (Part B) System A activity in rat liver plasma membrane vesicles from control or glucagon-injected (1 mg per 100 g body weight) animals. The freshly isolated vesicles were tested for Na^+-dependent AIB transport, as described above. The standard deviations, omitted for clarity, were generally less than 10%. (From Schenerman and Kilberg, 1986).

System A activity by starvation (Quinlan et al., 1982) or diabetes (Rosenthal et al., 1985) is also retained in the isolated vesicles from those animals. As shown in Fig. 3, most hepatic plasma membrane proteins, including the System A carrier, can be solubilized in detergent/urea, precipitated with polyethylene glycol, and reconstituted into artificial proteoliposomes using a freeze-thaw technique (Bracy et al., 1987).

Using the reconstitution procedure, the System A activity recovered in the proteoliposomes is proportional to that in the plasma membrane vesicles from which the proteins were solubilized (Table 1). This relation holds true regardless of whether one measures the basal transport activity or that induced by prior treatment of the animal with either glucagon alone or the combination of glucagon and dexamethasone. These two hormones have been shown to act synergistically to induce System A activity to levels greater than the sum of the induction by either hormone alone (Kletzien et al., 1975).

We have evidence that the hormone-stimulated System A activity is associated with a membrane protein which is not solubilized by a variety of detergents when tested alone, but which is subject to solubilization by 2.5% cholate/4 M urea and is fully active following the reconstitution procedure outlined in Fig. 3. This and other unique properties have permitted us to partially purify the System A carrier protein. The data presented in Table 2 illustrate the degree of purification which can be achieved through a number of steps including selective solubilization and differential polyethylene glycol precipitation of the carrier.

To achieve the indicated enrichment of approximately 2600-fold the membrane vesicles were first treated with 1% octanoyl-N-methylglucamide. This detergent solubilizes approximately 40% of the total membrane protein, while leaving nearly all of the System A carrier associated with the insoluble membrane remnants. The transport activity is then solubilized

membranes + 2.5% cholate/4M urea
30 min ice bath
centrifuge 105K g for 30 min

pellet —

supernatant + 30% PEG (1:2)
15 min ice bath
centrifuge 30K g for 30 min

supernatant —

pellet - wash 3 x with KMB

resuspend in KMB with asolectin (20:1)

freeze/thaw in dry ice/EtOH bath

dilute with KMB (10:1)

sonicate until clear (10 to 30 sec)

centrifuge 105 K g for 30 min

transport proteoliposomes

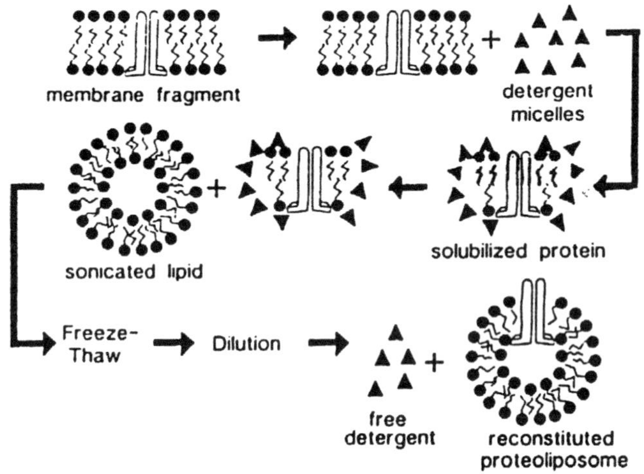

Figure 3: Reconstitution procedure for the hepatic System A carrier protein. Plasma membrane vesicles containing System A activity were solubilized in 2.5% cholate/4 M urea, as outlined in the diagram. Following precipitation of the membrane proteins by polyethylene glycol, the protein pellet was washed and resuspended in a buffer containing 200 mM KCl, 1 mM $MgCl_2$, and 10 mM HEPES-KOH, pH 7.5 (KMB). After a freeze/ thaw step, the mixture was diluted to lower the detergent concentration below its critical micellar concentration. The mixture was then sonicated until clear, and the resulting proteoliposomes collected by centrifugation. Amino acid transport via System A was measured as described for membrane vesicles in Fig. 2.

TABLE 1.

RECONSTITUTION OF HORMONE-INDUCED SYSTEM A TRANSPORT ACTIVITY

Treatment	Plasma Membrane Vesicles	Reconstituted Liposomes
	$pmol \cdot mg^{-1} protein \cdot min^{-1}$	
Control	114	147
Glucagon	514	422
Glucagon + Dexamethasone	1267	1147

TABLE 2.

PARTIAL PURIFICATION OF THE HEPATIC SYSTEM A CARRIER PROTEIN

Step	Protein (mg)	Total Activity	% Activity recovered	Specific Activity	Enrichment
Homogenate*	260	1560	---	6	---
Vesicles	4.7	1149	74	243±15	40
Proteoliposomes	0.07	1091	70	16000±1500	2600

*Estimated values, the transport activity in the homogenate is difficult to measure with any degree of precision.

by incubation of the membrane remnants in 2.5% cholate/4 M urea for 30 min at 4°C. The solubilized proteins, including the System A carrier, are subjected to differential precipitation. Instead of using 30% PEG (Fig. 3), which will precipitate all proteins in the solubilized fraction, the addition of 3% PEG selectively precipitates only 10% of the total protein, but approximately 75% of the System A carrier. The PEG-precipitated proteins are resuspended in 10 mM HEPES buffer containing 0.05% Triton X-100 and 0.25 mg/ml asolectin (crude soybean phospholipid mixture), sonicated for 10 seconds, and then the insoluble material is removed by centrifugation. The System A activity remains associated with the resulting supernatant and is reconstituted into asolectin liposomes as described in Fig. 3. The resulting proteoliposomes yield a specific activity for Na^{+}-dependent AIB transport of 16,000-20,000 pmoles·min^{-1}, an enrichment of greater than 2,600-fold (Table 2). This preparation contains only 0.03% of the initial protein, but approximately 70% of the total System A activity. Following one-dimensional polyacrylamide gel electrophoresis, silver staining reveals 6-8 prominant protein bands (some of which are known to be cytoskeletal components) and a number of proteins present in lesser amounts.

The hormonal and substrate-dependent regulation of the hepatic System A carrier has been well described at the cellular level, yet numerous questions remain concerning the molecular mechanisms responsible for these regulatory phenomenon. Answers to these questions will come only when we have the ability to monitor the absolute levels of the System A carrier protein and its corresponding mRNA. Identification of the System A carrier protein will allow the development of both immunological and molecular biological approaches in the study of hepatic amino acid transport and its impact on metabolism.

References

Bracy DS, Handlogten ME, Barber EF, Han H-P, Kilberg MS (1986) Cis-Inhibition, trans-inhibition, and repression of hepatic amino acid transport mediated by System A. J Biol Chem 261:1514-1520

Bracy DS, Schenerman MA, Kilberg MS (1987) Solubilization and reconstitution of hepatic System A-mediated amino acid transport. Preparation of proteoliposomes containing glucagon-stimulated transport activity. Biochim Biophys Acta 899:51-58

Christensen HN (1982) Interorgan amino acid nutrition. Physiol Rev 62:1193-1233

Christensen HN (1983) Hypothesis: Control of hepatic utilization of alanine by membrane transport or by cellular metabolism. Biosci Rep 3:905-913

Edmondson JW, Lumeng L (1980) Biphasic stimulation of amino acid uptake by glucagon in hepatocytes. Biochem Biophys Res Commun 96:61-68

Exton JH, Mallette LE, Jefferson LS, Wong EHA, Friedmann N, Miller TB, Jr, Park CR (1970) The hormonal control of hepatic gluconeogenesis. Recent Prog Horm Res 26:411-457

Gazzola GC, Dall'Asta V, Guidotti GG (1981) Adaptive regulation of amino acid transport in cultured human fibroblasts. J Biol Chem 256:3191-3198

Kilberg MS (1986) System A-mediated amino acid transport: metabolic control at the plasma membrane. Trends Biochem Sci 11:183-186

Kilberg MS, Barber EF, Handlogten ME (1985a) Characteristics and hormonal regulation of amino acid transport System A in isolated rat hepatocytes. Curr Top Cell Regul 25:133-163

Kilberg MS, Han H-P, Barber E, Chiles TC (1985b) Adaptive regulation of neutral amino acid transport System A in rat H4 hepatoma cells. J Cell Physiol 122:290-298

Kletzien RF, Pariza MW, Becker JE, Potter VR (1975) A "permissive" effect of dexamethasone on the glucagon induction of amino acid transport in cultured hepatocytes. Nature 256:46-47

Oxender DL, Christensen HN (1963) Distinct mediating systems for the transport of neutral amino acids by the Ehrlich cell. J Biol Chem 238:3686-3699

Prpic V, Green KC, Blackmore PF, Exton JH (1984) Vasopressin, angiotensin II-, and α_1-adrenergic-induced inhibition of Ca^{2+} transport by rat liver plasma membrane vesicles. J Biol Chem 259:1382-1385

Quinlan DC, Todderud CG, Kelley DS, Kletzien RF (1982) Sodium-gradient-stimulated transport of L-alanine by plasma-membrane vesicles isolated from liver parenchymal cells of fed and starved rats. Biochem J 208:685-693

Rosenthal NR, Jacob R, Barrett E (1985) Diabetes enhances activity of alanine transport in liver plasma membrane vesicles. Am J Physiol 248:E581-E587

Schenerman MA, Kilberg MS (1986) Maintenance of glucagon-stimulated System A amino acid transport acitivity in rat liver plasma membrane vesicles. Biochim Biophys Acta 856:428-436

Hormonal Control of Amino Acid Transport Systems in Cultured Periportal and Perivenous Hepatocytes

R. Gebhardt
Physiologisch-chemisches Institut
Universität Tübingen
Hoppe-Seyler-Str. 4
D-7400 Tübingen
Federal Republic of Germany

Introduction

The liver plays an important role in the metabolism of amino acids and the homeostasis of their concentrations in blood (Schimassek and Gerok, 1965). This is accomplished by uptake of amino acids in the absorptive phase followed by their metabolic conversion and by release of amino acids liberated by intracellular proteolysis in the postabsorptive phase or during starvation. The transport across the plasma membrane is mediated by several transport systems with overlapping specificity which are either shared with other cell types (e.g. systems A, ASC, L) or seem to be unique to the hepatocyte (e.g. system N) (for review see Kilberg, 1982). Several studies have shown that alanine transport across the plasma membrane is rate-controlling for its metabolism (Sips et al.,1980; Fafournoux et al.,1983a). Similar conclusions have been drawn for the uptake of glutamine via system N (Häussinger et al.,1985). If so, the transport step should be a major point of regulation of amino acid metabolism by the liver. Indeed, some of these transport systems have been found to be regulated by various mechanisms including long-term regulation by hormones (for adaptive and short-term hormonal control, not reviewed here, see Kilberg et al., 1985; Moule et al., 1987). Moreover, since many metabolic functions of the liver are known to be heterogeneously distributed among

Hepatic Transport in Organic Substances
Ed. by E. Petzinger, R. K.-H. Kinne, H. Sies

the three zones of the liver acinus (Jungermann and Katz, 1982), one would expect that amino acid transport, too, is zonated. The present paper summarizes recent studies from this as well as other laboratories concerning long-term hormonal regulation and zonation of amino acid transport in the liver.

Hormonal Regulation of Amino Acid Transport

System A was the first amino acid transport system recognized to respond to hormones by increasing the number and insertion of new carrier molecules into the plasma membrane. The wide variety of hormones as well as the individual features of this regulation have been reviewed in great detail (Kilberg, 1982; Kilberg et al., 1985). Worth to be presented here is the fact that both insulin and glucagon act as potent inducers under the permissive action of dexamethasone (dex) (Kletzien et al., 1975) and the rapid time course of induction leading to a 4 to 8-fold increase within 3 to 6 h (see Kilberg, 1982; Gebhardt and Kleemann, 1987).

The second example of a transport system showing long-term hormonal regulation was the Na^{+}-dependent transport of glutamate (Gebhardt and Mecke, 1983a) and soon thereafter, system N was found to be influenced by hormones, too (Gebhardt and Mecke, 1984a; Gebhardt and Kleemann, 1987). Whereas glutamate uptake, via system G^{-} (Gebhardt and Williams, 1986; Gebhardt and Burger, 1987), is stimulated more than 15-fold by dex alone, with some further enhancement by glucagon (Gebhardt and Mecke, 1984a), the hormonal control of system N, qualitatively, is much similar to that of system A (Gebhardt and Kleemann, 1987). Thus, insulin in combination of dex clearly stimulated the uptake via system N, while it strongly depressed the induction of system G^{-} by dex. Perhaps the most pronounced difference between the hormonal induction of systems N and A is the very slow time course of the former, which reached maximal stimulation only between 24 and 48 h (Gebhardt and Kleemann, 1987). A similar slow induction is found for system G^{-} (Gebhardt and Mecke, 1983a). Thus, these two amino acid transport systems, which seem to be important for controlling the nitrogen balance as well as pH homeostasis (Gebhardt and Mecke, 1984b; Häussinger et al., 1984), may adapt only during profound changes of the metabolic situation, e.g. during long-term starvation, high protein diet or diabetes.

The fact that the Na^+-dependent system G^- shows a relatively low capacity in isolated hepatocytes (Gebhardt and Mecke,1983a), membrane vesicles (Sips et al., 1982) or perfused rat liver (Taylor and Rennie, 1987) compared to the Na^+-independent uptake has favored the interpretation that mainly the latter might be responsible for the uptake of glutamate from the blood or perfusate (Häussinger, 1986; Ballatori et al., 1986). This view was supported by the finding that a Na^+-dependent component might be localized at the bile canalicular rather than at the sinusoidal membrane (Ballatori et al., 1986). However, the failure to detect such a transport in sinusoidal membrane vesicles may be due to the striking heterogeneity of this transport system (see below). In fact, the existence of a sinusoidal Na^+-dependent transport system for glutamate has been clearly demonstrated in the perfused rat liver (Taylor and Rennie, 1987).

Acinar Heterogeneity of Amino Acid Transport

Although in recent years accumulating evidence suggested that many metabolic functions are preferentially carried out in either the periportal or the perivenous zone of the liver acinus, nothing was known about a possible zonation of amino acid transport for a long time. However, when it was realized by enzyme immunohistochemistry (Gebhardt and Mecke, 1983b) as well as on metabolic grounds (Häussinger, 1983) that hepatic nitrogen metabolism, in particular the metabolism of glutamate and glutamine, showed an unexpected and rather extreme heterogeneity, people became aware that the transport functions related to this aspect of liver metabolism might be heterogeneous, too. Thus, it was concluded from metabolic studies that glutamate uptake might be located predominantly in the perivenous zone (Häussinger and Gerok, 1983). Similar conclusions were drawn from uptake measurements performed with liver preparations from rats treated with CCl_4 (Gebhardt and Burger, 1987; Taylor and Rennie, 1987). However, direct evidence for a heterogeneous zonal distribution of amino acid transport activities was still lacking.

A direct approach for studying the acinar distribution of hepatic amino acid transport activities and their hormonal regulation is to isolate and cultivate subpopulations of parenchymal liver cells from the perivenous and the periportal zone. Various methods have been proposed in the past which are based mainly on assumed zonal differences with respect to size, density, or surface properties of the cells (for review see Ugele et al., 1987). Often these methods have led to conflicting results as to the correlation of the cells to their acinar origin due to the lack of suitable marker enzymes. Using the striking heterogeneity of glutamine synthetase (GS) as a marker for perivenous hepatocytes (Ugele et al., 1987) we were not able to confirm claims in the literature that separation of hepatocytes on Percoll gradients would be successful. On the other hand, we could demonstrate the successful isolation and cultivation of hepatocytes from acinar zones 1 and 2 by selectively destroying cells of zone 3 by CCl_4 in vivo (Gebhardt and Burger, 1987). An excellent method for the isolation of both, periportal or perivenous hepatocytes, proved to be the so-called digitonin/collagenase perfusion technique, which was developed by Quistorff (1985) and Lindros & Penttilä (1985). The efficiency of this separation technique was recently demonstrated by us using again GS as a marker enzyme. The periportal/perivenous ratio of GS activity amounted to 56 or more (Ugele et al., 1987; Burger et al., 1988) indicating the almost complete absence of this enzyme from the periportal subfraction and a 2 to 3-fold enrichment of GS activity in the perivenous subfraction. The immunocytochemical demonstration of this enzyme and of carbamoylphosphate synthetase (CPS), the localization of which appears mutually exclusive in liver sections (Gaasbeek Janzen et al., 1984; Gebhardt and Mecke, 1984b), led to similar results and demonstrated that both enzymes were expressed by different hepatocytes with the exception of a very small fraction which co-expressed both enzymes (Table 1).

Table 1. Incidence of hepatocytes expressing GS, CPS or both enzymes in isolated periportal or perivenous subfractions.

Cell Preparation	Relative Incidence (%)		
	GS^+	CPS^+	GS^+/CPS^+
Total Susp.	95.4 ± 0.5[a]	7.8 ± 0.6	3.8 ± 0.7
Periportal[b]	99.6 ± 0.3	0.1 ± 0.1	n.d.[c]
Perivenous[b]	89.3 ± 1.8	21.7 ± 0.7	11.2 ± 0.4

[a] Means ± SD calculated from 150-200 cell countings each of 3-5 preparations; [b] isolated according to Burger et al. (1988); [c] n.d., not detectable in significant amounts.

Using these subpopulations we have investigated the possible heterogeneity of amino acid transport systems shortly after isolation (1 h) and during cultivation of the hepatocytes for up to 48 h (Burger et al., 1987; Burger et al., 1988). α-Amino-isobutyric acid (AIB), histidine (HIS), glutamine (GLN), and glutamate (GLU) were used to determine the distribution pattern of systems A, N, and G^-, as well as of the Na^+-independent uptake of these compounds. Specificity of the uptake measurements was guaranteed by the use of specific inhibitors (see Burger et al., 1988). In hepatocytes cultured for 1 h only (which reflect best the status of hepatocytes in vivo) a heterogeneity of the Na^+-independent uptake was found only for HIS and GLN characterized by a higher rate of uptake in pv-cells. On the other hand, the Na^+-dependent uptake via systems A and N showed no significant difference between pp and pv cells, whereas the uptake via system G^- was significantly higher in pv-cells ((pv/pp-ratio:-6.6). Moreover, a linear correlation (r = 0.896) was found between the Na^+-dependent transport for glutamate and the relative GS-activity of the respective hepatocyte preparation (Fig. 1) indicating that under physiological conditions system G^- is tightly associated with the GS^+ hepatocytes.

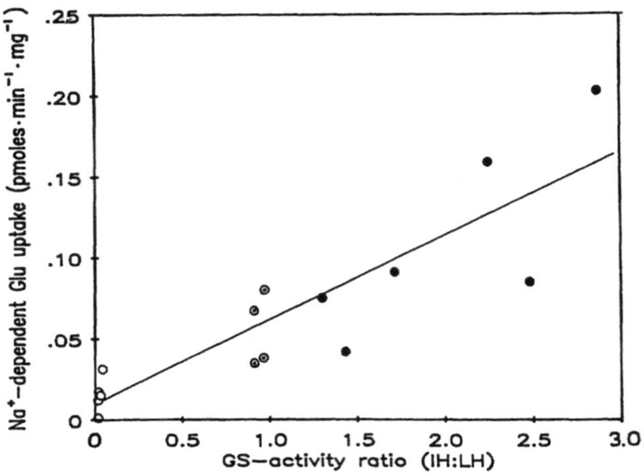

Fig. 1. Correlation of the ratio of GS-activity in isolated hepatocytes (IH) and in corresponding liver homogenate (LH) with Na^+-dependent uptake of glutamate measured in normal (⊙), periportal (○), and perivenous (●) hepatocyte preparations.

During cultivation with or without hormones for 48 h the general pattern of this heterogeneity of amino acid transport was largely maintained, but some interesting exceptions could be noted. With respect to the Na^+-independent uptake of the test substrates AIB and GLU no profound changes concerning their distribution were noted. The heterogeneity found with HIS as substrate was maintained, too. However, while there was no doubt that exposure to dex and insulin enhanced the V_{max} of HIS uptake particularly in the perivenous subpopulations (Burger et al., 1988), it remains uncertain whether the saturable Na^+-independent component for HIS has the same or a slightly lower K_m in the periportal hepatocytes. Even more interesting, the Na^+-independent uptake of GLN appeared to be strikingly different for pp and pv hepatocytes. In pv cells, a hormonal inducible agency with K_m-values between 4 and 5 mM could be easily distinguished (Table 2), whereas in pp cells a K_m of about 2 mM was apparent which most likely, according to preliminary experiments, reflects the contribution of the component mediating HIS uptake. Thus, there is probably an agency specific for GLN in pv cells

Table 2. Kinetic constants of the uptake of glutamate and glutamine by periportal and perivenous cultured hepatocytes

Substrate / Hormonal Treatment	Periportal K_m[a]	Periportal V_{max}[b]	Perivenous K_m	Perivenous V_{max}
	Na^+-dependent uptake[c]			
glutamate				
dex	0.057	0.37	0.066	0.86
glutamine				
control	0.82	0.63	0.79	0.73
dex+ins	1.15	2.81	1.01	2.53
	Na^+-independent uptake[c]			
glutamate				
dex	0.53	0.66	0.52	0.73
glutamine				
control	(2.06)[d]	(1.87)	4.45	3.10
dex/ins	(2.31)	(7.27)	4.81	15.2

[a] mM; [b] $nmoles \cdot min^{-1} \cdot mg\ protein^{-1}$; [c] determined after 48 h according to Burger et al. (1988); [d] for parenthesis see text.

but not in pp cells (indicated by parenthesis (Table 2) or question mark (Fig. 2)). It is not known at present whether these different agencies are also present in the isolated hepatocytes. It is tempting to speculate, however, that the high-K_m component (GLN) determined for the perivenous hepatocytes corresponds to the high K_m found by Fafournoux et al. (1983b) for glutamine efflux. In extension of the asumption of Burger et al. (1988) it is therefore hypothesized that glutamine efflux from the hepatocytes of the "afferent glutamine-synthesizing compartment" (Gebhardt and Mecke, 1984b) is not just mediated by a perivenously predominating Na^+-dependent component with broad substrate specificity but rather by a very specific transport agency which may be localized in the GS^+ hepatocytes only.

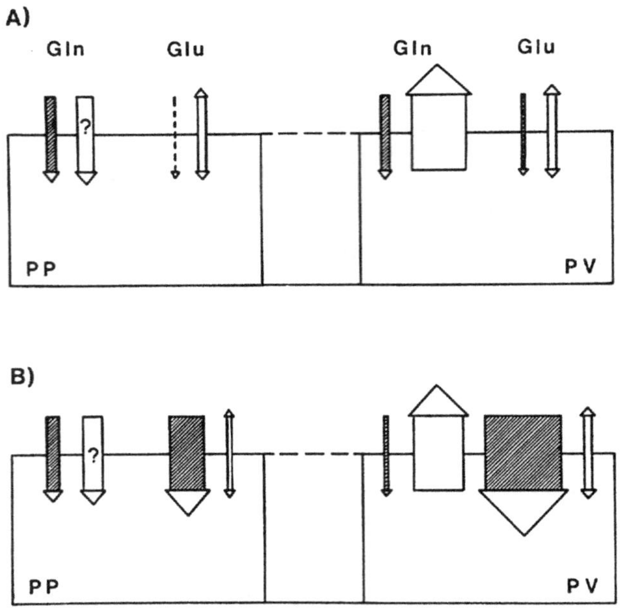

Fig. 2. Schematic illustration of the zonal distribution of the Na^+-dependent (hatched) and Na^+-independent (open arrows) transport agencies for glutamine (Gln) and glutamate (Glu) in periportal (pp) and perivenous (pv) hepatocytes cultured for 48 h without hormones (A) or in the presence of dex (B). The width of the arrow base is proportional to the transport rate (pmoles·min^{-1}·mg $protein^{-1}$) determined with 0.1 mM substrate. Arrows indicate assumed direction of fluxes. (?, see text)

For the Na^+-dependent transport, a surprising finding was that a heterogeneous distribution of system A could be induced by dex and glucagon (pv/pp ratio: 3.1). The individual hormones were ineffective. System N was equally induced in pp and pv cells by dex and insulin without any change in the K_m which was similar in pp and pv cells (Table 2). System G^-, which was barely detectable in pp control cultures, was strongly induced by dex in both hepatocyte populations, but always prevailed in the pv cells (pv/pp ratio: 2.4). In contrast to previous preliminary findings (Gebhardt and Mecke, 1983a) only one Na^+-dependent component could be discriminated by computer analysis even under induced conditions (Burger et al., 1988).

Conclusions

As apparent from the findings presented herein hepatic amino acid transport systems show a considerable zonal heterogeneity, particularly those agencies that mediate the uptake and/or release of glutamine and glutamate. Although the definite proof has to await for the isolation of the GS^+ hepatocyte population, several lines of evidence strongly suggest that the heterogeneity of system G^- and of the Na^+-independent transport components mediating the exchange of histidine and glutamine is tightly coupled to the unique distribution of GS. However, whereas the zonal distribution of GS is rather inflexible (Gebhardt et al., 1988), the data on the hormonal stimulation of amino acid transport show that an induction of the respective transport components in other zones of the acinus is possible, while the zonal preference is still maintained. A possible exception may be a high K_m component involved in glutamine efflux from the GS^+ hepatocytes. The flexibility of the distribution of the amino acid transport systems may partly compensate for the strict cellular expression of GS and may participate in controlling the fluxes through different metabolic pathways by directing the various amino acid in variable amounts to different acinar zones.

References

Ballatori N, Moseley RH, Boyer JL (1986) Sodium gradient-dependent L-glutamate transport is localized to the canalicular domain of liver plasma membranes. J Biol Chem 261:6216-6221

Burger HJ, Gebhardt R, Mayer C, Mecke D (1987) Hormonelle Induktion von Aminosäuretransportsystemen in kultivierten periportalen und perivenösen Rattenhepatocyten.Z Gastroenterol XXV:66

Burger HJ, Gebhardt R, Mayer C, Mecke D (1988) Different capacities for amino acid transport in periportal and perivenous hepatocytes isolated by digitonin-collagenase perfusion. Hepatology, in press

Fafournoux P, Remesy C, Demigne C (1983a) Control of alanine metabolism in rat liver by transport processes or cellular metabolism. Biochem J 210:645-652

Fafournoux P, Demigne C, Remesy C, LeCam A (1983b) Bidirectional transport of glutamine across the cell membrane in rat liver. Biochem J 216:401-408

Gaasbeek Janzen JW, Lamers WH, Moorman AFM, DeGraaf A, Los JA, Charles R (1984) Immunohistochemical localization of carbamoylphosphate synthetase (ammonia) in adult rat liver. Evidence for a heterogeneous distribution. J Histochem Cytochem 32:557-564
Gebhardt R, Mecke D (1983a) Glutamate uptake by cultured rat hepatocytes is mediated by hormonally inducible, sodium dependent transport systems. FEBS Lett 161:275-278
Gebhardt R, Mecke D (1983b) Heterogeneous distribution of glutamine synthetase among rat liver parenchymal cells in situ and in primary culture. EMBO J 2:567-570
Gebhardt R, Mecke D (1984a) Intercellular compartmentation of hepatic ammonia metabolism. In: Kleinberger G, Ferenci P, Riederer P, Thaler H (eds) Advances in hepatic encephalopathy and urea cycle diseases. Karger, Basel, p 126-131
Gebhardt R, Mecke D (1984b) Cellular distribution and regulation of glutamine synthetase in liver. In: Häussinger D, Sies H (eds) Glutamine metabolism in mammalian tissues. Springer, Berlin Heidelberg New York, p 98-121
Gebhardt R, Williams GM (1986) Amino acid transport in adult rat epithelial cells. Cell Biol Toxicol 2:9-20
Gebhardt R, Kleemann E (1987) Hormonal regulation of amino acid transport system N in primary culture of rat hepatocytes. Eur J Biochem 166:339-344
Gebhardt R, Burger HJ (1987) Selective loss of hormonal induction of glutamate transport in primary cultures of hepatocytes from rats treated with CCl_4. J Hepatology 4:381-389
Gebhardt R, Burger HJ, Heini H, Schreiber KL, Mecke D (1988) Alterations of hepatic enzyme levels and of the acinar distribution of glutamine synthetase in response to experimental liver injury in the rat. Hepatology, in press
Häussinger D (1983) Hepatocyte heterogeneity in glutamine and ammonia metabolism and the role of an intercellular glutamine cycle during ureogenesis in perfused rat liver. Eur J Biochem 133:269-275
Häussinger D, Gerok W (1983) Hepatocyte heterogeneity in glutamate uptake by isolated perfused rat liver. Eur J Biochem 136:421-425
Häussinger D, Gerok W, Sies H (1984) Hepatic role in pH regulation - Role of intercellular glutamine cycle. TIBS 9:300-302
Häussinger D, Soboll S, Meijer AJ, Tager JM, Sies H (1985) Role of plasma membrane transport on hepatic glutamine metabolism. Eur J Biochem 152:597-603
Häussinger D (1986) Regulation of hepatic ammonia metabolism: The intercellular glutamine cycle.Adv Enzyme Reg 25:159-180
Jungermann K, Katz N (1982) Functional heterogeneity of liver parenchyma. Hepatology 2:385-395
Kilberg MS (1982) Amino acid transport in isolated rat hepatocytes. J Membrane Biol 69:1-12
Kilberg MS, Barber EF, Handlogten ME (1985) Characteristics and hormonal regulation of amino acid transport system A in isolated rat hepatocytes. Curr Top Cell Regul 25:133-163
Kletzien RF, Pariza MW, Becker JE, Van Potter R (1975) A "permissive" effect of dexamethasone on the glucagon induction of amino acid transpot in cultured hepatocytes.Nature 256: 46-47

Lindros KO, Penttilä KE (1985) Digitonin-collagenase perfusion for efficient separation of periportal or perivenous hepatocytes. Biochem J 228:757-760

Moule SK, Bradford NM, McGivan JD (1987) Short-term stimulation of Na^+-dependent amino acid transport by dibutyryl cyclic AMP in hepatocytes. Biochem J 241:737-743

Quistorff B (1985) Gluconeogenesis in periportal and perivenous hepatocytes of rat liver, isolated by a new high yield digitonin/collagenase perfusion. Biochem J 229:221-226

Schimassek H, Gerok W (1965) Control of the levels of free amino acids in plasma by the liver.Biochem Z 343:407-415

Sips HJ, Groen AK, Tager JM (1980) Plasma membrane transport of alanine is rate-limiting for its metabolism in rat liver parenchymal cells. FEBS Lett 119:271-274

Sips HJ, DeGraaf PA, Van Dam K (1982) Transport of L-aspartate and L-glutamate in plasma membrane vesicles from rat liver. Eur J Biochem 122:259-264

Taylor PM, Rennie MJ (1987) Perivenous localization of Na-dependent glutamate transport in perfused rat liver. FEBS Lett 221:370-374

Ugele B, Locher M, Burger HJ, Gebhardt R (1987) Is there a heterogeneity of liver parenchyma in taurocholate uptake? In: Paumgartner G, Stiehl A, Gerok W (eds) Bile acids and the liver. MTP Press, Lancaster, p 153-160

Differential Inactivation of System A in Normal and Transformed Liver Cells

M.S. Kilberg
Department of Biochemistry and Molecular Biology
University of Florida
School of Medicine
Box J-245, JHMHC
Gainesville, FL 32610
U.S.A.

System A-mediated neutral amino acid transport into normal rat hepatocytes and several hepatoma cell lines was tested for inactivation by sulfhydryl-preferring protein-modifying reagents such as N-ethylmaleimide (NEM) and p-chloromercuribenzene sulfonate (PCMBS). System A activity in hepatocytes and four different hepatoma cell lines was equally sensitive to inhibition by the organic mercurial PCMBS. Inactivation by NEM was substantially different, normal hepatocytes showed the greatest sensitivity, while the hepatoma cells varied in their responsiveness from moderate to no inhibition. The inhibition by PCMBS was rapidly reversed by treatment of the cells with dithiothreitol. Amino acids with a high affinity for System A protected the transporter from inactivation in a stereospecific and systemspecific manner, whereas non-substrates afforded little or no protection. The substrate-dependent protection from PCMBS inactivation was observed in hepatoma cells, but not in normal rat hepatocytes. The differential sensitivity to NEM was also observed when plasma membrane vesicles were prepared from hepatocytes and hepatoma cells and then treated with PCMBS or NEM. As in whole cells, PCMBS inactivated System A in both membrane populations, whereas NEM inactivated the System A carrier in hepatocyte vesicles, but not those from the rat hepatoma

Hepatic Transport in Organic Substances
Ed. by E. Petzinger, R. K.-H. Kinne, H. Sies

cells. The amino acid-dependent protection of PCMBS inactivation was observed only in the membrane vesicles from the hepatoma cell lines. To eliminate influences of cellular structure or membrane lipid composition and/or packing, System A carrier activity was solubilized from membrane vesicles of hepatocytes and hepatoma cells and then treated with NEM. Subsequent reconstruction into artificial proteoliposomes demonstrated that the solubilized carrier from the hepatocytes was sensitive to NEM, whereas that from the hepatoma cells was completely resistant. Collectively, the data suggest an inherent difference in the System A carrier protein present in normal rat hepatocytes and several hepatoma cell lines of both rat and human origin.

Mechanism of Renal Transport of L-Alanine in Luminal Membrane Vesicles

M.I. Sheikh, K.E. Jørgensen
Institute of Medical Biochemistry, University of Åarhus, 8000 Århus C, Denmark

The characteristics of L- and D-alanine transport in luminal membrane vesicles isolated either from proximal convoluted (pars convoluta) or proximal straight (pars recta) tubules were studied. Renal L- and D-alanine transport systems, displaying the following properties, were shown: In vesicles from pars convoluta, the uptake of both L- and D-alanine was mediated by both Na^+-dependent and Na^+-independent transport processes. It was found that an inwardly directed H^+ gradient could drive the transport of alanine into the vesicles both in the presence and absence of Na^+. Thus, in addition to Na^+, the transport of L- and D-isomers of this amino acid is influenced by the H^+ gradient. In vesicles from pars recta, the transient accumulation of L- and D-alanine was strictly dependent of Na^+, since no 'overshoot' was ever observed in the absence of Na^+. Although the N^+-dependent uptake of alanine was stimulated at acid pH, H^+ did not substitute for Na^+, as it apparently does in pars convoluta, but instead potentiated the Na^+ effect. Kinetic analysis of the saturation experimental data showed the existence of two electrogenic and Na^+-requiring systems with high and low affinity in vesicles from pars recta for the transport of L-alanine, whereas D-alanine is taken up by a single Na^+-dependent electrogenic process in this region of nephron. Modification of histidyl

Hepatic Transport in Organic Substances
Ed. by E. Petzinger, R. K.-H. Kinne, H. Sies

residues of the intact luminal membrane vesicles from pars recta by diethylpyrocarbonate (DEP) completely abolished the transient renal accumulation of L-alanine. Substrate and Na^+ protection experiments suggest that histidyl residues may be at or close to the active site of the L-alanine transporter. Addition of L-alanine to vesicle preparations, both from pars convoluta and pars recta, specifically inhibited renal uptake of D-alanine. A comparison between the transport characteristics of L-alanine and D-alanine indicated that these two isomers of alanine probably share common transport systems located along the proximal tubule of rabbit kidney. Studies on the stoichiometric properties of Na^+- and H^+-dependent L-alanine transporters indicate that one Na^+ and one H^+ ion are involved in the translocation of L-alanine in vesicles from pars convoluta. Furthermore, our experimental data suggest that both the high affinity and the low affinity Na^+-dependent L-alanine transport system of pars recta vesicles operate with a 1:1 stoichiometry.

Hepatic Transport of Organic Anions and Bile Acids

Characteristics of Organic Anion Binding Proteins from Rat Liver Sinusoidal Plasma Membranes

Paul D. Berk, Barry J. Potter, Dario Sorrentino, Wolfgang Stremmel*, Decherd Stump, Chih-Li Kiang and Sheng-Li Zhou

Polly Annenoerg Levee Hematology Center and Hepatic Research Group, Departments of Medicine and Biochemistry, Mount Sinai School of Medicine of the City University of New York, New York, NY 10029 USA and * Medizinische Klinik D, Universität Düsseldorf, Federal Republic of Germany

Introduction

Within the past decade it has become widely accepted that the entry into the hepatocyte of low molecular weight amphipathic organic compounds which are albumin-bound within the circulation is mediated by membrane proteins on the sinusoidal (basolateral) surface of the hepatocyte (Berk and Stremmel, 1986; Berk et al., 1987; Frimmer and Ziegler, 1988). These proteins are presumed to be components of one or more specific transport systems. Precisely how these proteins participate in cellular uptake is, in most instances, unclear, and there is a considerable debate, between "lumpers" and "splitters", as to how many such proteins there are, and how many different transport systems they serve. "Lumpers" argue for the existence of a very limited number of sinusoidal membrane transport systems of broad specificity; perhaps as few as one mediating sodium-

† Supported by Grants AM-26438 from the National Institutes of Health, AA-06860 from the National Institute of Alcohol Abuse and Alcoholism, SRR 216/2-2 from the Deutsche Forschungsgemeinschaft, and a grant to Dario Sorrentino from Regione Sardegna Assessorato PI, Italy, as well as generous gifts from the Jack Martin Fund, The Monique Weil-Caulier Bequest and the Polly Annenberg Levee Charitable Trust.

Hepatic Transport in Organic Substances
Ed. by E. Petzinger, R. K.-H. Kinne, H. Sies

dependent and another sodium-independent uptake of a wide spectrum of compounds which includes anions, cations and neutral compounds (Buscher et al, 1986). "Lumpers" base their conclusions heavily on studies employing photoaffinity labeling of membrane components with photolabile analogs of the compounds of interest. "Splitters", on the other hand, argue for the existence of multiple transport systems, with considerable (although probably overlapping) substrate specificity. Their approach has utilized a wide variety of techniques for studying uptake, as outlined below.

The authors number themselves among the "splitters". Specifically, they believe that the hepatocyte contains, among others, three distinct transport systems for the uptake of sulfobromophthalein (BSP), bilirubin, and a number of other anionic drugs and dyes; conjugated trihydroxy bile acids such as taurocholate; and long chain fatty acids such as oleate and palmitate. Evidence for the separate nature of these three systems derives from: (a) studies of uptake kinetics in a variety of systems, from intact animals to isolated perfused livers to isolated hepatocytes to, most recently, sinusoidal plasma membrane vesicles; (b) studies of the binding of representative ligands of each class to cells or derived plasma membrane preparations, and of the effects of proteases and protein denaturing conditions on binding; (c) the isolation and partial characterization of distinct BSP/bilirubin-, bile acid-, and free fatty acid-binding proteins from sinusoidal liver plasma membranes; (d) the demonstration of specific functional roles for these proteins in the uptake of their respective ligands by means of antibody inhibition studies (BSP/bilirubin- and fatty acid-binding proteins) or liposome reconstitution studies (bile acid-binding proteins); and (e) the identification of distinct, class-specific cytosolic binding or "acceptor" proteins for each of the 3 classes of organic anions in question.

Space restrictions preclude our acknowledging all of the studies relevant to this issue, and the bibliography cited in this brief review is highly selective. However, we have reviewed elsewhere (Berk and Stremmel, 1986; Berk et al.,

1987), with extensive bibliographies, the evidence in favor of the two positions in this debate. Moreover, the principal protagonists will present their own viewpoints in other chapters in this volume. At this point, we would merely like to highlight two aspects of the data which, in our estimate, strongly argue in favor of separate transport systems for the 3 principal classes of albumin-bound organic anions.

The first deals with kinetics. If uptake is carrier-mediated, binding between substrate and carrier must be, perforce, reversible. If two substrates share a carrier system, they will compete for binding to the carrier, and will demonstrate mutually competitive inhibition of uptake. Non-competitive inhibition may occur under a variety of circumstances, but only the demonstration of competitive inhibition of uptake at physiological concentrations of both ligands has any relevance for the argument about shared versus separate carrier systems. In isolated hepatocytes, BSP and bilirubin show mutually competitive inhibition of uptake, which is not altered by physiologic concentrations of taurocholate or fatty acids (Stremmel and Berk, 1986a). These data suggest that BSP and bilirubin share a common uptake mechanism, distinct from that for taurocholate and oleate. Many ligands inhibit the uptake of taurocholate into plasma membrane vesicles. In the case of other bile acids, or of toxic mushroom peptides, the inhibition is competitive, suggesting a common transport mechanism (see Frimmer and Ziegler, 1988 for review). In the cases of BSP and indocyanine green, the inhibition of taurocholate uptake is non-competitive and therefore not suggestive of a shared transport system (Zimmerli et al., 1987).

The second line of evidence deals with the nature of the sinusoidal plasma membrane-binding proteins for organic anions, which have been reported from many laboratories. Three distinct proteins, with high affinities for BSP/bilirubin (54-55 kDa), long chain free fatty acids (40 kDa), and conjugated bile acids (54 kDa), respectively, have been isolated from sinusoidal rat liver plasma membranes in our laboratory since 1979. The former two have been purified to homogeneity and will be described in detail in the sections which follow. The 54 kDa bile acid-binding protein, which we isolated by a technique initially described by Levy and von Dippe (1983), has thus far only partially been evaluated in our hands. However, it has been shown immunologically to be entirely distinct from the BSP/bilirubin-binding protein of the same molecular weight (Berk et al., 1987), and to have little affinity for BSP. Conversely, the BSP/bilirubin-binding protein has a low affinity for conjugated bile acids. These data suggest that the 54 kDa BSP/bilirubin- and bile acid-binding proteins are, in fact, different. The functional role for the latter remains to be established.

Membrane Proteins that Bind BSP and Bilirubin

BSP/Bilirubin-Binding Protein

Stimulated by kinetic studies and studies of the binding of ligands to sinusoidal liver plasma membrane (LPM) preparations, all of which suggested that an LPM protein participated in the hepatocellular uptake of BSP and bilirubin, 3 laboratories have reported the isolation of putative BSP/bilirubin carrier proteins. We solubilized the integral membrane proteins of sinusoidal enriched rat LPM preparations with triton X-100, and isolated two proteins of interest by affinity chromatography over BSP-agarose or bilirubin-agarose, respectively (Reichen and Berk, 1979; Stremmel et al, 1983). Proteins isolated by this affinity procedure were further purified and characterized, and used to prepare monospecific polyclonal antibodies by intradermal injection into rabbits in complete Freund's adjuvant. Monoclonal antibodies have also been prepared.

Bilirubin-agarose and BSP-agarose each retained a single principal protein from the detergent solubilized LPM protein mixture. In each case the protein migrated as a 54-55 kDa band on SDS-polyacrylamide gel electrophoresis (PAGE); on gradient PAGE without SDS the proteins migrated as apparent dimers of ~105 kDa. Dimerization was also observed in studies in which the isolated BSP-binding protein was chromatographed in the presence of [^{35}S]BSP over Sephadex G-150. In the absence of SDS a single protein with an estimated molecular mass of 103 kDa was observed; addition of 3.3% SDS to the sample led to the elution of two protein peaks of ~103 and ~ 55 kDa which co-chromatographed with [^{35}S]BSP.

Prior studies had shown competition between BSP and bilirubin both for binding to LPM (Reichen et al., 1981) and for hepatocellular uptake (Scharschmidt et al., 1975; Stremmel and Berk, 1986a). Not surprisingly, the protein isolated by BSP-agarose affinity chromatography was shown, in co-chromatography studies, to have a high affinity for bilirubin; conversely, the protein isolated over bilirubin-agarose co-chromatographed with BSP. A monospecific rabbit antibody directed against the rat LPM BSP-binding protein demonstrated immunologic identity between this protein, a single component in the triton X-100 solubilized LPM protein mixture, and the LPM-bilirubin-binding protein. We have subsequently designated this presumed single protein as LPM BSP/bilirubin-binding protein (BSP/BR-BP). Both the BSP- and bilirubin- agarose matrices intermittently retained traces of a 48 kDa protein as well as the more abundant 55 kDa species. Its recovery from this procedure is in quantities too small to isolate, and we have made no further effort to characterize it except to show that it does not react with the rabbit anti BSP/BR-BP.

BSP/BR-BP has proved to be a glycoprotein, with 2 sialic residues per polypeptide chain. It has a Kd (BSP) ~ 10×10^{-6} M. Co-chromatography studies suggest little affinity for taurocholate. By the same token, the rabbit antiserum against BSP/BR-BP does not cross react with another 54 kDa LPM-binding protein isolated by affinity chromatography over glycocholate agarose. We conclude that BSP/BR-BP and the 54 kDa bile acid-binding protein isolated as described above are different. This would appear to be consistent with the finding that in most but not all studies, <u>physiologic</u> concentrations of bile acids do not competitively inhibit either binding of BSP or bilirubin to LPM or their hepatocellular uptake. Furthermore, the rabbit anti-LPM BSP/BR-BP selectively inhibits both the membrane binding of BSP and bilirubin and their hepatocellular uptake, while having no effect on either on the binding or hepatocellular uptake of long chain fatty acids or bile acids (Stremmel et al, 1983; Stremmel and Berk, 1986a).

<u>Organic Anion Binding Protein</u>

Virtually simultaneously with the initial description of BSP/BR-BP, Wolkoff and Chung (1980) were able to photoaffinity label a single 55 kDa protein in LPM with [^{35}S]BSP. Using affinity chromatography of deoxycholate solubilized LPM proteins over BSP-GSH-coupled agarose, they isolated a 55 kDa protein with a high affinity for both BSP and bilirubin, which they designated organic anion-binding protein (OA-BP). Similarities between BSP/BR-BP and OABP are indicated in Table 1 and suggest that the proteins are either closely related or identical. The only apparent difference, in isoelectric points, may relate to methodology: the acidic pI of 3.5 for BSP/BR-BP was determined on samples to which BSP - used to elute it from BSP-agarose - was still bound (Stremmel et al.,1983); the pI of 7.0 for OA-BP was determined in 6 M urea on samples eluted from BSP-GSH-agarose with a pH gradient, and therefore presumably free of BSP (Wolkoff and Chung, 1980). The latter figure thus more likely represents the pI of the native protein.

Although some immunologic cross reactivity between BSP/BR-BP and OA-BP has been reported (Wolkoff et al, 1985) difficulties in isolating and preparing truly monospecific antibodies against these trace proteins has hindered proving conclusively the nature of the relationship. The recent successful cloning of the gene for OA-BP by Wolkoff et al. (1987) has suggested a new approach. Rabbit anti BSP/BR-BP was used to screen a rat liver cDNA library constructed in bacteriophage λgt11 (Clontech, Palo Alto, CA). Primary screening of $4x10^5$ recombinants in E. coli strain Y 1090 resulted in recovery of 3 positive isolates, which were plaque purified. We plated 0.1 ml aliquots of a $1:10^5$ dilution of each of the purified positives in 0.7% soft agarose onto a

Table 1. Comparison of Two Liver Plasma Membrane Proteins That Bind BSP and Bilirubin.

	BSP/BR-BP*	OA-BP†
MW (kDa)	55	55
Sialic acid (mol/mol)	2	1-2
pI	3.5	7.0
Kd (BSP)	10×10^{-6} M	4×10^{-6} M
Kd (BR)	---	20×10^{-6} M

* Reichen and Berk, 1979; Stremmel et al, 1983.
† Wolkoff and Chung, 1980.

lawn of Y-1090. These dilutions were sufficient to yield 5-20 discrete bacteriophage plaques per 90 mm plate. We then incubated the mature plates for 4 hours each with 3 successive nitrocellulose filters which had previously been saturated with isopropyl thio-β-galactoside. After blocking with 3% BSA/0.25% dried milk, each filter was incubated with a 1:250 dilution of one of 3 rabbit antisera obtained from (a) AW Wolkoff (anti OA-BP); (b) our own laboratory (BJ Potter: anti BSP/BR-BP) and (c) the University of Düsseldorf (W. Stremmel: anti BSP/BR-BP), respectively. Filters were washed, and re-incubated with peroxidase-conjugated goat anti rabbit immunoglobin, re-washed, and developed with chloro-1-naphthol/H_2O_2. As illustrated in figure 1, all bacteriophage plaques seen in the agar plate (panel A) reacted with each of the 3 antisera, clearly demonstrating a strong immunologic relationship between BSP/BR-BP and OA-BP. This cross reactivity was demonstrable for each of the 3 separate phage isolates. The cDNA inserts from each of the 3 positive λgt11 isolates have been cut out with the bacterial restriction endonuclease Eco-R1, re-cloned into the plasmid vector pGEM-1 (Melton et al, 1984), and used to transform the rec A^+ E. coli strain M 1060. Electrophoresis of the 3 inserts in 1% agarose reveals them to have molecular weights of about 550,

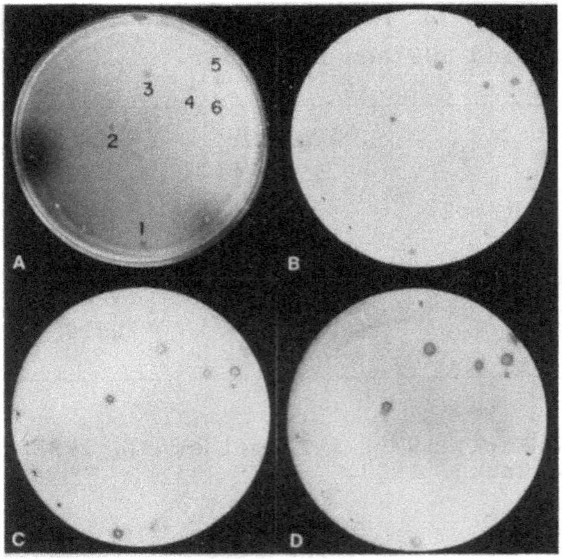

Figure 1. Secondary screening of λgt11 rat liver cDNA library with various antisera to rat LPM membrane proteins. A) Agar plate: 0.1 ml of 1:10^5 dilute of plaque purified phage isolate. (B-D) Consecutive nitrocellulose filter blots of plates probed with the listed rabbit antisera, followed by peroxidase conjugated goat anti-rabbit immunoglobin. B) Anti OA-BP (Wolkoff, New York); C) Anti BSP/BR-BP (Potter, New York); D) Anti BSP/BR-BP (Stremmel, Düsseldorf).

625 and 1400 bp, respectively (Figure 2). These inserts are currently being sequenced. In sequencing the largest of the 3 fragments we are making use of the fact that it can be further cleaved at an internal HindIII site.

By immunofluorescence, BSP/BR-BP has been detected almost exclusively in sinusoidal LPM, and in the plasma membranes of rat bile duct epithelial cells following duct ligation (Stremmel et al, 1983). By contrast, using an ELISA assay which detects total cellular antigen, not merely that expressed on the cell surface, OA-BP has been found in numerous tissues, with highest concentration in the heart (Wolkoff et al, 1985). We remain concerned about the possibility that the ELISA assay is detecting an intracellular antigen - possibly a distinct but cross-reacting protein - with little or no role in organic anion transport.

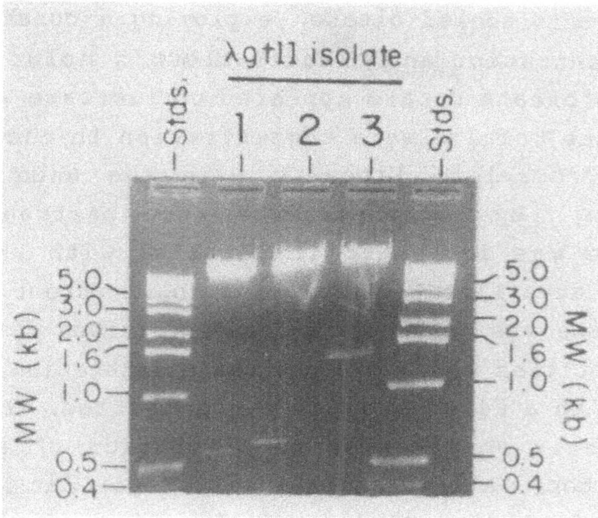

Figure 2. Eco-R1 inserts from λgt11 bacteriophage which code for BSP/BR-BP and OA-BP. Inserts were cut out with ECO-R1 and electrophoresed in a 1% agarose gel containing 1 μg/ml ethidium bromide.

Bilitranslocase

A third membrane BSP/bilirubin-binding protein, bilitranslocase, has been isolated from an acetone powder extract of rat liver (Tiribelli et al, 1978). Most recent reports describe it as a 100 kDa trimer of 2 non-identical subunits (α~ 37 kDa; β 35.5 kDa), with a subunit composition of $\alpha_2\beta$ (Lunazzi et al, 1982). Insertion of bilitranslocase into liposomes has been reported to reconstitute BSP transport, and anti-bilitranslocase antibodies inhibit bilirubin transport in the perfused rat liver (Sottocasa et al, 1982). The structural relationship, if any, of bilitranslocase to BSP-BR-BP and OA-BP is unknown.

Plasma Membrane Fatty Acid Binding Proteins

It has long been believed that long chain fatty acids enter cells passively, perhaps by simply partitioning into the lipid bilayer (Noy et al., 1986). In 1981, Weisiger, Gollan, and Ockner reported that, when perfusing rat livers with radiolabeled oleate, employing a constant 150 μM albumin concentration and oleate: albumin molar ratios of 0.25:1 to 3:1, oleate uptake appeared to increase as a linear function of the total oleate concentration in the perfusate and did not correlate linearly with the unbound oleate concentration. By contrast, when the perfusate oleate concentration was increased in parallel with the albumin concentration at a fixed 1:1 molar ratio, apparent saturation of oleate uptake was observed, at concentrations of oleate and at uptake velocities which did not cause saturation during the perfusions with a fixed albumin concentration. The authors attributed their observation to saturation of a putative albumin receptor. A re-examination of their data, involving re-plotting of oleate uptake as a function of the unbound rather than total oleate concentration in the experiments with a fixed 150 μM albumin concentration suggested that, in these experiments, oleate uptake was a saturable function of the unbound oleate level.

We confirmed in isolated hepatocyte suspensions that [^{3}H]oleate uptake velocity was, indeed, a saturable function of the unbound oleate concentration (Stremmel and Berk, 1986b; Stremmel et al., 1986), and that oleate binding to isolated LPM had the characteristics of saturable, specific, high affinity binding to a single class of binding sites (Stremmel et al., 1983). Both cellular uptake and membrane binding were inhibited by trypsin, suggesting a role for a membrane protein. Using oleate-agarose affinity chromatography, a basic (pI=9.1) 40 kDa fatty acid-binding protein was isolated from hepatocyte plasma membranes (Stremmel et al, 1985). By contrast to BSP/BR-BP, which requires detergent solubilization to extract it from membranes, the hepatic plasma membrane fatty acid binding protein (designated h-$FABP_{PM}$) can be extracted from membranes with 2M salt (Potter et al, 1987), suggesting that it may be a peripheral rather than an integral membrane protein, perhaps analogous to the 46 kDa fatty acid acceptor protein of E. coli (Nunn et al., 1986). The h-$FABP_{PM}$ is unrelated either structurally or immunologically to the smaller, cytosolic fatty acid-binding protein, designated h-$FABP_C$ (Ockner, Manning and Kane, 1982).

Oleate uptake has now been shown to have the kinetic characteristics of a saturable, facilitated process in jejunal enterocytes (Stremmel, 1987), adipocytes (Abumrad et al, 1981; 1984; Schwieterman et al., 1988) and cardiac myocytes (Stremmel, 1988; Sorrentino et al, 1988) as well as in the liver, and basic 40 kDa plasma membrane FABP's have been isolated from each of these sites (Potter et al, 1987; Stremmel, 1988; Sorrentino et al, 1988); all react on Western blot with anti-h-$FABP_{PM}$. Anti h-$FABP_{PM}$ also selectively inhibits fatty acid uptake in adipocytes (Schwieterman et al., 1988), enterocytes (Stremmel, 1987) and cardiac myocytes (Sorrentino et al, 1988) as well as in the liver (Stremmel, Strohmeyer and Berk, 1986) suggesting that these proteins are closely related, if not identical.

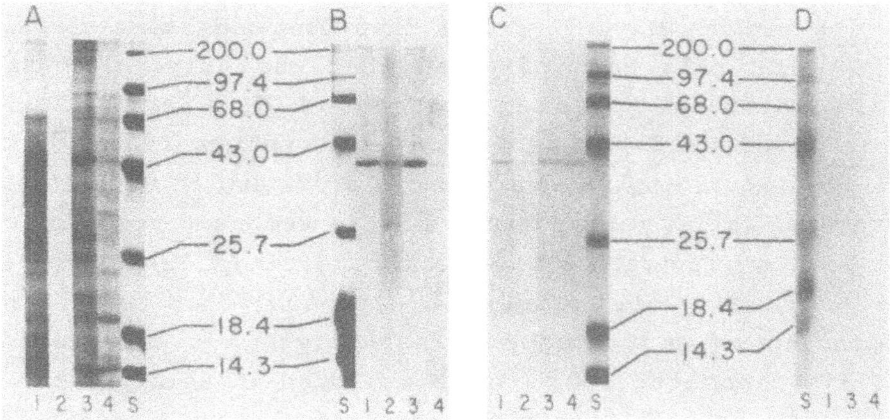

Figure 3. SDS-PAGE analysis of plasma membrane fatty acid-binding proteins. At each purification step the proteins were examined by electrophoresis in sodium dodecyl sulfate under reducing conditions in slab gels. A: Crude protein extracts from membranes, B: Affinity purified membrane fatty acid-binding proteins, C: Membrane fatty acid-binding proteins purified by isoelectric focusing followed by oleate-agarose affinity chromatography or gel permeation HPLC, D: Western blots of the purified proteins using a monospecific antiserum to the liver fatty acid-binding protein. 1: liver, 2: jejunum; 3: adipocyte; 4: myocardium; S: standards. Reproduced from Potter et al (1987) with permission.

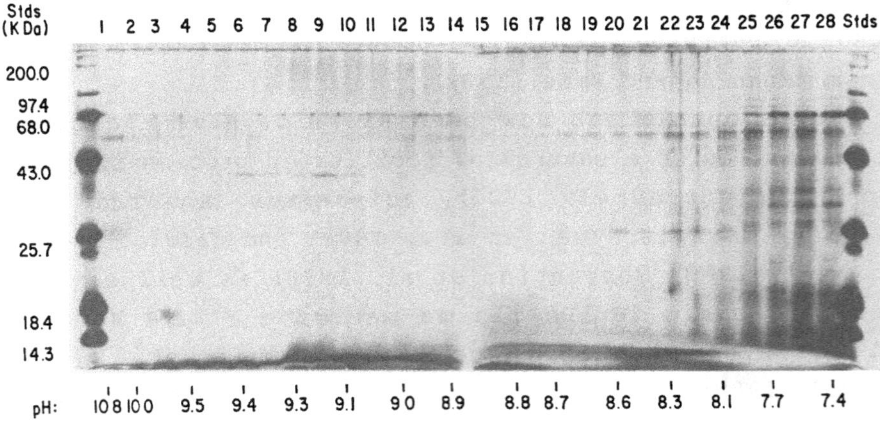

Figure 4. SDS-PAGE analysis of the fractions obtained from preparative isoelectric focusing of 350 mg of solubilized adipocyte plasma membrane proteins. The proteins were focused for 17 hours at 1500 V and 10 W (constant power) in a 7-11 pH gradient before elution in phosphate buffered saline, pH 7.4, and subsequent electrophoretic analysis. Reproduced from Potter et al, (1987) with permission.

The use of silver- rather than Coomassie blue-staining of SDS-PAGE gels had indicated that the original 1 step purification of plasma membrane FABP's by affinity chromatography did not yield a pure preparation (Figure 3, see panel B). Accordingly, we now use a procedure in which, after extraction from membranes in 2M NaCl, membrane proteins are separated by preparative isoelectric focusing. Fractions with pI's $\geq$ 9.0 are pooled (figure 4) and subjected to HPLC gel permeation chromatography; alternatively, oleate-agarose affinity chromatography may be employed at this point. In most instances this 2 step procedure results in a single band on silver-stained SDS-PAGE gels (Potter et al, 1987). Occasionally, especially with canine cardiac tissue, a discrete contaminant of either slightly greater or slightly lesser molecular weight remains. This is removed by reverse phase HPLC, or by hydrophobic interaction HPLC employing an ammonium sulfate gradient (figure 5) (Sorrentino et al, 1988).

We recently isolated sufficient quantities of purified h-$FABP_{PM}$ to obtain a partial sequence. To our surprise, the sequence of the first 24 N-terminal amino acids was identical

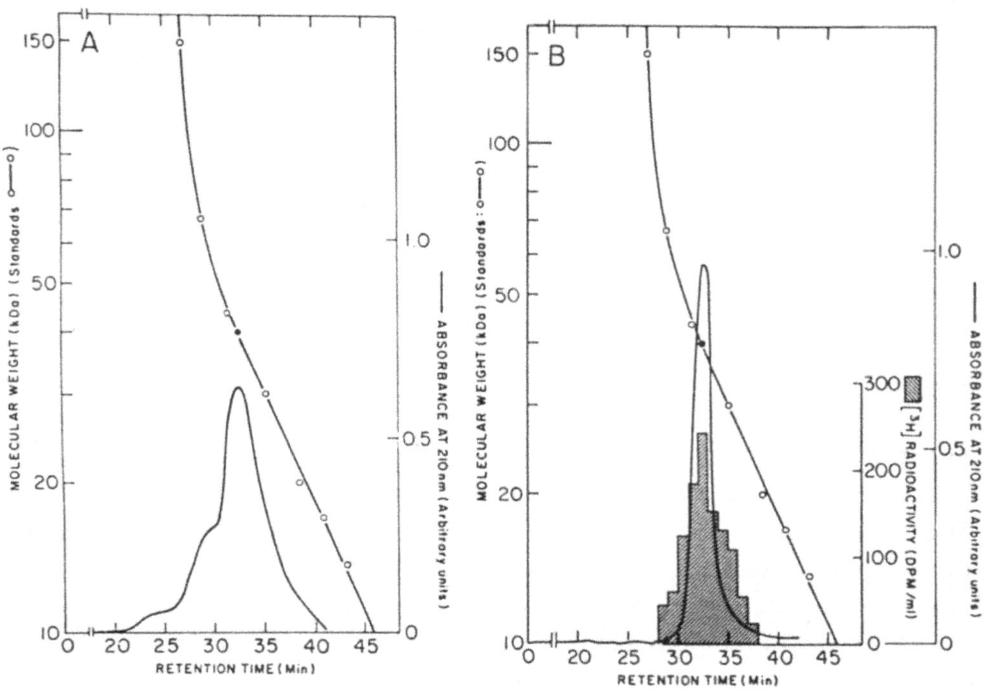

Figure 5. HPLC gel permeation elution profile of the canine myocardial $FABP_{PM}$ isolated by affinity chromatography before (panel A) and after (panel B) further purification by hydrophobic interaction chromatography. Panel B also shows co-chromatography of the protein with [^{3}H]-oleate at a retention time corresponding to a molecular weight of 40 kDa. Reproduced from Sorrentino et al (1988) with permission.

to that previously reported for mitochondrial glutamate-oxaloacetate transaminase (mGOT) (EC 2.6.1.1) (Huynh et al, 1981). Furthermore, we found h-$FABP_{PM}$ to have GOT enzymatic activity at a specific activity virtually identical to that reported for purified mGOT. Authentic samples of pure mGOT and anti mGOT (courtesy of Professor H. Wada and Dr. Y. Horio) were obtained: anti h-$FABP_{PM}$ reacted strongly with mGOT, which was found to have a high affinity for long chain fatty acids, and to be retained on oleate-agarose columns. Anti mGOT identified a single 40 kDa band on Western blots of LPM extracts, and also reacted with purified h-$FABP_{PM}$. Furthermore anti mGOT inhibited fatty acid uptake by isolated rat hepatocytes. Finally, in immunofluorescence studies, both anti h-$FABP_{PM}$ and anti mGOT stained hepatocyte plasma membranes, as well as showing punctate cytoplasmic staining presumed to be in mitochondria. While still preliminary, these data suggest that the recently described liver plasma membrane fatty acid binding protein is closely related to a well known mitochondrial protein, mGOT. Since many plasma membrane and mitochondrial proteins are believed to be synthesized on different populations of ribosomes and to have distinctly different signal sequences, this finding - if confirmed - will pose interesting issues in cell and molecular biology.

The isolation from a rat liver cDNA library of λgt11 bacteriophage containing inserts encoding for h-$FABP_{PM}$ will greatly facilitate the cloning and sequencing of this protein. The consequent availability of appropriate cDNA probes will permit us to address such questions as the number of genes and the number of messages in rat liver which encode for this and closely related proteins. Cloning of the gene for h-$FABP_{PM}$ and its expression in a prokaryotic system should also provide sufficient protein to probe both its structure and its function more definitively.

References

Abumrad NA, Perkins RC, Park JH, Park CR (1981) Mechanism of long chain fatty acid permeation in the isolated adipocyte. J Biol Chem 256:9183-8181

Abumrad NA, Park JH, Park CR (1984) Permeation of long chain fatty acid into adipocytes. J Biol Chem 259:8945-8953

Berk PD, Stremmel W (1986) Hepatocellular uptake of organic anions. In: Popper H and F Schaffner (eds) Progress in Liver Disease, vol VIII, Grune and Stratton, New York, p 125-144

Berk PD, Potter BJ, Stremmel W (1987) Role of plasma membrane ligand-binding proteins in the hepatocellular uptake of albumin-bound organic anions. Hepatology 7:165-76

Buscher HP, Fricker G, Gerok W, Kramer W, Kurz G, Müller M, Schneider S (1986) Membrane transport of amphiphilic compounds by hepatocytes. In: Greten H, Windler E, Beisiegel U (eds). Receptor mediated uptake in the liver. Springer, New York, p 189-199

Frimmer M, Ziegler K (1988) The transport of bile acids in liver cells. Biochim Biophys Acta 947:75-99

Huynh Q-K, Sakakibara R, Watanabe T, Wada H (1981) The complete amino acid sequence of mitochondrial glutamic oxaloacetic transaminase from rat liver. J Biochem 90:863-875

Levy D, von Dippe P (1983) Reconstitution of the bile acid transport system derived from hepatocyte sinusoidal membranes (abstract). Hepatology 3:837

Lunazzi G, Tiribelli C, Gazzin B, Sottocasa G (1982) Further studies on bilitranslocase, a plasma membrane protein involved in hepatic organic anion uptake. Biochim Biophys Acta 685:117-122

Melton DA, Krieg PA, Rebagliati MR, Maniatis T, Zinn K, Green MR (1984) Efficient in vitro synthesis of biologically active RNA and RNA hybridization probes from plasmids containing a bacteriophage SP6 promoter. Nucleic Acids Res 12:7035-7056.

Noy N, Donnely TM, Zakim D (1986) Physical-chemical model for the entry of water insoluble compounds into cells. Studies of fatty acid uptake by the liver. Biochemistry 25:2013-2021

Nunn WD, Colburn RW, Black PN (1986) Transport of long chain fatty acids in Escherichia coli: evidence for role of fadL gene product as long chain fatty acid receptor. J Biol Chem 261:167-171

Ockner RK, Manning JA, Kane JP (1982) Fatty acid binding protein: isolation from rat liver, characterization and immunochemical quantification. J Biol Chem 257:7872-7878

Potter BJ, Stump D, Schwieterman W, Sorrentino D, Jacobs LN, Kiang C-L, Rand J, Berk PD (1987) Isolation and partial characterization of plasma membrane fatty acid binding proteins from myocardium and adipose tissue and their relationship to analogous proteins in liver and gut. Biochem Biophys Res Commun 148:1370-1376

Reichen J, Berk PD (1979) Isolation of an organic anion binding protein from rat liver plasma membrane fractions by affinity chromatography. Biochem Biophys Res Commun 91:484-489

Reichen J, Blitzer BL, Berk PD (1981) Binding of unconjugated and conjugated sulfobromophthalein to rat liver plasma membrane fractions _in vitro_. Biochim Biophys Acta 640:298-312

Scharschmidt BF, Waggoner JG, Berk PD (1975) Hepatic organic anion uptake in the rat. J Clin Invest 56:1280-1292

Schwieterman W, Sorrentino D, Potter BJ, Rand J, Kiang C-L, Stump D, Berk PD (1988) Uptake of oleate by isolated rat adipocytes is mediated by a 40 kDa plasma membrane fatty acid binding protein closely related to that in liver and gut. Proc Natl Acad Sci USA 85:359-363

Sorrentino D, Stump D, Potter BJ, Robinson R, White R, Kiang C-L, Berk PD (1988) Oleate uptake by cardiac myocytes is carrier mediated and involves a 40 kDa plasma membrane fatty acid binding protein similar to that in liver, adipose tissue and gut. J Clin Invest (in press)

Sottocasa GL, Baldini G, Sandri G, Lunazzi G, Tiribelli C (1982) Reconstitution in vitro of sulfobromophthalein transport by bilitranslocase. Biochim Biophys Acta 685:123-128

Stremmel W (1987) Uptake of fatty acids by the jejunal mucosa is mediated by a specific fatty acid binding membrane protein. Gastroenterology 92:1656

Stremmel W (1987) Translocation of fatty acids across the basolateral rat liver plasma membrane is driven by an active potential-sensitive sodium-dependent transport system. J Biol Chem 262:6284-6289

Stremmel W (1988) Fatty acid uptake by isolated rat heart myocytes represents a carrier-mediated transport process. J Clin Invest 81:844-852

Stremmel W, Berk (1986a) Hepatocellular uptake of sulfobromophthalein (BSP) and bilirubin is selectively inhibited by an antibody to the liver plasma membrane BSP/bilirubin binding protein. J Clin Invest 78:822-826

Stremmel W, Berk PD (1986b) Hepatocellular influx of ^{14}C-oleate reflects membrane transport rather than intracellular metabolism or binding. PNAS 83:3086-3090

Stremmel W, Kochwa S, Berk PD (1983) Studies of oleate binding to rat liver plasma membranes. BBRC 112:88-95

Stremmel W, Strohmeyer G, Berk PD (1986) Hepatocellular uptake of oleate is energy dependent, sodium linked, and inhibited by an antibody to a hepatocyte plasma membrane fatty acid binding protein. PNAS 83:3584-3588

Stremmel W, Gerber MD, Glezerov V, Thung SN, Kochwa S, Berk PD (1983) Physicochemical and immunological studies of a sulfobromophthalein- and bilirubin- binding protein from rat liver plasma membranes. J Clin Invest 71:1796-1805

Stremmel W, Strohmeyer G, Borchard F, Kochwa S, Berk PD (1985) Isolation and partial characterization of a fatty acid binding protein in rat liver plasma membranes. PNAS 82:4-8

Tiribelli C, Lunazzi GC, Luciani M, Panfili E, Gazzin B, Liut GF, Sandri G, Sottocasa GL (1978) Isolation of a sulfobromophthalein-binding protein from hepatocyte plasma membrane. Biochim Biophys Acta 532:105-112

Weisiger R, Gollan J, Ockner R (1981) Receptor for albumin on the liver cell surface may mediate uptake of fatty acids and other albumin-bound substances. Science 211:1048-1051

Wolkoff AW, Burk RD, Sosiak A, Nakata B (1987) Cloning of a rat liver cell surface membrane organic anion binding protein (OABP) (abstract). Hepatology 7:1035

Wolkoff AW, Sosiak A, Greenblatt HC, van Renswoude J, Stockert RJ (1985) Immunological studies of an organic anion binding protein (OABP) isolated from rat liver cell plasma membrane. J Clin Invest 76:454-459

Wolkoff AW, Chung CT (1980) Identification, purification and partial characterization of an organic anion binding protein from rat liver cell plasma membranes. J Clin Invest 65:1152-1161

Zimmerli B, Valantinas J, Meier PJ (1987) Multispecificity of Na^+ dependent taurocholate uptake in basolateral (sinusoidal) rat liver plasma membrane (blLPM) vesicles (abstract). Hepatology 7:1036

Mechanism of Hepatic Fatty Acid Uptake

Wolfgang Stremmel
Division of Gastroenterology
Department of Internal Medicine
University Clinics of Düsseldorf
Moorenstr. 5
4000 Düsseldorf
West-Germany

Introduction

In the past few years, a new concept of the cellular uptake mechanism of long-chain fatty acids, in particular by the liver, has been described. It was shown that the cellular influx of this major class of energy-yielding substrates is mediated by a specific membrane carrier protein (Stremmel and Berk, 1986; Stremmel et al., 1985; 1986; Stremmel and Theilmann, 1986). This is of physiologic significance, since such a carrier-mediated uptake process might represent a site of metabolic and hormonal control of fatty acid metabolism.

Arguments against Uptake by Simple Diffusion

It was long assumed that fatty acids, due to their lipophilic character, permeate the phospholipid bilayer of the plasma membrane by simple diffusion (Heimberg et al., 1978; Kuhl and Spector, 1970; Spector and Fletcher, 1978). However, due to various unfavorable structural and thermodynamic conditions it seems unlikely that they are taken up as a linear function of the transmembrane concentration gradient. One argument is the fact that fatty acids are presented to the liver cell plasma

Hepatic Transport in Organic Substances
Ed. by E. Petzinger, R. K.-H. Kinne, H. Sies

membrane as albumin complexes within the plasma. Albumin, however, is not taken up by hepatocytes and the non-albumin-bound fraction is far too low to explain the high cellular influx rates observed in vivo (Weisiger et al., 1987). Therefore, accelerated dissociation of the fatty acid-albumin complex at the liver cell plasma membrane has to occur (Berk and Stremmel, 1986). Although this mechanism is still a matter of debate (Berk and Stremmel, 1986), the high concentration of albumin with its high affinity for fatty acids makes it unlikely that the efficient cellular uptake of fatty acids is due to a free exchange of unbound fatty acids between the extra- and intracellular space.

Another consideration relates to the movement of fatty acids across the tight structure of the phospholipid bilayer of the plasma membrane itself. The polar groups facing the extra- and intracellular space may hinder fatty acid permeation. Whether, in addition, the high concentration of free fatty acids in the plasma membrane (Simpson and Peters, 1987) may interfere with the partition of these hydrophobic compounds into the lipid phase remains speculative.

Moreover, at physiologic pH the fatty acid molecules in the plasma are present as anions. Therefore, they have to be taken up against an unfavorable electrical gradient due to the negative charge at the cytosolic site of the plasma membrane.

All of the above mechanisms may, in fact, protect the cells from uncontrolled influx of fatty acids and are not compatible with a simple diffusion process. Therefore, it was suggested that cellular fatty acid uptake requires certain characteristics of the plasma membrane and does not occur in any cell. This was confirmed by an experimental model in which fatty acid uptake was examined in cells, which cannot metabolize fatty acids. Those cells were spheroplasts kept in a minimal medium with 1.2 M sorbitol (Veenhuis et al., 1987). They were prepared from yeast cells after digestion of the surrounding cell wall with helicase, leaving the plasma membrane intact.

When those cells were incubated with fatty acid-albumin complexes under various conditions no binding and no uptake was observed. It indicates that at least in those cells passive diffusion is of no relevance. In contrast, under the same experimental conditions fatty acid uptake by liver cells was very efficient (Stremmel et al., 1986; Stremmel and Theilmann, 1986).

Hepatocellular Fatty Acid Uptake Represents a Carrier-Mediated Uptake Process

For evaluation of the transport mechanism of fatty acids by hepatocytes, kinetic studies with a representative long-chain fatty acid, oleate, were performed in isolated rat hepatocytes and short-term cultured hepatocytes (Stremmel and Berk, 1986; Stremmel et al., 1986; Stremmel and Theilmann, 1986). It was shown that with increasing oleate concentrations incubated, uptake revealed saturation kinetics (K_m = 83 nM; V_{max} = 394 pmol x min^{-1} per 10^5 cells) (Stremmel et al., 1986). This uptake mechanism was specific for long-chain fatty acids but not for other classes of organic anions; temperature-dependent with an optimum at 37°C; and inhibited by trypsin pretreatment of the cells. These observations are criteria for a carrier-mediated transport process.

Since the first step in any carrier-mediated uptake process represents the binding of ligand to the carrier, the interaction of oleate with isolated rat liver plasma membranes was examined (Stremmel et al., 1985). It was shown that membrane binding was saturable with increasing oleate concentrations incubated and revealed a high affinity with a K_D = 20 nM. Furthermore, binding was specific for long-chain fatty acids and inhibited by heat denaturation or trypsin pretreatment of the plasma membranes, suggesting interaction with a membrane fatty acid binding protein.

In fact, a single 40 kDa membrane protein with high affinity for long-chain fatty acids was identified and isolated by affinity chromatography from the total mixture of rat liver plasma membrane proteins (Stremmel et al., 1985). For evaluation of its biologic significance an antibody to this protein was raised in rabbits. By Western blot analysis the antibody reacted with a single corresponding membrane protein as well as with an identical single protein in whole human liver homogenate. By immunofluorescence and immune electron microscopy this protein was localized in the plasma membrane of hepatocytes. It was shown that this antibody selectively inhibited binding of long-chain fatty acids to isolated plasma membranes (Stremmel et al., 1985). Its carrier function was documented when, in presence of the antibody to this membrane protein, fatty acid influx into hepatocytes was significantly inhibited, whereas uptake of other classes of organic anions was not affected (Stremmel et al., 1986; Stremmel and Theilmann, 1986). According to these results it was concluded that the translocation of long-chain fatty acids across the sinusoidal membrane of hepatocytes involves accelerated dissociation of the albumin-fatty acid complex at the cell surface and binding to a specific membrane protein which functions as transmembrane carrier protein.

Driving Forces for Hepatocellular Fatty Acid Uptake

The efficiency by which the liver takes up fatty acids is very high with a single pass extraction capacity of 30-40% under physiologic conditions (Weisiger et al., 1987). It seems unlikely, however, that this high fatty acid uptake rate is driven by a large transmembrane concentration gradient. Although the concentration of unbound fatty acids in the cytoplasmatic compartment close to the plasma membrane is not known, the chemical gradient is assumed to be small due to the extensive binding of fatty acids to plasma and cytosolic proteins. Moreover, the driving forces for cellular fatty acid uptake have to overcome an unfavorable electrical gradient, since at physiologic pH, fatty acids are negatively charged and the cell interior is also electrically negative with respect to the exterior (s. above). Despite these unfavorable conditions uptake is very efficient and might be explained if influx of fatty acids is coupled to the movement of a second solute down its electrochemical gradient. Active transport of small molecules by mammalian cells is typically coupled to influx of Na^+, which is favored both electrically and chemically. For determination of the responsible driving forces, basolateral rat liver plasma membrane vesicles were used as experimental model (Stremmel, 1987). They were prepared such that the outside of the vesicles represented the site facing the extracellular space of Disse. The advantage of vesicle studies is that they allow a direct evaluation of the membrane translocation process independent of intracellular metabolism. Furthermore, it is possible to determine influx after any modulation of the ion composition in the extra- and intravesicular space as well as after imposition of various potential gradients across the plasma membranes.

When vesicle uptake of oleate was examined in the presence of an inwardly directed Na^+ gradient with 100 mM NaCl in the medium, a rapid influx phase was observed with maximal accumulation of fatty acids after 20 sec. This was followed by a slow decline, as oleate effluxed from the vesicles until equilibrium was reached after 20 min (Fig. 1) (Stremmel, 1987).

This typical overshoot phenomenon which is characteristic for active transport, was not observed in the presence of KCl, LiCl, or choline chloride. That the effect of Na^+ in stimulating oleate influx was not due to a vesicle inside-positive diffusion potential but to a Na^+-oleate cotransport system was demonstrated under ion equilibrated conditions. Preloading of vesicles with 100 mM NaCl or KCl in order to equilibrate the ions across the membrane revealed that influx of oleate proceeded more rapidly in the presence of Na^+ than in the

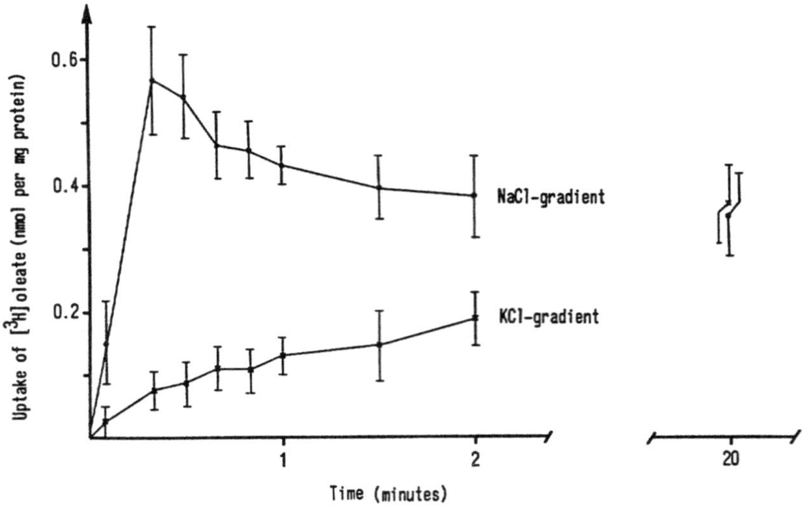

Figure 1: Time course of Na^+-dependent and Na^+-independent [^3H]oleate uptake by basolateral rat liver plasma membrane vesicles under cation gradient conditions (reprinted with permission from Stremmel (1987)).

presence of K^+ (Fig. 2) (Stremmel, 1987). Since there was no Na^+ gradient as driving force, an overshoot of uptake was not observed.

Uptake via a Na^+-fatty acid cotransport system implies that the transport system has a distinct affinity for Na^+. This was indeed shown when the influx velocity was examined as a function of the Na^+ concentration in the medium revealing a K_m value for Na^+ of 52 mM (Stremmel, 1987).

Furthermore, the potential sensitivity of uptake was examined. Transmembrane charge differences in the presence of an inwardly directed Na^+ gradient were created by substitution of various accompanying anions covering a range of lipid permeabilities. Anions which diffuse into vesicles more rapidly than Na^+ (e.g. SCN^-) produce a transient more negative intravesicular membrane potential than anions which permeate the plasma membrane slower than Na^+ (e.g. gluconate$^-$). Since

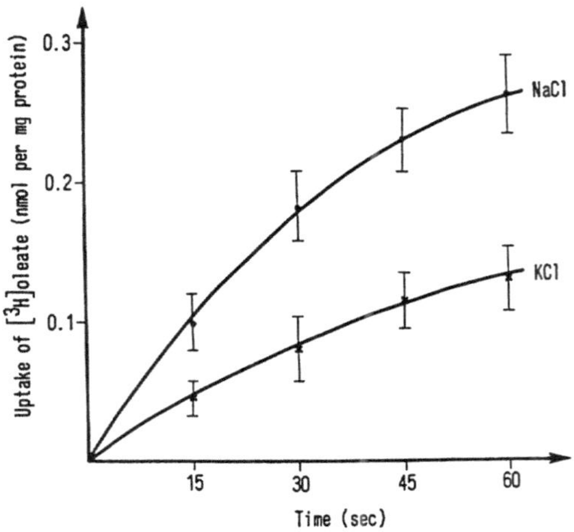

Figure 2: Time course of Na^+-dependent and Na^+-independent [^3H]oleate uptake by basolateral rat liver plasma membrane vesicles in the absence of ion gradients (reprinted with permission from Stremmel (1987).

initial uptake of oleate is accelerated in the presence of more permeable accompanying anions (Fig. 3, left), it was suggested that the translocation of fatty acids across the plasma membrane is stimulated by a relatively more negative intravesicular charge. This was confirmed by another experiment. Vesicles were preloaded with 100 mM K^+ gluconate and exposed to the K^+ ionophore valinomycin. In the absence of extravesicular K^+, valinomycin permits a rapid out-diffusion of K^+, thereby creating a transient negative charge within the vesicles. In those valinomycin-pretreated vesicles uptake of fatty acids is in fact accelerated, supporting the hypothesis that cellular influx is stimulated by a negative intracytoplasmatic potential (Fig. 3, right) (Stremmel, 1987).

Complementary results were obtained by Weisiger, Fitz and Scharschmidt in the experimental model of the isolated perfused rat liver using microelectrodes to continuously monitor the electrical potential difference across the plasma membrane, while simultaneously monitoring the rate of oleate uptake (Weisiger et al., 1987). Isoosmotic cation or anion substitution was used to modulate the membrane potential. It

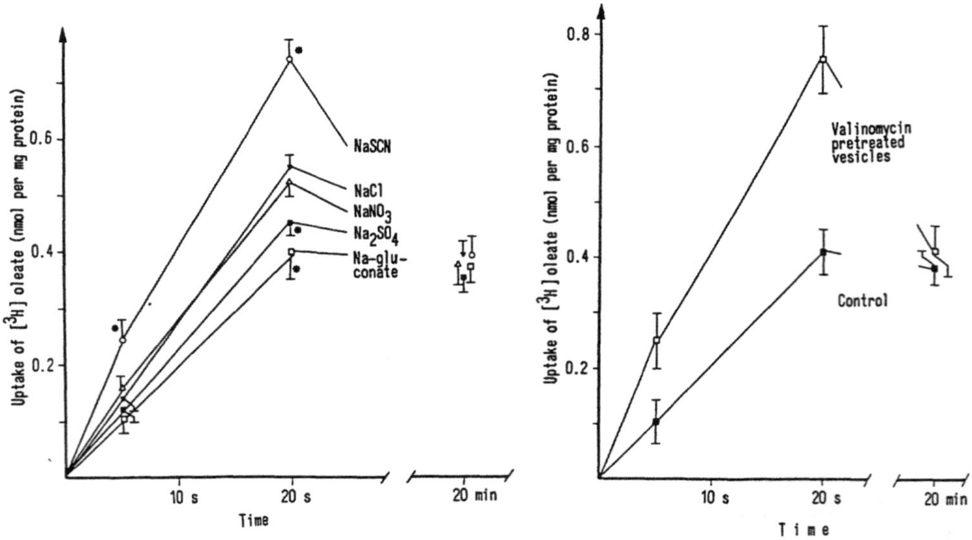

Figure 3: Potential sensitivity of oleate uptake.
Left: Transmembrane charge differences in the presence of an inwardly directed Na^+ gradient were created by substitution of various accompanying anions (increasing negative intravesicular membrane potential with SCN^- > Cl^- = NO_3 > SO_4^{2-} > gluconate$^-$).
Right: Creation of a negative intravesicular charge by a valinomycin-induced K^+ diffusion potential.
(Reprinted with permission from Stremmel (1987).

was shown that a more negative membrane potential is associated with a more rapid uptake rate of fatty acids and a more positive potential with less rapid uptake. This was suggestive for an electrogenic transport mechanism in which fatty acids enter the cells as positively charged complexes. Since removal of Na^+ from the medium reduced uptake to a higher degree than expected from the degree of depolarization, it was concluded that the cotransported positive charge is provided by Na^+. Direct evidence for an electrogenic Na^+-fatty acid cotransport was the observation that in the presence of Na^+ in the perfusion medium, the abrupt exposure of the liver to a 1.5 mM oleate solution resulted in a significant depolarization of the membrane potential due to the associated influx of positive charge, whereas in Na^+-free medium no depolarization was seen (Weisiger et al., 1987). These results confirm that depolarization was caused by a specific Na^+-dependent process rather than a nonspecific effect of oleate on the potential difference.

Conclusion

The carrier-mediated uptake of fatty acids is driven by an electrogenic transport mechanism in which fatty acids enter the cell as positively charged complexes. It appears to represent a Na^+-fatty acid cotransport system. The physiologic transcellular Na^+ gradient is maintained by the Na^+/K^+-ATPase which transports Na^+ again in the extracellular space in exchange for K^+.

References

Berk PD, Stremmel W (1986) Hepatocellular uptake of organic anions. In: Popper H, Schaffner F (eds) Progress in liver disease, Vol VIII. Grune & Stratton, New York, p 125-144

Heimberg M, Goh EH, Klausner HJ, Soler-Argilaga C, Weinstein I, Wilcox HG (1978) Regulation of hepatic metabolism of free fatty acids: Interrelationships among secretion of very low density lipoproteins, ketogenesis, and cholesterogenesis. In: Dietschy JM, Gotto AM, Ontko JA (eds) Disturbances in lipid and lipoprotein metabolism. Williams and Wilkins, Baltimore, p 251-267

Kuhl WE, Spector AA (1970) Uptake of long-chain fatty acid methyl esters by mammalian cells. J Lipid Res 11: 458-465

Simpson RJ, Peters TJ (1987) Iron-binding lipids of rabbit duodenal brush-border membrane. Biochim Biophys Acta 898: 181-186

Spector AA, Fletcher JE (1978) Transport of free fatty acids in the circulation. In: Dietschy JM, Gotto AM, Ontko JA (eds) Disturbances in lipid and lipoprotein metabolism. Williams and Wilkins, Baltimore, p 229-249

Stremmel W, Strohmeyer G, Borchard F, Kochwa S, Berk PD (1985) Isolation and partial characterization of a fatty acid binding protein in rat liver plasma membranes. Proc Natl Acad Sci USA 82:4-8

Stremmel W, Strohmeyer G, Berk PD (1986) Hepatocellular uptake of oleate is energy dependent, sodium linked, and inhibited by an antibody to a hepatocyte plasma membrane fatty acid binding protein. Proc Natl Acad Sci USA 83:3584-3588

Stremmel W, Berk PD (1986) Hepatocellular influx of ^{14}C oleate reflects membrane transport rather than intracellular metabolism or binding. Proc Natl Acad Sci USA 83: 3086-3090

Stremmel W, Theilmann L (1986) Selective inhibition of long-chain fatty acid uptake in short-term cultured rat hepatocytes by an antibody to the rat liver plasma membrane fatty acid-binding protein. Biochim Biophys Acta 877: 191-197

Stremmel W (1987) Translocation of fatty acids across the basolateral rat liver plasma membrane is driven by an active potential-sensitive sodium-dependent transport system. J Biol Chem 262:6284-6289

Veenhuis M, Mateblowske M, Kunau WH, Harder W (1987) Proliferation of microbodies in saccharomyces cerevisiae. Yeast 3:77-84

Weisiger RA, Fitz JG, Scharschmidt BF (1987) Hepatic oleate uptake: Electrochemical driving forces. Clin Res 35:416A

Studies on the Mechanism of Organic Anion Transport by the Liver

Allan W. Wolkoff
Liver Research Center
Albert Einstein College of Medicine
1300 Morris Park Avenue
Bronx, NY 10461
U.S.A.

Introduction

The organic anions bilirubin and sulfobromophthalein (BSP) circulate tightly bound to albumin, but are rapidly and efficiently extracted from their albumin carrier by hepatocytes (Scharschmidt et al., 1975; Stollman et al., 1983). These organic anions are transported into cell cytosol where they bind to soluble proteins, primarily glutathione (GSH)-S-transferases before conjugation and biliary excretion (Wolkoff et al., 1978). Previous studies in intact animals and isolated perfused liver revealed that uptake of bilirubin and BSP is mutually competitive and saturable (Goresky, 1964; Scharschmidt et al., 1975). Countertransport and a preloading effect have been reported (Scharschmidt et al., 1975) and the uptake mechanism has been suggested to be one of facilitated diffusion. Although several candidate transport proteins have been proposed, the identity of the putative carrier is not clear (Lunazzi et al., 1982; Stremmel et al., 1983; Wolkoff and Chung, 1980; Wolkoff et al., 1985).

Recently, we have extended our studies on the mechanism of organic anion transport to short-term cultured rat hepatocytes. The ability to easily manipulate culture conditions

Hepatic Transport in Organic Substances
Ed. by E. Petzinger, R. K.-H. Kinne, H. Sies

permits studies to be performed that would be difficult in other systems. This cell system has been valuable in other studies in which the pathway of receptor-mediated endocytosis has been dissected (Samuelson et al., 1988; Wolkoff et al., 1984). Results obtained in these cultured cells have been in agreement with studies performed in vivo and in other in vitro systems.

Methods

Preparation of short-term cultured rat hepatocytes

Hepatocytes are isolated from Sprague-Dawley rats after perfusion of the liver with collagenase. Cells are suspended in medium consisting of Waymouth's 752/1 (Gibco), 25 mM Hepes, pH 7.2, 5% heat-inactivated fetal bovine serum, 100 U/ml penicillin, and 0.1 mg/ml streptomycin. Approximately 1.5 x 10^6 cells in 3 ml are placed in 60 mm Lux culture dishes and cultured in a 5% CO_2 atmosphere at 37°C. Approximately 2 hours later, the medium is changed and cells are cultured for 16-18 hours until they are used. During this time, the cells spread, distribute into a cord-like arrangement, and form tight junctions and bile canaliculi. Morphologically they resemble closely normal rat hepatocytes.

Transport of organic anions by cultured hepatocytes

Transport of ^{35}S-BSP and ^{3}H-bilirubin have been studied in these cells in the presence of a 20-fold molar excess of bovine serum albumin (BSA) (Wolkoff et al., 1987). Prior to a transport study, cells are washed twice with 1.5 ml of modified serum-free medium. This consists of 135 mM NaCl, 1.2 mM $MgCl_2$, 0.81 mM $MgSO_4$, 27.8 mM glucose, 2.5 mM $CaCl_2$, and 25 mM Hepes adjusted to pH 7.2 with solid Tris base. 1 ml of 0.1% BSA in SFM is then added to each plate which is incubated at 4°C or 37°C for 15 min. 10 μl of a stock solution of ^{35}S-BSP or ^{3}H-bilirubin (80 nmole/ml) is added to the 1 ml of medium on each plate and incubation is continued for varied times.

Plates are then washed twice at 4°C with 1.5 ml of SFM and incubated in 5% BSA or a 100-fold molar excess of BSP for 5 min at 4°C, to displace surface-bound ligand. Following three washes in SFM at 4°C, plates are scraped into 1 ml of SFM and radioactivity was determined.

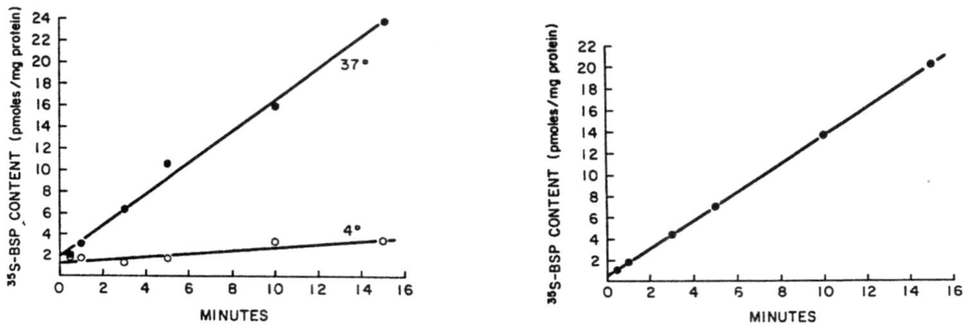

Figure 1: Initial uptake of ^{35}S-BSP by short-term cultured rat hepatocytes.
Cells were incubated at 4°C or 37°C for 15 min in 1 ml of SFM. 10 μl (0.8 nmoles) of a solution of ^{35}S-BSP was added to each culture dish and incubation was continued at 4°C or 37°C. Cells were then washed at 4°C in unlabelled BSP (80 nmole/ml) to displace surface-bound radioactivity. After washing, cells were scraped and radioactivity was determined.
Left: Cell accumulation of ^{35}S-BSP was linear for at least 15 min at 4°C and 37°C. As compared with results at 37°C, accumulation of ligand at 4°C was markedly reduced.
Right: The small amount of residual cell-associated radioactivity at 4°C was subtracted from uptake data at 37°C. This revealed linear kinetics with a y-intercept close to zero.
(Reprinted from Wolkoff et al., 1987, with permission).

Results

Cell uptake of ^{35}S-BSP

As seen in Fig. 1 on the left, cultured hepatocytes accumulated ^{35}S-BSP linearly for 15 min at 4°C and 37°C. As compared with results at 37°C, accumulation of ligand at 4°C was markedly reduced. Subtraction of this 4°C blank from uptake data at 37°C revealed linear kinetics with a y-intercept close

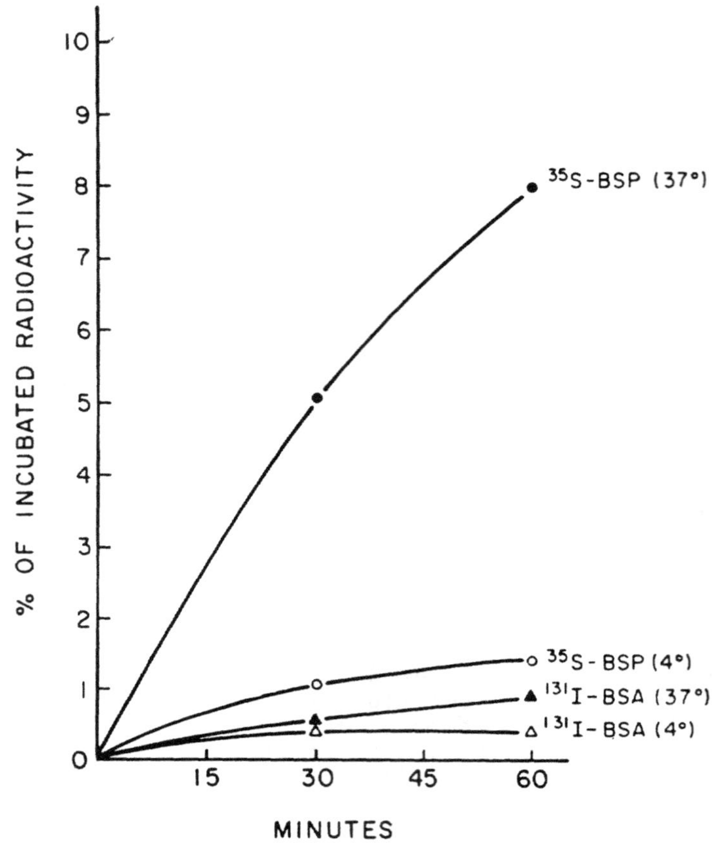

Figure 2: Comparison of uptake of ^{35}S-BSP and ^{131}I-BSA by short-term cultured rat hepatocytes.
Following 60 min of incubation, less than 1% of ^{131}I-BSA was cell-associated while over 8% of ^{35}S-BSP was in cells, indicating extraction of ligand from its albumin carrier. (Reprinted from Wolkoff et al., 1987, with permission).

to 0 (Fig. 1, right). Similar results were found in studies of ^{3}H-bilirubin transport. Previous studies performed in the isolated perfused rat liver (Stollman et al., 1983) revealed that ^{35}S-BSP was extracted from albumin prior to uptake by the liver. Similar results were found in studies in cultured rat hepatocytes (Fig. 2), in which cell association of ^{131}I-albumin and ^{35}S-BSP was determined. After 60 min at 37°C, less than 1% of labelled albumin was cell-associated as compared to

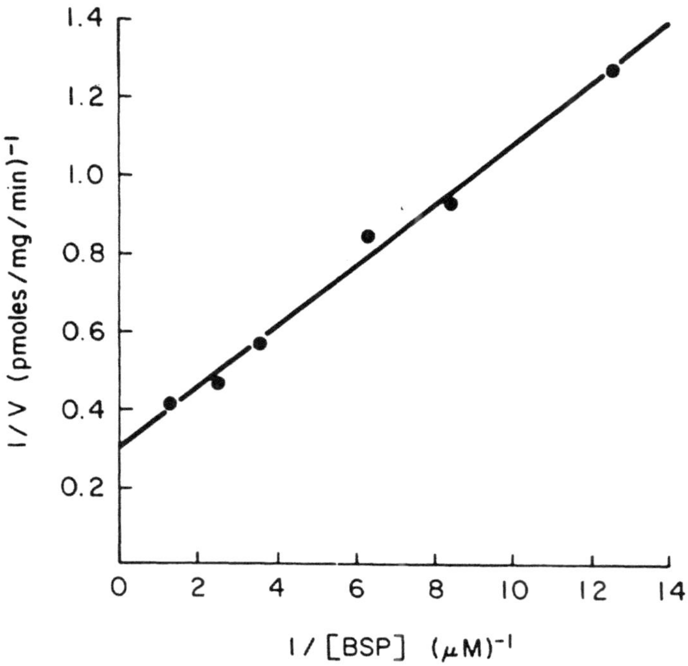

Figure 3: Saturation of initial ^{35}S-BSP uptake by short-term cultured rat hepatocytes.
Initial uptake of varied concentrations of ^{35}S-BSP (0.08-1.2 μM) was determined keeping the proportion of BSA constant. In this representative study, K_m was 0.28 μM and V_{max} was 3.3 pmole/mg protein/min. (Reprinted from Wolkoff et al., 1987, with permission).

over 8% of BSP. Previous studies have also revealed that hepatic organic anion uptake is saturable (Scharschmidt et al., 1975). As seen in Fig. 3, uptake of ^{35}S-BSP by cultured hepatocytes is also saturable and inhibited by bilirubin (data not shown).

From these features, then, transport of BSP and bilirubin by cultured rat hepatocytes is similar to that in intact liver. The influence of temperature on ^{35}S-BSP uptake by cultured cells was also studied (Fig. 4) and revealed little uptake between 4°C and 27°C. Between 27°C and 37°C, there was a large increase in BSP uptake, suggesting a high energy of activation for transport. Consistent with this finding was a marked

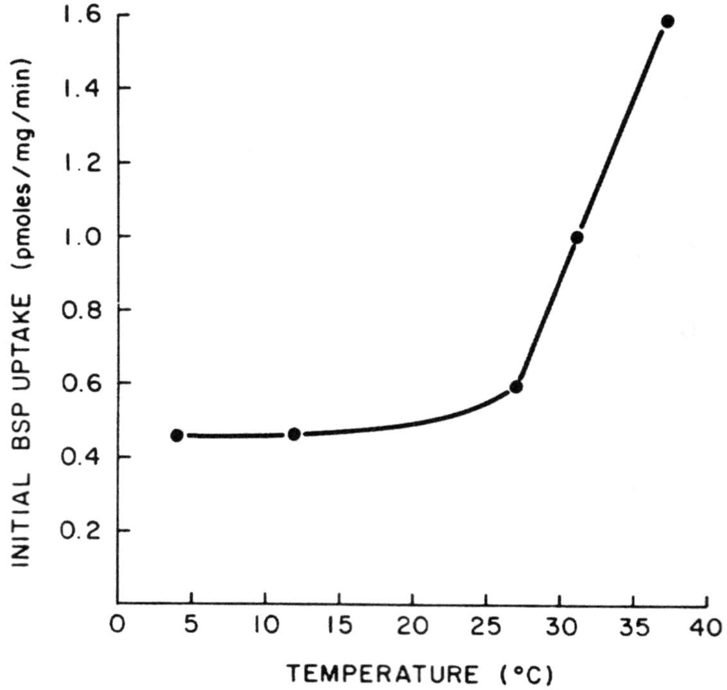

Figure 4: Temperature dependence of ^{35}S-BSP uptake by short-term cultured rat hepatocytes.
The increase in BSP uptake between 27°C and 37°C suggests a large energy of activation for transport. (Reprinted from Wolkoff et al., 1987, with permission).

reduction in BSP transport by cells following a 30 min preincubation in 0.1% sodium azide and 50 mM 2-deoxyglucose. This manipulation reduces cell content of ATP by approximately 50% and in previous studies (Wolkoff et al., 1984) has been found to inhibit ATP-dependent endosomal acidification.

Salt requirements for cultured cell uptake of ^{35}S-BSP

Although Na^+-coupled transport has been described for bile acids, little is known about ion requirements for transport of bilirubin and BSP. Whether NaCl in media was required for cell uptake of ^{35}S-BSP and bilirubin was determined in media in which NaCl was isosmotically replaced by sucrose. As seen in Fig. 5, uptake of ^{35}S-BSP was markedly reduced in this medium.

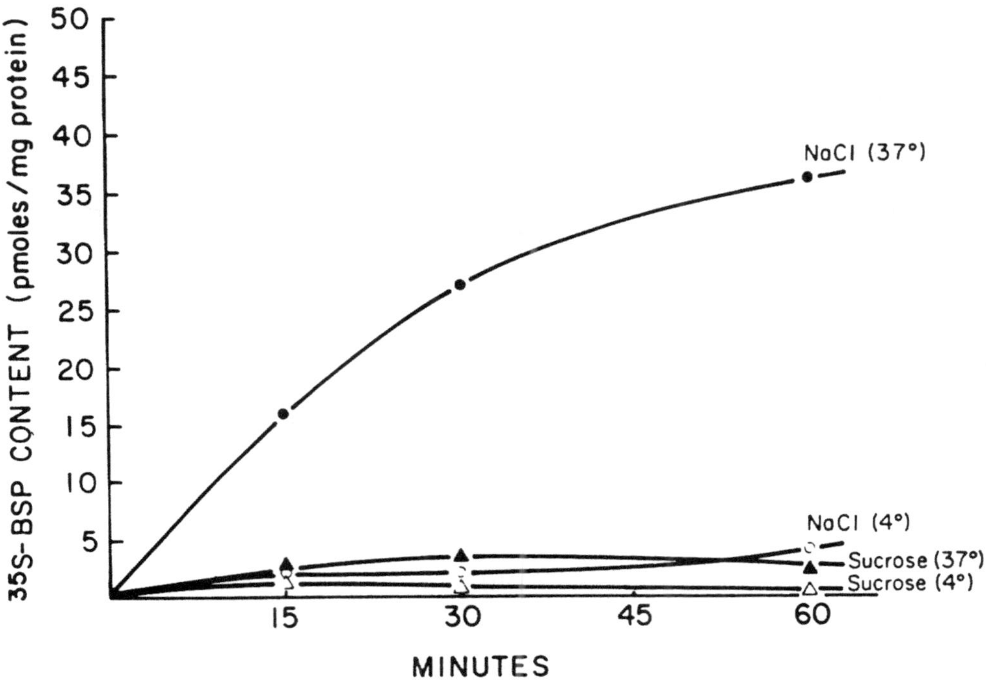

Figure 5: Effect of isosmotic substitution of NaCl in media by sucrose on uptake of ^{35}S-BSP by short-term cultured rat hepatocytes.
During 60 min in the absence of NaCl, there was little uptake of ligand. Quantitation of initial uptake revealed an 80% reduction as compared with control. (Reprinted from Wolkoff et al., 1987, with permission).

Cells still retained the ability, however, to bind and internalize asialoglycoproteins (Wolkoff et al., 1984). To determine whether this effect was due to a requirement for extracellular Na^+, NaCl in medium was substituted isosmotically by KCl, LiCl, or choline Cl. These manipulations had no effect on either initial uptake of bilirubin or BSP or on 30 min cell accumulation of BSP. However, substitution of Cl^- by HCO_3 or gluconate markedly reduced cell accumulation and initial uptake of bilirubin and BSP (Fig. 6). Substitution of Cl^- by the more permeant NO_3^- enhanced initial BSP uptake by 30%, while substitution by I^- had no effect on initial uptake.

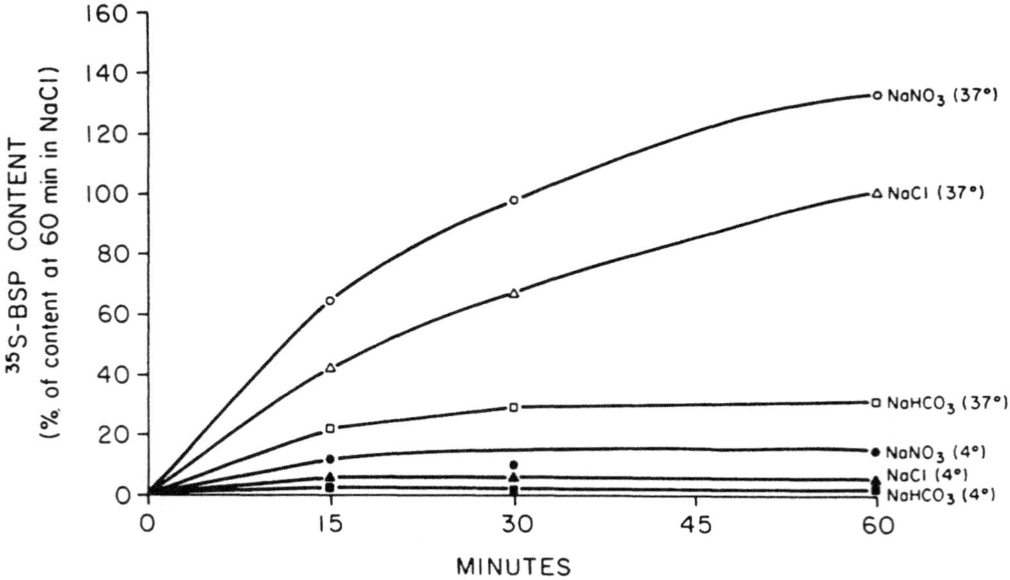

Figure 6: Effects of anion substitution in media on uptake of ^{35}S-BSP by short-term cultured rat hepatocytes.
Although substitution of NaCl by $NaNO_3$ augmented BSP uptake, substitution by $NaHCO_3$ markedly reduced uptake. Similar results were obtained for initital uptake of BSP. (Reprinted from Wolkoff et al., 1987, with permission).

As would be expected from the lack of effect of KCl substitution, BSP uptake was unaffected by inclusion of 10 μM valinomycin in the medium. These findings suggest that the uptake process in not electrogenic.

Studies of the effect of Cl^- replacement by gluconate were also conducted in the isolated perfused rat liver (Wolkoff et al., 1987). This system permits manipulation of the cellular milieu while keeping the organ intact. Livers were perfused with control perfusate or in perfusate in which NaCl was replaced by Na gluconate. All parameters of viability during gluconate perfusion were identical to that in control perfusion. Single-pass indicator dilution studies were performed and data was analyzed by computer-fit to the model of Goresky

(1964). These studies revealed approximately 50% reductions in both influx and efflux of ^{3}H-bilirubin during gluconate perfusion.

Salt requirements for cultured cell uptake of other organic anions

Uptake of ^{3}H-taurocholate by cultured hepatocytes was studied by methods similar to that used for bilirubin and BSP. Elimination of NaCl from medium and isosmotic substitution with sucrose markedly reduced taurocholate uptake. Unlike the findings for BSP and bilirubin, substitution of NaCl by Na gluconate had no effect on taurocholate uptake. However, substitution of NaCl by KCl or K gluconate reduced taurocholate uptake by over 60%.

^{99m}Tc-DISIDA is another organic anionic compound that is taken up rapidly by hepatocytes. Its uptake was unaffected by sucrose substitution, indicating a third pattern of liver cell organic anion transport.

Influence of chloride gradients on BSP transport

Uptake and efflux of ^{36}Cl by short-term cultured rat hepatocytes were also studied. Incubation of cells in 135 mM Na ^{36}Cl resulted in an equilibrium Cl^- content of 176±13.9 nmole/mg protein (mean ±SEM) by 30 min with over 95% located in cell cytosol. Subsequent incubation in Cl^--free medium depleted cellular Cl^- by 80% after 30 min; however, over 75% of total Cl^- remained in cells at 1 min. Initial ^{36}Cl influx into Cl^--depleted cells was 3.9±0.2% of equilibrium Cl^- content/min. This indicated that Cl^- gradients could be maintained in cells for at least 1 min. Initial BSP uptake with extracellular Cl^- (Cl_e) > intracellular Cl^- (Cl_i) was three-fold that with $Cl_i > Cl_e$.

The influence of pH gradients on BSP transport was also determined. Previous studies (Samuelson et al., 1988) revealed that in Hepes buffer, pH gradients between cells and medium could be maintained for several minutes. Although Cl^-/OH^- exchange occurs in other cells, we found no effect of inwardly or outwardly directed pH gradients on Cl^- uptake. However, BSP uptake with $Cl_e > Cl_i$ was almost 2-fold higher in cells with intracellular pH ($pH_i = 8$) > extracellular pH ($pH_e = 6$) as compared to $pH_e > pH_i$.

Discussion

Previous studies have revealed that hepatic uptake of organic anions such as bilirubin and BSP has carrier-mediated kinetics. Several investigators have studied BSP transport by isolated hepatocytes (Laperche et al., 1981; Schwarz et al., 1980; Schwenk et al., 1976; van Bezooijen et al., 1976). As distinct from the studies described above, these investigations were performed in the absence of albumin and suggested temperature-dependent, sodium-independent uptake. Quantitation of ligand actually transported into cells was difficult due to the rapid binding of a significant proportion of ligand to the cell surface. Incubation of cells in the present study with a molar excess of BSA minimizes this low affinity interaction of organic anion with the liver cell surface.

Our studies in short-term cultured rat hepatocytes reveal linear uptake of BSP over at least 15 min. This contrasts with a more rapid process that has been described _in vivo_ in freshly isolated hepatocytes, and in perfused liver. Similar slowing of cellular events in cultured hepatocytes as compared with these other systems has been noted in studies of receptor-mediated endocytosis of asialoorosomucoid (Wolkoff et al., 1984). However, this has facilitated quantitation and characterization of several otherwise very rapid steps of the endocytic pathway (Wolkoff et al., 1984; Samuelson et al., 1988).

Our studies reveal temperature-dependent extraction of BSP and bilirubin from their albumin carrier. This marked temperature dependence suggests that uptake of these compounds requires energy, and this is confirmed by inhibition of initial uptake with ATP depletion. The site at which organic anion transport may be coupled to energy utilization is not known.

Our studies also indicate that liver cell transport of bilirubin and BSP requires Cl^- or other specific inorganic anions in the medium. This finding has been confirmed in the isolated perfused rat liver. Although others have described organic anion/Cl^- exchange in kidney and intestine (Kahn and Aronson, 1983; Kahn et al., 1983; 1985; Weinberg et al., 1986), studies in which unidirectional Cl^- gradients have been established suggest that this is not the case in hepatocytes. Our data are consistent with Cl^-/organic anion cotransport by hepatocytes, and suggest that this electroneutral uptake process may also have a component of OH^-/organic anion exchange. Future studies performed in sinusoidal liver cell membrane vesicles may help to further elucidate this complex uptake mechanism.

References

Goresky CA (1964) Initial distribution and rate of uptake of sulfobromophthalein in the liver. Am J Physiol 207:13-26

Kahn AM, Aronson PS (1983) Urate transport via anion exchange in dog renal microvillus membrane vesicles. Am J Physiol 244:F56-F63

Kahn AM, Branham S, Weinman EJ (1983) Mechanism of urate and p-aminohippurate transport in rat renal microvillus membrane vesicles. Am J Physiol 245:F151-F158

Kahn AM, Harnath S, Weinman EJ (1985) Urate and p-aminohippurate transport in rat renal basolateral vesicles. Am J Physiol 249:F654-F661

Laperche Y, Preaux AM, Berthelot P (1981) Two systems are involved in the sulfobromophthalein uptake by rat liver cells: one is shared with bile salts. Biochem Pharmacol 30: 1333-1336

Lunazzi G, Tiribelli C, Gazzin B, Sottocasa G (1982) Further studies on bilitranslocase, a plasma membrane protein involved in hepatic organic anion uptake. Biochim Biophys Acta 685:117-122

Samuelson AC, Stockert RJ, Novikoff PM, Spray D, Saez J, Wolkoff AW (1988) Influence of cytosolic pH on receptor-mediated endocytosis of asialoorosomucoid. Am J Physiol (Cell Physiol), in press

Scharschmidt BF, Waggoner JG, Berk PD (1975) Hepatic organic anion uptake in the rat. J Clin Invest 56:1280-1292

Schwarz LR, Gotz R, Klaassen CD (1980) Uptake of sulfobromophthalein-glutathione conjugate by isolated hepatocytes. Am J Physiol 239:C118-C123

Schwenk M, Burr R, Schwarz L, Pfaff E (1976) Uptake of bromosulfophthalein by isolated liver cells. Eur J Biochem 64:189-197

Stollman YR, Gartner U, Theilmann L, Ohmi N, Wolkoff AW (1983) Hepatic bilirubin uptake in the isolated perfused rat liver is not facilitated by albumin binding. J Clin Invest 72:718-723

Stremmel W, Gerber MD, Glezerov V, Thung SN, Kochwa S, Berk PD (1983) Physiochemical and immunohistological studies of a sulfobromophthalein and bilirubin-binding protein from rat liver plasma membranes. J Clin Invest 71:1796-1805

van Bezooijen CFA, Grell T, Knock DL (1976) Bromosulfophthalein uptake by isolated liver parenchymal cells. Biochem Biophys Res Commun 69:354-361

Weinberg SL, Burckhardt G, Wilson FA (1986) Taurocholate transport by rat intestinal basolateral membrane vesicles: evidence for the presence of an anion exchange transport system. J Clin Invest 78:44-50

Wolkoff AW, Chung CT (1980) Identification, purification, and partial characterization of an organic anion binding protein from rat liver cell plasma membrane. J Clin Invest 65:1152-1161

Wolkoff AW, Ketley JN, Waggoner JG, Berk PD, Jacoby WB (1978) Hepatic accumulation and intracellular binding of conjugated bilirubin. J Clin Invest 61:142-149

Wolkoff AW, Klausner RD, Ashwell G, Harford J (1984) Intracellular segregation of asialglycoproteins and their receptor: a prelysosomal event subsequent to dissociation of the ligand-receptor complex. J Cell Biol 98:375-381

Wolkoff AW, Samuelson AC, Johansen KL, Nakata R, Withers DM, Sosiak A (1987) Influence of Cl^- on organic anion transport in short-term cultured rat hepatocytes and isolated perfused rat liver. J Clin Invest 79:1259-1268

Wolkoff AW, Sosiak A, Greenblatt HC, van Renswoude J, Stockert RJ (1985) Immunological studies of an organic anion-binding protein isolated from rat liver cell plasma membrane. J Clin Invest 76:454-459

Mechanisms of Sinusoidal Bile Acid Transport

B.L. Blitzer, M.D.
Liver Study Unit
University of Cincinnati College of Medicine
Cincinnati, Ohio 45267
USA

Na^+-Dependent Uptake of Conjugated Bile Acids

During the past decade, the Na^+-dependence of hepatic conjugated bile acid uptake has been well established. Initial studies in the isolated perfused rat liver, isolated rat hepatocytes, and hepatocyte monolayer culture first demonstrated Na^+-dependence and ouabain inhibition of taurocholate uptake (for a review, see Blitzer and Boyer, 1982).

More recently, the availability of techniques for isolating highly purified basolateral liver plasma membrane vesicles has permitted the direct demonstration of Na^+ gradient-dependent "uphill" transport of taurocholate (Inoue et al., 1982; Blitzer and Donovan, 1984; Meier et al., 1984). In these studies, both the energetic and kinetic effects of Na^+ gradients are readily apparent. In the presence of an inwardly directed Na^+ gradient, the bile acid is transiently accumulated at a concentration exceeding that found at equilibrium ("overshoot"). With other cation (e.g. K^+) gradients, entry of taurocholate into the vesicle is relatively slow (manifested by a decrease in initial velocity) and no accumulation above equilibrium concentrations is observed. Even in the absence of

Hepatic Transport in Organic Substances
Ed. by E. Petzinger, R. K.-H. Kinne, H. Sies

a gradient ($Na^+_{out} = Na^+_{in}$), Na^+ increases the initial velocity of uptake and rate of approach to equilibrium, implying a kinetic effect of the cation. These kinetic effects on bile acid uptake observed under Na^+-equilibrated conditions strongly support a direct cotransport (symport) mechanism for the uptake of taurocholate and Na^+. Presumably, in the absence of a gradient, bile acid uptake occurs via carrier-mediated facilitated transport, a process expected to be more rapid than simple diffusion in the absence of Na^+. An overshoot is not observed under these conditions due to the lack of a Na^+ gradient to provide the energy for uphill transport. In contrast, if taurocholate entry was only indirectly coupled to Na^+ transport via the parallel operation of two ion exchangers (e.g. Na^+/H^+ exchange and OH^-/taurocholate exchange), bile acid transport rates would not increase under Na^+-equilibrated conditions.

Kinetic studies of Na^+-dependent taurocholate uptake in both vesicles and isolated hepatocytes suggest that the K_m for this process is approximately 40 μM (Blitzer and Donovan, 1984; Blitzer et al., 1982). Studies of the structural determinants for Na^+-dependent bile acid transport by basolateral vesicles (Dember and Blitzer, 1984) have shown that while the uptakes of conjugated bile acids (taurocholate, glycocholate, taurochenodeoxycholate, and cholylglycyltyrosine) were markedly stimulated by an inwardly directed Na^+ gradient, the uptakes of unconjugated bile acids (cholate, deoxycholate, chenodeoxycholate) showed similar velocities in the presence of Na^+ or K^+ gradients. Further, taurine conjugates appeared to be the "preferred" substrate compared to glycine conjugates. As shown in the isolated perfused liver (Anwer et al., 1985) and in isolated rat hepatocytes (Hardison et al., 1984), a negative charge on the bile acid side chain is required for efficient transport.

The stoichiometry of Na^+-dependent bile acid transport has been the subject of debate. Vesicle studies have suggested electroneutral (Meier et al., 1984) or electrogenic (Inoue et al., 1982) transport. However, the finding that hyperpolarization of the hepatocyte with glucagon stimulates taurocholate uptake (Edmondson et al., 1985) and documentation of a small depolarization of the liver cell during Na^+-dependent taurocholate uptake by the intact liver (Fitz and Scharschmidt, 1987) both suggest a Na^+:bile acid stoichiometry greater than 1:1. Uptake of more than one Na^+ per bile acid transported theoretically would permit transport of the bile acid against a steeper concentration gradient than would a 1:1 stoichiometry.

The following model summarizes our current concepts of the mechanisms of conjugated bile acid uptake by the intact hepatocyte. The anionic moieties of conjugated bile acids bind to a carrier site on the basolateral membrane and enter the cell coupled to the movement of Na^+. No convincing evidence for ordered binding has been obtained thus far. Transport of the bile acid is uphill as the bile acid enters the cell against unfavorable electrical (negative transmembrane potential) and chemical (bile acids are concentrated in the cell) gradients. The energy for this process is derived from the entry of Na^+ which moves "downhill" along favorable electrical and chemical gradients. The ultimate driving force for this coupled transport mechanism is provided by Na^+,K^+-ATPase which maintains the transmembrane Na^+ gradient and contributes to the inside negative potential.

One might predict from this model that brisk uptake of other Na^+-dependent substrates might inhibit Na^+-dependent bile acid uptake by dissipating the transmembrane Na^+ gradient. Such inhibition of taurocholate uptake by Na^+-dependent amino acids has been observed in both isolated hepatocytes (Blitzer et al., 1983) and basolateral membrane vesicles (Blitzer and Bueler, 1985).

The role of albumin in the uptake of albumin-bound bile acids is controversial. While it has generally been thought that uptake is driven solely by the free (unbound) concentration of bile acid, studies in both the isolated perfused rat liver (Forker and Luxon, 1981) and in basolateral vesicles (Blitzer and Lyons, 1985) suggest that albumin may actually enhance uptake at the membrane surface. Although there is no definitive evidence for an albumin "receptor" on the hepatocyte surface, interactions of albumin with the plasma membrane may occur which could account for an increase in affinity of the bile acid receptor for taurocholate observed in vesicles (Blitzer and Lyons, 1985). A recent electron spin resonance study (Horie et al., 1988) suggests that the albumin molecule undergoes a conformational change during interaction with the membrane surface and speculates that this alteration may accelerate the dissociation of organic anions from albumin.

The specific basolateral membrane polypeptide mediating hepatic Na^+-dependent bile acid uptake has not been definitively identified. In several studies, photoaffinity derivatives of bile acids labelled up to 5 polypeptide bands on SDS-polyacrylamide gels (Kramer et al., 1982; Von Dippe and Levy, 1983). Moreover, Na^+-stimulated bile acid uptake by liposomes was observed after incorporation of a group of membrane proteins with molecular weights ranging from 49 to 54,000 (Von Dippe et al., 1986). In a preliminary report (Ledfrod and Blitzer, 1988), using substrate protection and sulfhydryl labelling techniques, we have provided evidence that a band of 50,000 M_r mediates Na^+-dependent bile acid transport.

Uptake of Unconjugated Bile Acids

The mechanisms underlying the uptake of unconjugated bile acids are less well established than those for conjugated bile acids. We (Blitzer et al., 1986) have shown that outwardly directed OH^- gradients drive uphill transport of the unconjugated bile acid cholate in basolateral membrane vesicles.

Transport is saturable and furosemide-inhibitable. Moreover, these effects of an outwardly directed OH^- gradient are not due to induced electrical potentials. This process is apparently specific for cholate as pH gradients did not drive uphill transport of other unconjugated bile acids (deoxycholate, chenodeoxycholate, or ursodeoxycholate). Since the pK_a of the latter three bile acids (5.0) is identical to that of cholate (Roda and Fini, 1984), a non-ionic diffusion mechanism cannot fully account for uphill transport of cholate. In addition, this process is domain-specific as significant uphill transport of cholate was not observed in canalicular membranes in the presence of pH gradients. These findings are consistent with the following model for cholate uptake. Cholate enters the cell in exchange for OH^- at a carrier site on the basolateral membrane (this process is phenomenologically indistinguishable from H^+/cholate symport). Although the pH of the hepatocyte interior is actually slightly more acidic than that of extracellular fluid, the outward movement of OH^- is strongly favored by the markedly negative transmembrane potential. In addition, the parallel operation of the basolateral Na^+/H^+ exchanger (Moseley, 1986) would contribute to OH^- generation and may establish local outwardly directed OH^- gradients across the membrane. Thus in the intact cell, cholate uptake may arise from the parallel operation of three transporters and represent an example of "tertiary active transport" (Aronson, 1983). According to this view, Na^+,K^+-ATPase, by establishing the transmembrane Na^+ gradient, would drive Na^+/H^+ exchange which in turn would drive OH^-/cholate exchange.

A preliminary report (Caflisch et al., 1987) questioned the existence of this putative carrier because of the finding of a lack of DIDS sensitivity. In addition, furosemide-sensitive uphill transport of cholate was found in experiments with artificial membranes which the authors interpreted as consistent with non-ionic diffusion. However, in renal membranes, various relatively DIDS-insensitive anion exchangers have now been described (Karniski and Aronson, 1987). While a small component of pH gradient-driven basolateral cholate uptake probably arises from non-ionic diffusion, the following features suggest that non-ionic diffusion cannot fully account for the stimulation of uptake by OH^- gradients: (1) the lack of significant uphill transport in control biological (canalicular) membrane vesicles; (2) specificity for cholate; and (3) predominance of the anionic moiety at pH 6 at which the studies were performed.

References

Anwer MS, O'Maille ERL, Hofman AF, Di Pietro RA, Michelotti E (1985) Influence of side-chain charge on hepatic transport of bile acids and bile acid analogues. Am J Physiol 249:G479-G488

Aronson PS (1983) Mechanisms of active H^+ secretion in the proximal tubule. Am J Physiol 245:F647-F659

Blitzer BL, Boyer JL (1982) Cellular mechanisms of bile formation. Gastroenterology 82:346-357

Blitzer BL, Bueler RL (1985) Kinetic and energetic aspects of the inhibition of taurocholate uptake by Na^+-dependent amino acids: studies in rat liver plasma membrane vesicles. Am J Physiol 249:G120-G124

Blitzer BL, Donovan CB (1984) A new method for the rapid isolation of basolateral plasma membrane vesicles from rat liver: characterization, validation, and bile acid transport studies. J Biol Chem 259:9295-9301

Blitzer BL, Lyons L (1985) Enhancement of Na^+-dependent bile acid uptake by albumin: direct demonstration in rat basolateral liver plasma membrane vesicles. Am J Physiol 249:G34-G38

Blitzer BL, Ratoosh SL, Donovan CB (1983) Amino acid inhibition of bile acid uptake by isolated rat hepatocytes: relationship to dissipation of transmembrane Na^+ gradient. Am J Physiol 245:G399-G403

Blitzer BL, Ratoosh SL, Donovan CB, Boyer JL (1982) Effects of inhibitors of Na^+-coupled ion transport on bile acid uptake by isolated rat hepatocytes. Am J Physiol 243:G48-G53

Blitzer BL, Terzakis C, Scott KA (1986) Hydroxyl/bile acid exchange: a new mechanism for the uphill transport of cholate by basolateral liver plasma membrane vesicles. J Biol Chem 261:12042-12046

Caflisch C, Zimmerli B, Hugentobler G, Meier PJ (1987) pH gradient driven cholate uptake into rat liver plasma membrane vesicles represents nonionic diffusion rather than a carrier mediated process. Gastroenterology 92:1722

Dember L, Blitzer BL (1984) Conjugation is a requirement for Na^+-coupled bile acid transport by rat basolateral liver plasma membrane vesicles. Gastroenterology 86:1315
Edmondson JW, Miller BA, Lumeng L (1985) Effect of glucagon on hepatic taurocholate uptake: relationship to membrane potential. Am J Physiol 249:G427-G433
Fitz JG, Scharschmidt BF (1987) Regulation of transmembrane electrical potential in rat hepatocytes in situ. Am J Physiol 252:G56-G64
Forker EL, Luxon BA (1981) Albumin helps mediate removal of taurocholate by rat liver. J Clin Invest 67:1517-1522
Hardison WGM, Bellentani S, Heasley V, Shelhamer D (1984) Specificity of an Na^+-dependent taurocholate transport site in isolated rat hepatocytes. Am J Physiol 246:G477-G483
Horie T, Mizuma T, Kasai S, Awazu IM (1982) Taurocholate transport by rat liver sinusoidal membrane vesicles. Am J Physiol 254:G465-G470
Inoue M, Kinne R, Tran T, Arias IM (1982) Taurocholate transport by rat liver sinusoidal membrane vesicles. Hepatology 2:572-579
Karniski LP, Aronson PS (1987) Anion exchange pathways for Cl^- transport in rabbit renal microvillus membranes. Am J Physiol 253:F513-F521
Kramer W, Bickel U, Buscher HP, Gerok W, Kurz G (1982) Bile salt-binding polypeptides in plasma membranes of hepatocytes revealed by photoaffinity labelling. Eur J Biochem 129:13-24
Ledford CH, Blitzer BL (1988) Identification of the hepatic Na^+-dependent bile acid transporter through substrate protection and sulfhydryl labelling. Gastroenterology 94:A560
Meier PJ, Meier-Abt AS, Barrett C, Boyer JL (1984) Mechanisms of taurocholate transport in canalicular and basolateral rat liver plasma membrane vesicles. J Biol Chem 259:10614-10622
Moseley RH, Meier PJ, Aronson PS, Boyer JL (1986) Na-H exchange in rat liver basolateral but not canalicular plasma membrane vesicles. Am J Physiol 250:G35-G43
Roda A, Fini A (1984) Effect of nuclear hydroxy substituents on aqueous solubility and acidic strength of bile acids. Hepatology 4:72S-76S
Von Dippe P, Ananthanarayanan M, Drain P, Levy D (1986) Purification and reconstitution of bile acid transport system from hepatocyte sinusoidal plasma membranes. Biochim Biophys Acta 862:353-360
Von Dippe P, Levy D (1983) Characterization of the bile acid transport system in normal and transformed hepatocytes. J Biol Chem 258:8896-8901

Binding and Uptake of Taurocholate in Subcellular Organelles from Rat Liver

B. Fleischer, F.A. Simion[1] and S. Fleischer
Department of Molecular Biology
Vanderbilt University
Nashville, TN 37235
USA

Introduction

As part of the enterohepatic circulation, bile acids and bile salts are taken up by hepatocytes on the sinusoidal surface, rapidly pass through the cell and are secreted into bile via the canalicular surface. In rat liver, transepithelial transport from blood to bile has been shown to be the major pathway of bile secretion. Early studies on both perfused liver and isolated hepatocytes suggested that uptake of bile salts is a carrier-mediated processes which is stimulated by a sodium gradient, outside > inside (for a review, see Boyer, 1980). Little is known concerning the intracellular pathway by which bile acids move from the sinusoidal surface to the bile canalicular surface. Morphologic evidence has suggested that Golgi plays a role in this process. Thus, increased taurocholate secretion brought about by feeding taurocholate results in proliferation of the Golgi complex around the bile canaliculus (Jones et al., 1979). Furthermore, photoactivatable cholate derivatives have been shown to associate with the Golgi complex and smooth endoplasmic reticulum *in situ* (Suchy et al., 1983; Goldsmith et al., 1983).

[1]Present address, Colgate-Palmolive Research Center, Piscataway, NJ 08854.

Hepatic Transport in Organic Substances
Ed. by E. Petzinger, R. K.-H. Kinne, H. Sies

Subcellular Distribution of Bile Acids and Bile Salts

In order to gain some insight into the possible role of the Golgi apparatus in the intracellular transport of bile acids, we analyzed and quantitated the bile acids and salts present in subcellular fractions of rat liver. Greater than 95% of the bile acids found in rat liver homogenates consist of about equal amounts of taurocholate and tauromuricholate (Table I). Simple differential centrifugation showed that most of the bile salts are recovered in the soluble fraction. Since bile salts are rapidly washed out of cell fractions, only low levels of bile salts were found in purified cell fractions such as rough endoplasmic reticulum (RER) or Golgi.

TABLE I - Distribution of Bile Salts in Rat Liver

	taurocholate		tauromuricholate	
fraction	µg/g liver	µg/mg protein	µg/g liver	µg/mg protein
homogenate	35.8 ± 14.1	0.176	20.0 ± 5.7	0.157
cell supernatant	21.9 ± 1.9	0.369	17.2 ± 1.4	0.285
total microsomes	3.1 ± 1.6	0.184	0.8 ± 0.3	0.060
rough microsomes	<0.1	0.030	ND	
D_2O Golgi	<0.2	0.060	ND	

No taurodeoxycholate, taurochenodeoxycholate, glycine-conjugated bile salts, or unconjugated bile acids were detected (limit of detectability in homogenate, 3 µg/g of liver). ND, not determined (from Simion et al., 1984a).

High Affinity Binding Sites in Subcellular Organelles

Since subcellular fractions involved in intrahepatic bile acid transport might be expected to contain high affinity binding sites for bile acids, the subcellular distribution of taurocholate binding sites was also studied. Taurocholate (TC) binding to purified subcellular fractions from rat liver was determined using equilibrium dialysis in a TC concentration range from 0.1 to 100 µM. This is well below the critical micellar concentration of taurocholate (3 mM). The results are summarized in Table II.

TABLE II - Taurocholate Binding to Subcellular Fractions[a]

fraction	N_1 (nmol/mg prot.)	A_1 (µM)	N_2 (nmol/mg prot.)	A_2 (µM)
cell supernatant	0.35 ± 0.06	4.8 ± 0.8	3.0 ± 0.3	79 ± 7
rough microsomes	<0.02[b]		8.9 ± 1.6	102 ± 19
smooth microsomes	<0.02[b]		9.2 ± 1.2	94 ± 16
Golgi complex	0.15 ± 0.03	1.0 ± 0.3	8.9 ± 1.9	99 ± 21
mitochondria	<0.02[b]		9.0 ± 1.9	238 ± 51
PM (total)	0.15 ± 0.02	1.3 ± 0.2	5.4 ± 1.6	91 ± 28
PM (bile canalicular)	<0.02[b]		6.8 ± 3.2	116 ± 55

[a]Binding of TC to the fractions was determined by equilibrium dialysis in 0.1 M Tris-HCl and 0.25 M sucrose, pH 8.0 at 4°C. To estimate binding constants, binding data were fitted to a model which assumed two groups of binding sites. A_1 and A_2 are mean dissociation constants for the two groups of binding sites and N_1 and N_2 represent the total binding capacity of each group of binding sites. [b]These fractions appeared to possess only one class of binding sites. The binding constants were estimated by fitting the Scatchard plot data to a straight line (from Simion et al., 1984a).

All of the fractions investigated exhibited low affinity binding with dissociation constants from 80 to 240 µM as did membrane lipid vesicles. Therefore, low affinity binding appears referable to taurocholate non-specifically partitioning into the lipid bilayer. High affinity binding is present in total plasma membranes, Golgi, and cell supernatant. The high affinity binding sites in Golgi have a mean dissociation constant (A_1) of 1.0 µM and bind 0.15 nmoles TC per mg protein. Similarly, the high affinity binding sites of total PM have an

A_1 of 1.3 µM and bind 0.15 nmoles TC were bound per mg protein. Mitochondria, smooth and rough microsomes, bile canalicular-enriched PM and Golgi liposomes showed no detectable amounts of high affinity binding. These results are compatible with a role for the Golgi complex, cytoplasmic component(s) and basolateral PM in transhepatic bile acid transport.

Transport in Subcellular Organelles

In order to further assess whether intracellular organelles are involved in the transcellular secretion of bile acids, we measured directly the ability of purified subcellular fractions of rat liver to take up taurocholate using a Millipore filtration assay. Two distinct uptake mechanisms were found, one localized in the plasma membranes, the other in the Golgi and smooth microsomal fractions. Total plasma membranes (PM) prepared according to Fleischer and Kervina

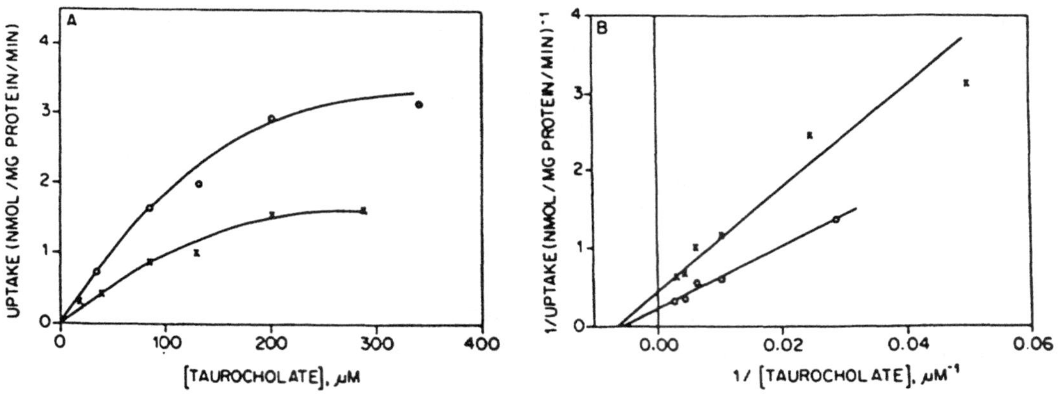

Fig. 1. Stimulation of taurocholate (TC) uptake with a Na^+ gradient as a function of TC concentration. A, dependence of uptake rate in total plasma membranes on TC concentration. B, Lineweaver-Burk plot of data in A. Initial uptake rates were measured after 20 s. For measuring uptake in the presence of a 100 mM sodium gradient (O), the fraction was preincubated for 3 min in 0.1 M potassium phosphate, 0.25 M sucrose, pH 7.0. In the absence of a sodium gradient (basal), preincubation was with 0.1 M Tris-HCl, 0.25 M sucrose, pH 7.0 (X). The use of 0.1 M potassium phosphate instead of 0.1 M KCl does not affect the TC uptake rate (from Simion et al., 1984b).

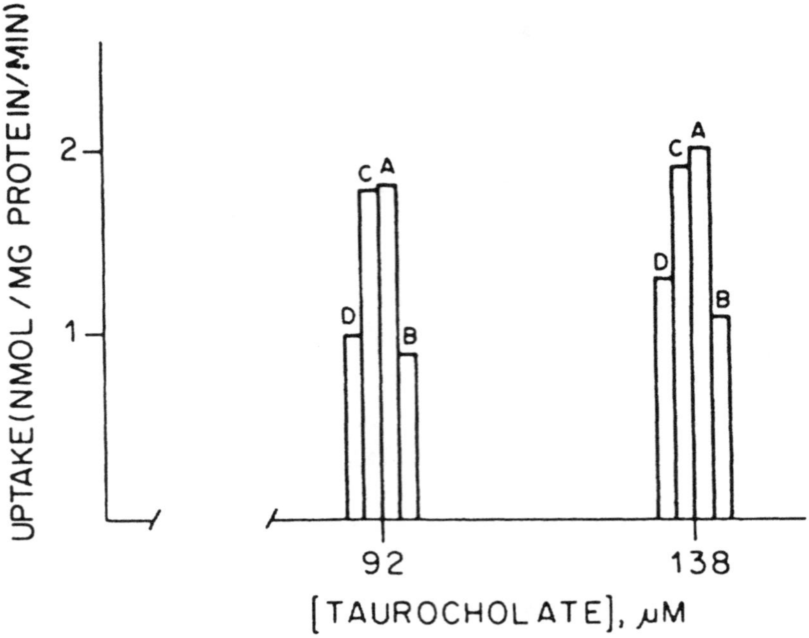

Fig. 2. Differential effects of phalloidin and bromosulfophthalein (BSP) on sodium gradient-stimulated taurocholate uptake by total plasma membrane preparations. After preincubating the plasma membranes in 0.1 M potassium phosphate, 0.25 M sucrose, pH 7.0, for 3 min at 37°C, uptake was initiated by the addition of TC at the indicated concentrations and 100 mM NaSCN (A), together with phalloidin (10 µM) (B), or BSP (120 µM) (C), respectively. The basal TC uptake rate (D) was determined as in Fig. 1. Addition of phalloidin does not affect the basal rate of TC uptake (from Simion et al., 1984b).

(1974) take up taurocholate in a saturable manner with an apparent V_{max} of 2.4 nmoles min^{-1} mg $protein^{-1}$ and a K_m of 190 µM at 37°C. After preincubation of the membranes with K^+ ions, a sodium gradient (100 mM outside) stimulates the uptake rate by 90% with the observed K_m unchanged (Fig. 1).

The stimulation by the sodium gradient is inhibited by phalloidin (Fig. 2), a known competitive inhibitor of TC uptake in hepatocytes (Frimmer et al., 1977), but not by bromosulfophthalein (BSP), a noncompetitive inhibitor of TC uptake by hepatocytes (Schwarz et al., 1975). TC uptake in the absence of a Na^+ gradient (Basal) is not affected.

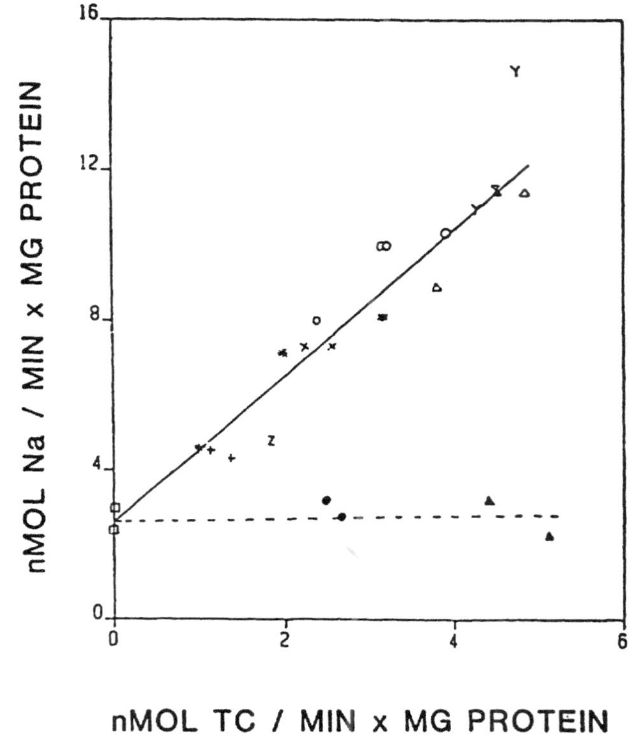

Fig. 3. Stoichiometry of sodium ion to taurocholate (TC) taken up during sodium gradient-stimulated taurocholate uptake by plasma membranes. Plasma membranes were preincubated with 0.1 M KCl, 0.05 M Tris · HCl, 0.25 M sucrose, pH 7.0, for 3 min before the [^{22}Na]-NaSCN and [^{3}H]TC were added simultaneously at varying concentrations. The uptake rate of ^{22}Na is plotted against that for [^{3}H]TC. The slope of the graph (2.0 ± 0.3) provides the Na^+/TC stoichiometry (from Simion et al., 1984c).

We also observed (Simion et al., 1984c) that TC uptake in isolated liver PM is specific for sodium and that there is cotransport of two Na^+ ions per TC taken up (Fig. 3). The presence of K^+ ions inside the vesicles was also essential for maximum Na^+ stimulation of TC uptake, although a K^+ gradient was not necessary. Mg^{2+} was almost as effective as K^+ in this regard. However, at physiological concentrations, potassium is much more effective than magnesium and is probably the cation involved <u>in vivo</u>. The symport of Na^+ and TC during uptake was

TABLE III - Subcellular distribution of taurocholate uptake and marker enzymes

Fraction	Taurocholate uptake rate			Marker enzymes	
	Basal	Total	Stimulation by Na^+ (%)	Galactosyl-transferase (Golgi complex)[a]	Cholic acid: CoA ligase (endoplasmic reticulum)[b]
Smooth microsomes	13.0	14.6	12	192	11.2
Rough microsomes	1.3	1.0	0	8	11.3
Golgi complex	9.8	11.1	13	741	1.6
PM (total)	2.4	4.6	93	18	1.5
PM (bile canalicular)	0.9	0.6	0	ND[c]	ND
Mitochondria	0.1	0.1	0	0	0.3

Activities expressed as [a]nmoles/mg protein x h and [b]µmoles/mg protein x min. Basal taurocholate uptake was measured after incubation in 0.1 M Tris-HCl, 0.25 M sucrose, pH 7.0, in the absence of a sodium gradient, and total taurocholate uptake was measured in the presence of a 100 mM Na gradient after preincubation in buffer containing a 0.1 M potassium salt. [c]Not determined. PM = plasma membranes (from Simion et al., 1984b).

shown to be electrogenic, so that K^+ may act as an exchange counterion preventing accumulation of positive charge within the vesicles.

Bile canaliculus-enriched plasma membrane vesicles prepared by the method of Kramer et al., 1982, as well as rough endoplasmic reticulum and mitochondria did not exhibit sodium-stimulated TC uptake (Table III).

Golgi-enriched fractions and smooth microsomes are also capable of taking up TC. This uptake differs from that in the total PM fraction in that it is not stimulated by a sodium gradient, has a V_{max} of 12 nmoles min^{-1} mg $protein^{-1}$ and a K_m of 440 µM at 37°C, and is inhibited by BSP but not by phalloidin (Fig. 4). The uptake rates observed were at least an order of magnitude higher than the rate observed with PM enriched in canalicular membranes, which we estimate to be about 30% pure.

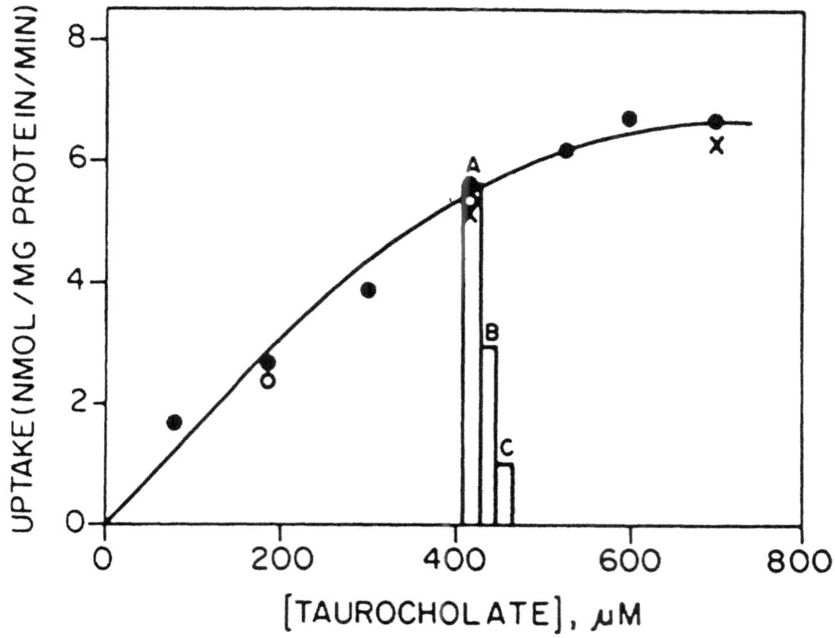

Fig. 4. Taurocholate (TC) uptake by the Golgi complex as a function of TC concentration. Golgi (2.0 mg/ml) was preincubated in either 0.1 M potassium phosphate, 0.25 M sucrose, pH 7.0 (O) or 0.1 M Tris-HCl, 0.25 M sucrose, pH 7.0, for 3 min at 37°C (X). Uptake was started by the addition of different TC concentrations. When Golgi had been preincubated in potassium phosphate buffer, NaSCN was simultaneously added to produce a 100 mM transmembrane sodium gradient. The initial uptake rate was measured after 20 s. The effect of phalloidin was determined after preincubation in 0.1 M potassium phosphate, 0.25 M sucrose, pH 7.0, by adding NaSCN (100 mM) and phalloidin (20 μM) simultaneously with TC (O). Inhibition by bromosulfophthalein (BSP) was determined after preincubation in 0.1 M potassium phosphate, 0.25 M sucrose, pH 7.0, by adding TC (440 μM), NaSCN (100 mM), and BSP simultaneously [BSP = 0 μM (A); 60 μM (B); or 150 μM (C)] (from Simion et al., 1984b).

Taurocholate uptake into smooth microsomes is abolished by filipin, an antibiotic that complexes with cholesterol to disrupt the membrane. This suggests that taurocholate uptake occurs into a non-endoplasmic reticulum subfraction since endoplasmic reticulum membranes contain negligible amounts of cholesterol (Zambrano et al., 1975) and treatment with filipin does not affect the latency of glucose-6-phosphatase in the smooth microsome fraction (Fig. 5).

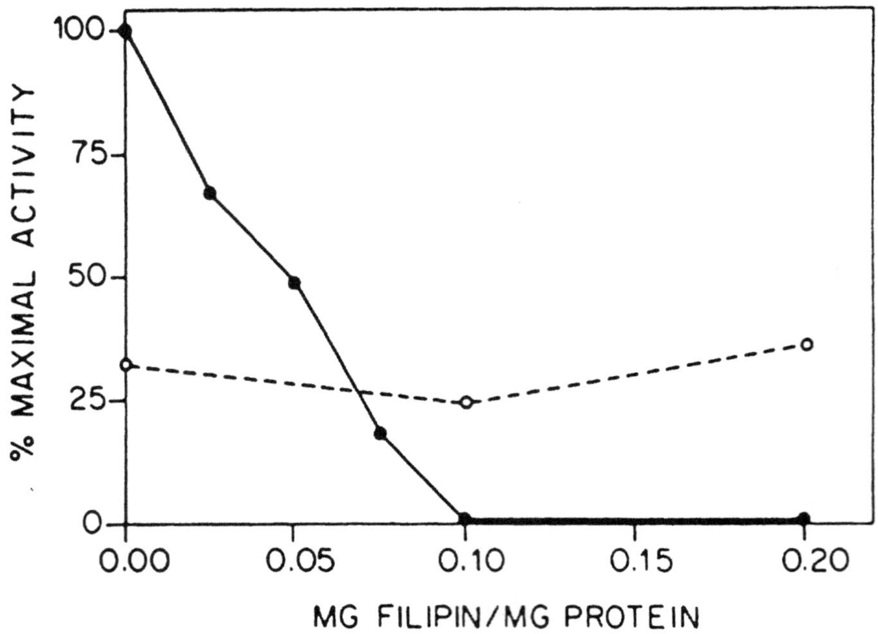

Fig. 5. The differential effect of filipin on taurocholate uptake and the latency of mannose 6-phosphatase activity in smooth microsomes. Uptake of TC, 700 µM (O) was measured after the smooth microsomes had been preincubated with various concentrations of filipin for 10 min at 37°C in 0.1 M Tris-HCl, 0.25 M sucrose, pH 7.0. Mannose 6-phosphatase (O) and glucose 6-phosphatase activities were determined using the same protocol after the smooth microsomes were preincubated with various concentrations of filipin in 0.1 M Tris-HCl, 0.25 M sucrose, pH 7.0, for 10 min at 37°C. The glucose 6-phosphatase activity determined under the same conditions but without filipin was used as the maximal phosphatase rate (from Simion et al., 1984b).

Previous studies have shown that the high affinity binding protein for BSP present in rat liver plasma membranes does not bind taurocholate (Reichen et al., 1981). Furthermore, BSP uptake into hepatocytes is not sodium-dependent nor is it inhibited by TC (Schwenk et al., 1976). Thus, it appears that BSP and TC are taken up into hepatocytes by different carriers in the plasma membrane. Once inside, BSP may compete for TC binding to an intracellular smooth membrane locus involved in transcellular transport. Our results with isolated fractions indicate that a BSP-sensitive taurocholate uptake is present in

TABLE IV - Comparison of taurocholate uptake by plasma membrane, Golgi complex, and smooth microsomes

Fraction	Total Uptake		Inhibition by		
	V_{max}	K_m	BSP	Phalloidin	Filipin
PM (total)	4.6 ± 1.0	190 ± 40	No	K_i ~ 10 μM	Yes
Golgi complex	11.1 ± 1.0	440 ± 40	K_i ~ 10 μM	No	Yes
Smooth microsomes	14.6 ± 1.0	400 ± 20	K_i ~ 4 μM	No	Yes

V_{max} expressed as nmoles TC/mg protein x min and K_m as μM TC. Total uptake rates measured in the presence of a 100 mM Na^+ gradient after preincubation for 3 min with 0.1 M potassium salt. Phalloidin or bromosulfophthalein (BSP) were added simultaneously with [^{3}H]taurocholate. However, disruption of the membranes by filipin requires 10 min of preincubation at 37° with 0.3 mg of filipin/mg of plasma membrane protein; 0.2 mg of filipin/mg of Golgi protein; 0.1 mg of filipin/mg of smooth microsomal protein (from Simion et al., 1984b).

both smooth microsomes and in Golgi fractions at about equal levels (Table IV). This activity does not colocalize with either endoplasmic reticulum or Golgi marker enzymes. Possible candidates for this uptake are endosomes, multivesicular bodies, or specialized secretory vesicles derived from the Golgi apparatus destined for secretion into bile.

Transport in Isolated Liver Cells

We have also used isolated liver cells to assess whether agents known to block secretion of serum proteins by the Golgi apparatus have any effect on secretion of TC. Rat hepatocytes prepared by a modification of the method of Berry and Friend, 1969, take up TC at a linear rate until saturation is reached (Fig. 6). The rate of uptake and the equilibrium level is dependent on the initial concentration of TC. The initial uptake rate is saturable with an apparent K_m of 45 μM and a V_{max} of 1.9 nmoles per 2 x 10^6 cells per min (Fig. 7).

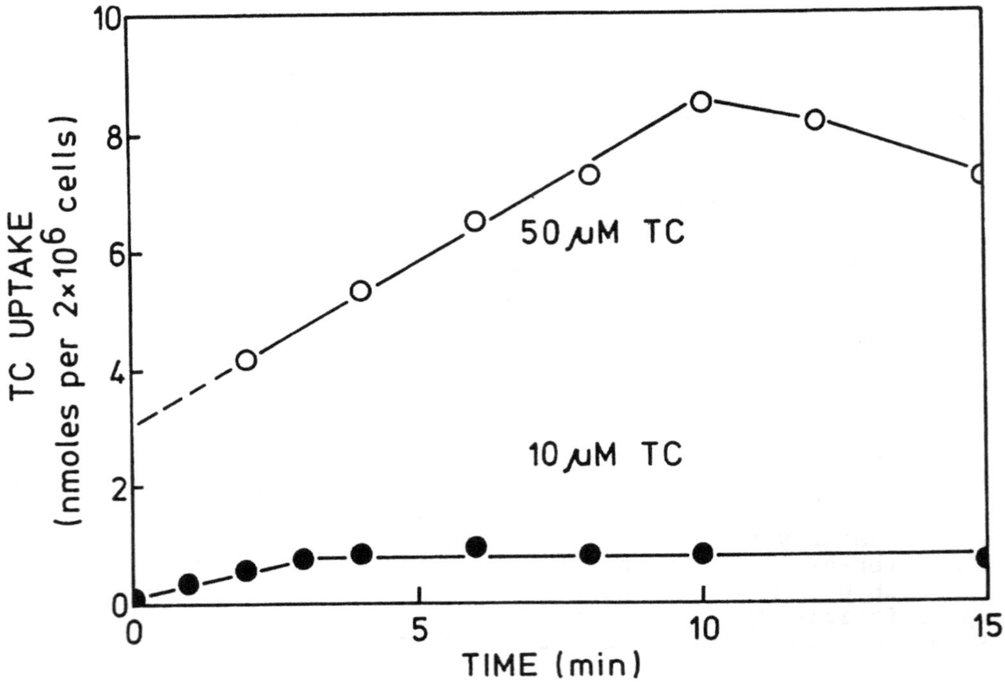

Fig. 6. Uptake of TC by isolated rat liver cells. Cells, 2×10^6 per ml, were preincubated in 1-2 ml of medium designed for optimum serum albumin synthesis and secretion (Schreiber and Schreiber, 1973). Cells were placed in wide cellulose nitrate tubes and shaken at 37°C using a reciprocal motion at 125 rpm for 30 min before addition of [^{3}H]taurocholate. To determine uptake, 0.1 ml aliquots were diluted into 2.0 ml cold medium and 1.0 ml filtered using 0.25 µm HA type Millipore filters. Filters were washed with 5 ml cold buffer, dissolved in 2 ml ethylene-glycol methyl ether, and counted after addition of 10 ml scintillation fluid.

Efflux of TC from preloaded cells can also be demonstrated after dilution of the external TC concentration (Fig. 8). The efflux is not due to simple washing out of externally bound TC since it is inhibited at 4°C. Efflux is linear for about 5 min with a rate of 0.2 nmoles per min per 2×10^6 cells when cells are loaded using 50 µM TC.

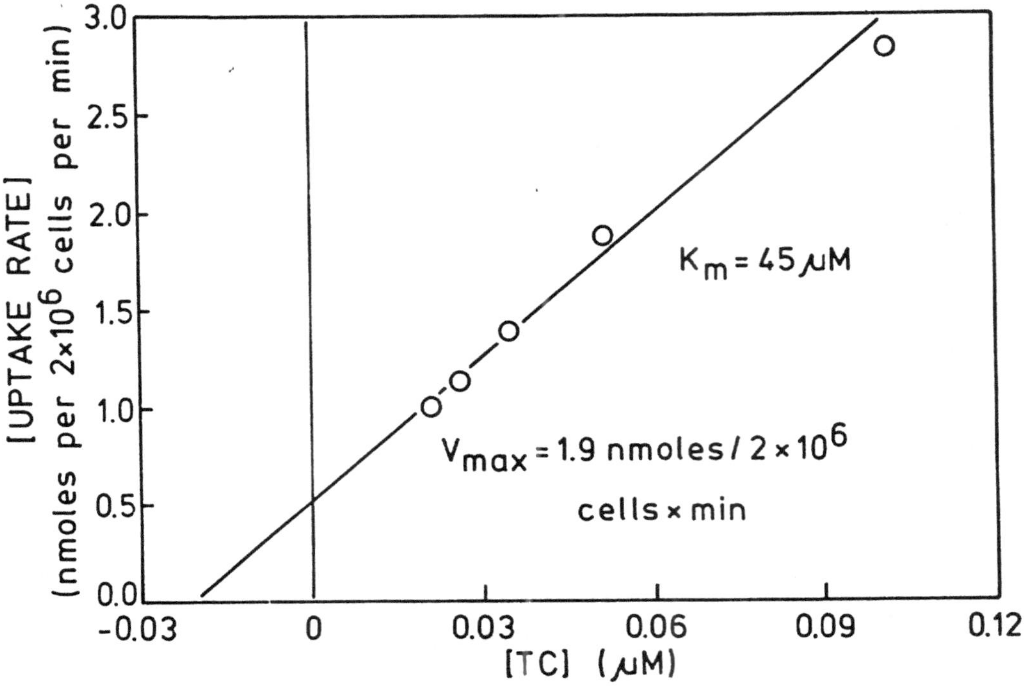

Fig. 7. Effect of TC concentrations on TC uptake by isolated liver cells. Uptake was measured after 2 min at 37°C as described in legend to Fig. 6.

The rate of efflux of TC can be increased by addition of cold TC to the resuspending medium after preloading the cells with ^{3}H-TC. Thus, the secretion of stored TC is stimulated by further addition of extracellular TC. This indicates that the transcellular pathway of TC secretion is operating in the isolated cells (Fig. 9).

The cells are capable of synthesizing and secreting newly formed albumin after a lag of about 15 min as described previously (Schreiber and Schreiber, 1973). Secretion is inhibited by preincubating the cells with 5 μM colchincine or by adding 5 μM monensin to the cells as indicated in Fig. 10.

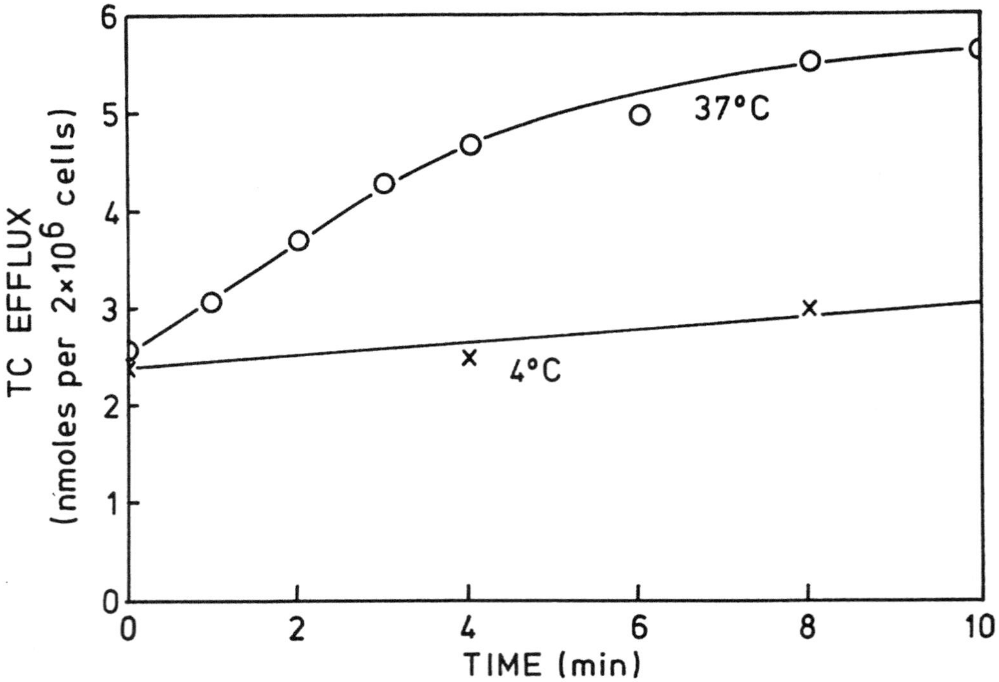

Fig. 8. Effect of temperature on efflux of TC from preloaded liver cells. Cells were incubated with 50 µM ^{3}H-TC for 15 min as described in the legend to Fig. 6. The cells were then diluted 10-fold with cold medium without TC, sedimented at 50-100xg at 4°C for 30 sec and 9 volumes of supernatant removed leaving cells in the original volume of medium containing about 5 µM TC. Cells were resuspended and reincubated at 37°C. Samples were taken at various times thereafter, diluted with 10 volumes cold buffer and centrifuged a few seconds in a microfuge to remove cells. Aliquots of the final supernatant were counted to determine TC efflux.

In contrast to the effects on serum albumin secretion, addition of colchicine during preincubation plus addition during TC secretion of the cells had no effect on the rate of TC secretion of the cells. Monensin present during TC secretion also showed no significant effect on the rate of TC secretion (Fig. 11).

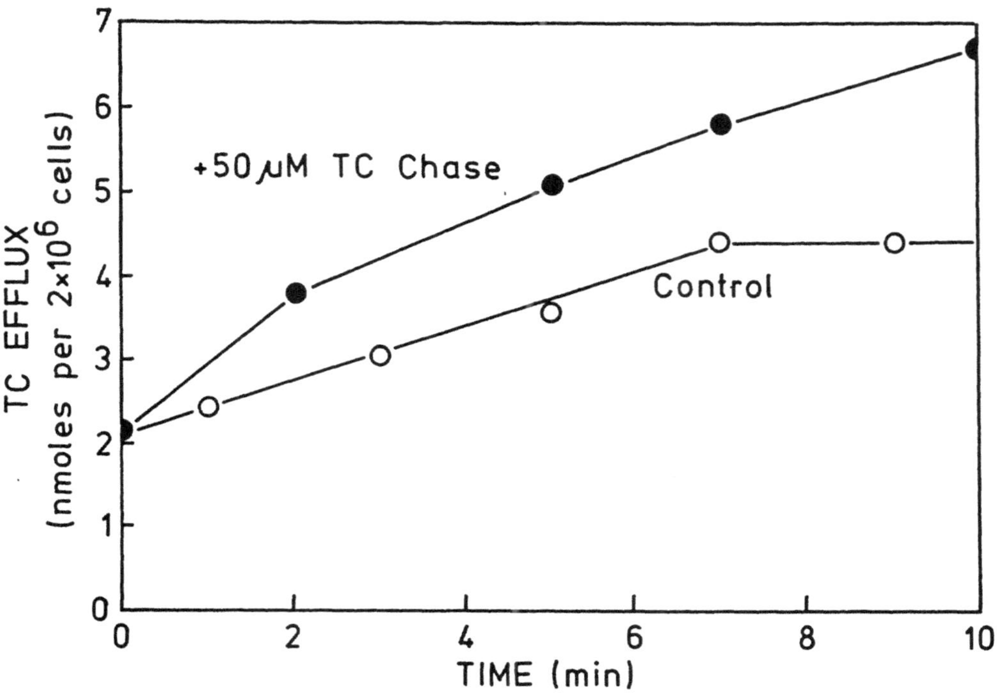

Fig. 9. Effect of added external cold TC on the efflux rate of ^{3}H-TC from preloaded cells. Cells were loaded, washed and reincubated ± 50 μM cold TC as described in the legend to Fig. 8.

The efflux of TC from isolated hepatocytes appears to be a process that is independent of the secretory pathway for newly formed serum proteins and therefore independent of Golgi secretory vesicles. Nonetheless, our results strongly suggest that an intracellular transport vesicle of some type is involved in TC transport in liver.

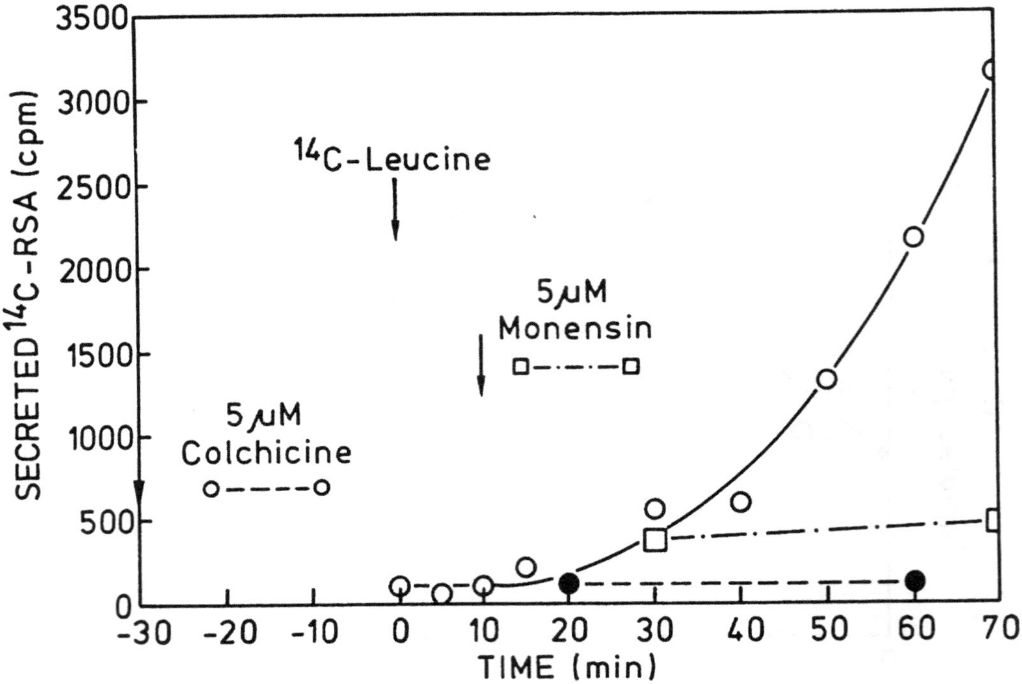

Fig. 10. Effect of monensin or colchicine on secretion of ^{14}C-rat serum albumin (RSA) from isolated liver cells. Newly formed albumin was labelled by addition of 4 µCi L-[^{14}C]-leucine (300 mCi/mmole) per ml of cells and incubating at 37°C. At given time points aliquots were removed, diluted 10-fold and supernatants recovered. Serum albumin was precipitated from 400 µl supernatant after addition of 5 µg carrier serum albumin, 1 mM cold leucine and 20 µl rabbit anti-rat albumin (Cappel Labs). After incubation at 37°C for 1 hr, then 18 hrs at 4°C, the immunoprecipitates were recovered by centrifugation, washed 3 times with 0.9% NaCl, dried, dissolved in 50 µl Folin A reagent, 10 ml scintillation fluid added, and counted.

Summary

Rat liver contains mainly taurocholate and tauromuricholate which are easily washed out during fractionation so that their subcellular distribution cannot be determined directly. High affinity binding sites for taurocholate can be demonstrated in the supernatant, the basolateral plasma membranes and the Golgi apparatus fractions. In addition, two distinct uptake mechanisms can be distinguished, one localized in basolateral PM, the other present in both Golgi and smooth microsome fractions. Uptake by basolateral PM is

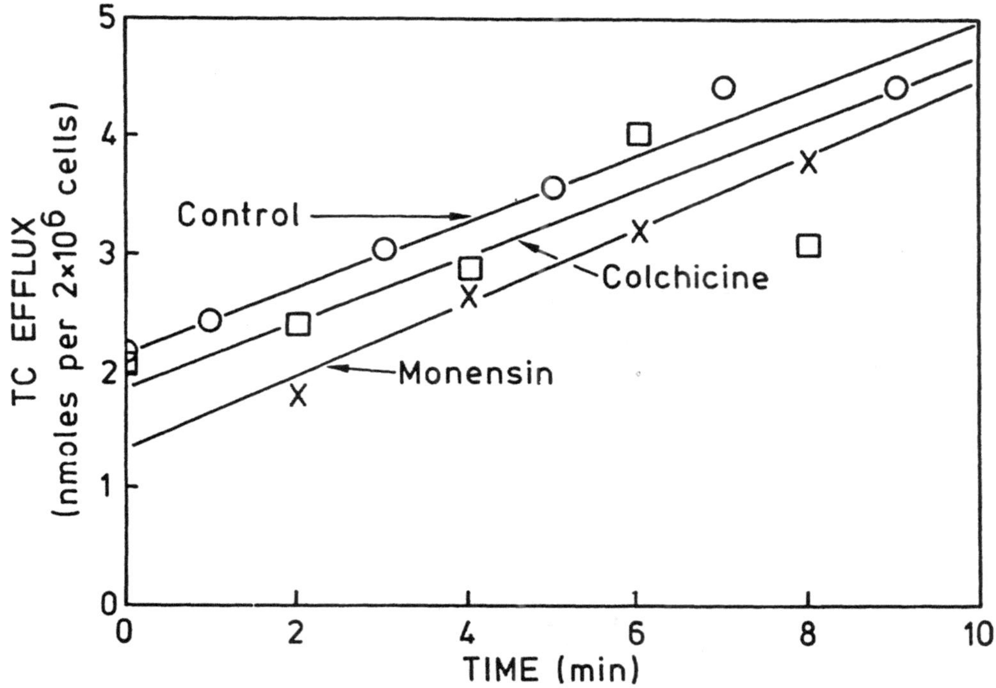

Fig. 11. Effects of monensin or colchicine on TC efflux. Cells were loaded with 50 μM TC as described in the legend to Fig. 6. Colchicine (5 μM) was added to one batch of cells at the start of the 30 min preincubation and was present throughout the incubations for uptake and secretion. Monensin, 5 μM, was added to another batch during incubation for secretion only.

stimulated by a Na^+ gradient outside > inside when vesicles are preloaded with K^+. A cotransport of 2 Na^+ ions per taurocholate can be demonstrated. This transport is inhibited by phalloidin but not by bromosulfophthalein (BSP). Transport by the Golgi and smooth microsome fractions, on the other hand, is not stimulated by a Na^+ gradient and is inhibited by BSP but not by phalloidin. We have also investigated whether agents known to block the secretion of serum proteins by the Golgi apparatus have any effect on secretion of TC in isolated liver cells. Levels of monensin or colchicine which could be shown to block secretion of newly formed albumin had no detectable effect on TC loading or efflux. These studies suggest that the mechanism of secretion of TC in liver cells involves intracellular smooth vesicles which are distinct from those involved in serum albumin secretion by the Golgi apparatus.

References

Berry MN, Friend DS (1969) A high yield preparation of isolated rat liver parenchymal cells. J Cell Biol 43:506-520

Boyer JL (1980) New concepts of mechanisms of hepatocyte bile formation. Physiol Rev 60:303-326

Fleischer S, Kervina M (1974) The subcellular fractionation of rat liver. Methods Enzymol 31:180-191

Goldsmith MA, Huling S, Jones AL (1983) Hepatic handling of bile salts and protein in the rat during intrahepatic cholestasis. Gastroenterology 84:978-986

Johnson E, Fleischer B (1986) Independence of taurocholate and albumin secretion in isolated liver cells. J Cell Biol 103:64a

Jones AL, Schmucker DL, Mooney JS, Ockner RK, Adler RD (1979) Alterations in hepatic pericanalicular cytoplasm during enhanced bile secretory activity. Lab Inv 40:512-517

Kramer W, Bickel U, Buscher HP, Gerok W, Kurz G (1982) Bile-salt-binding polypeptides in plasma membranes of hepatocytes revealed by photoaffinity labelling. Eur J Biochem 129:13-24

Reichen J, Blitzer BL, Berk PD (1981) Binding of unconjugated and conjugated sulfobromophthalein to rat liver plasma membrane fractions in vitro. Biochim Biophys Acta 640:298-312

Schreiber G, Schreiber M (1973) The preparation of single cell suspensions from liver and their use for the study of protein synthesis. Sub-Cell Biochem 2:321-383

Schwenk M, Burr R, Schwarz L, Pfaff R (1976) Uptake of bromosulfophthalein by isolated liver cells. Eur J Biochem 64:189-197

Simion FA, Fleischer B, Fleischer S (1984a) Subcellular distribution of bile acids, bile salts and taurocholate binding sites in rat liver. Biochem 23:6459-6466

Simion FA, Fleischer B, Fleischer S (1984b) Two distinct mechanisms for taurocholate uptake in subcellular fractions from rat liver. J Biol Chem 259:10814-10822

Simion FA, Fleischer B, Fleischer S (1984c) Ionic requirements for taurocholate transport in rat liver plasma membrane vesicles. J Bioenerg Biomemb 16:507-515

Suchy FJ, Balistreri WF, Hung J, Miller P, Garfield SA (1983) Intracellular bile acid transport in rat liver as visualized by electron microscopy autoradiography using a bile acid analogue. Am J Physiol 245(Gastrointest Liver Physiol 8):G681-G689

Zambrano F, Fleischer S, Fleischer B (1975) Lipid composition of the Golgi apparatus of rat kidney and liver in comparison with other subcellular organelles. Biochim Biophys Acta 380:357-369

The Ontogeny of Hepatic Bile Acid Transport

F.J. Suchy, M. Ananthanarayanan, and
J.C. Bucuvalas
Division of Pediatric Gastroenterology
Children's Hospital Medical Center
Elland and Bethesda Avenues, Cincinnati, Ohio 45229
USA

Introduction

Bile formation is an extraordinarily complex process that requires the participation of numerous transport mechanisms for organic and inorganic anions on the sinusoidal and canalicular domains of the hepatocyte plasma membrane. The vectorial transport of bile acids from sinusoidal blood into bile requires the polarized insertion of specific transport proteins into the basolateral and apical domains of the plasma membrane. Recent studies indicate that the carrier mechanisms for bile acids are ontogenically regulated and expressed at very specific times during perinatal development. These findings are consistent with observations in the intact animal showing that bile secretory function is immature at birth and develops during postnatal life (Suchy et al., 1987). In addition, hepatic uptake and elimination of bile acids and other organic anions are decreased in developing animals including the human compared with the adult. In the intact rabbit and dog and in the isolated perfused rat liver, basal and bile salt stimulated flow rates are significantly lower in the neonate when compared with the adult (Shaffer et al., 1985; Tavolini et al., 1988; Piccoli et al., 1986). Similarly, both

Hepatic Transport in Organic Substances
Ed. by E. Petzinger, R. K.-H. Kinne, H. Sies

bile acid-dependent and -independent fractions of bile flow are generally immature in the neonate and develop progressively with postnatal age. Canalicular bile flow as estimated by ^{14}C erythritol clearance is markedly reduced and matures near the time of weaning. In the full term human infant, biliary bile acid concentrations are extremely low in comparison with the older child and adult. Intraluminal bile acid concentrations of 1 to 2 mM have been found in several studies during meals and exhibit very little variation during the day (Ricour and Rey, 1972). Thus, bile secretion and bile acid output appear to function at a maximum during early life and cannot be stimulated further by the hormonal milieu of the postprandial period. The cumulative effect of immature hepatobiliary function in newborns is an exaggerated tendency towards cholestatic liver disease.

Cellular Mechanisms of Bile Acid Transport during Development

Bile acid transport by the developing liver is critically dependent upon the driving forces provided by the Na^+ pump as well as on the ontogenesis of specific membrane carriers for bile acids on the basolateral and canalicular domains of the hepatocyte plasma membrane (Blitzer and Boyer, 1982). Ontogenic studies of hepatic transport are difficult, particularly in intact experimental systems, and are often complicated by a number of age-related variables such as organ perfusion, lobular geometry, volume of distribution, cell size, cell metabolism, and intracellular compartmentation. Many of these problems can be overcome by performing studies in isolated hepatocytes and domain-specific plasma membrane vesicles. However, great care must be taken when isolating hepatocytes to insure that cell viability and the population of cells isolated are comparable at each age. Similarly, the characteristics of the plasma membrane vesicles must be comparable in terms of the yield, recovery and enrichment of marker enzyme activities for contaminating intracellular organelles and the specific plasma membrane domains. Our laboratory has been successful in employing the Percoll gradient technique of Blitzer and Donovan to prepare highly purified basolateral liver plasma membrane vesicles from fetal, neonatal, and adult rats (Blitzer and Donovan, 1984). The nitrogen cavitation and calcium precipitation method of Inoue et al. has been used to prepare canalicular plasma membrane vesicles from developing rat liver (Inoue et al., 1983). The main limitation of these methods is the relatively large amount of tissue (20-25 grams) required for these membrane preparations. The basolateral membranes are approximately 30-fold enriched at each age group in the basolateral marker enzyme Na^+-K^+-ATPase, while marker enzyme activities indicate only modest contamination with canalicular membranes and intracellular organelles. Approximately 98% of surface membranes present in the final fraction are derived from the basolateral domain. Canalicular membranes are 80-fold enriched in the apical markers alkaline phosphatase and leucine aminopeptidase, 1.5-fold enriched in Na^+,K^+-ATPase, and not enriched in marker enzymes for intracellular organelles. There is an approximately 10% contamination with basolateral membranes prepared in this fashion.

The ontogeny of bile acid uptake by the developing liver has been studied in hepatocytes isolated from rats between the ages of 7 and 56 days (Suchy and Balistreri, 1982). The rate of taurocholate uptake was found to be a saturable function of its concentration in the incubation media at each age, but uptake was markedly lower in hepatocytes isolated from suckling and weanling rats in comparison with the adult. Analysis of the kinetics of taurocholate uptake revealed a similar apparent K_m at each age but a 4-fold increase in V_{max} was observed between 7 and 56 days of age. These data suggested that during development there is a progressive increase in the number of bile acid carriers in the plasma membrane. The similar K_m indicated that the apparent affinity of the carrier for taurocholate remained constant during this period of development. However, the complexity of the intact cell prevented a conclusion about other factors which would limit uptake such as the development of Na^+,K^+-ATPase activity. In addition, the loss of polarity which occurs on cell isolation precluded an absolute identification of the carrier mechanism mediating taurocholate entry into the cell.

The ontogenesis of Na^+,K^+-ATPase activity which provides the driving force for bile acid uptake across the basolateral membrane and excretion across the canalicular membrane has been studied in rat liver basolateral plasma membranes (Suchy et al., 1986). Enzyme activity was significantly lower in membranes from late fetal (day 21-22) and neonatal (day 1) rat liver compared with membranes from the adult. Kinetic analysis of Na^+,K^+-ATPase activity at various concentrations of adenosine triphosphate revealed that the V_{max} of the enzyme reaction was 70 and 90% of adult activity in the term fetus and neonate, respectively. These differences in enzyme activity were statistically significant, but the question can be raised as to whether they are of biological importance because of the large reserve capacity of the Na^+ pump. Na^+ pump activity may acutely increase 4 to 5-fold in response to an elevation of the intracellular Na^+ concentration (Van Dyke and Scharschmidt, 1983). This increase appears to be related to an increased cycling of each pump unit rather than to insertion of new pumps into the plasma membrane. Although measurements of enzyme activity in membrane fractions in vitro may not directly correlate with the capacity of intact hepatocytes for cation pumping, several other lines of evidence would suggest that Na^+,K^+-ATPase matures relatively early in the perinatal period. A recent study demonstrated greater ouabain inhibitable uptake of ^{86}Rb (a substrate for the Na^+,K^+ pump) by hepatocytes isolated from neonatal compared with mature rats (Bellerman, 1981). Thus, available data would indicate that the Na^+,K^+-ATPase is well developed during early postnatal life and is unlikely to be rate limiting for transport of bile acids and other ions in the neonate. The maturation of Na^+,K^+-ATPase activity is probably required to sustain high rates of

Na^+-coupled transport processes such as those for amino acids during the phase of rapid hepatic growth.

Further studies of bile acid uptake have been performed in our laboratory using extensively characterized basolateral plasma membrane vesicles from fetal, neonatal, and adult rat livers (Suchy et al., 1986). As reported by several other laboratories using membrane vesicles from adult rat liver, an inwardly directed transmembrane Na^+ gradient was found to provide the driving force for uphill transport of taurocholate. In contrast, in membrane vesicles isolated from the 19 day fetal liver, taurocholate uptake was similar in the presence of an inwardly directed Na^+ or K^+ gradient and taurocholate accumulation above its equilibrium concentration (or an overshoot) was not observed, suggesting the absence of the Na^+-coupled transport mechanism for bile acids. Twenty-four hours later on day 20 of gestation, imposition of a Na^+ gradient stimulated taurocholate influx and energized concentrative transport of the bile acids against an electrochemical gradient. Kinetic studies demonstrated a markedly lower V_{max} for taurocholate uptake in late fetal and one day neonatal vesicles compared with the adult. There was a gradual increase in V_{max}, but rates of transport remained significantly less than the adult even at 14 days of age (Suchy et al., 1985). The apparent taurocholate K_m was 3-fold higher in the term fetus and neonate compared with the adult. These studies directly demonstrate that the specific carrier that mediates Na^+-bile acid cotransport appears to be functionally expressed on the basolateral membrane during late gestation. Following birth, the decreased number or translocation rate of this carrier limits bile acid uptake during postnatal development.

In recent studies a strategy has been developed for identification and isolation of the basolateral Na^+-dependent bile acid carrier protein based on its ontogenic expression in perinatal rat liver. Several studies suggest that the carrier for Na^+-coupled bile acid transport is a protein of an approximate M_r of 48,000. First, using photoreactive bile acid probes Kramer and associates have identified putative bile acid-binding peptides of approximate M_r 48 and 54 kDa (Kramer et al., 1982). These proteins are thought to represent a Na^+-dependent and a Na^+-independent carrier mechanism, respectively. Furthermore, in hepatocytes isolated from the small marine skate, taurocholate uptake occurs via a carrier-mediated but a Na^+-independent process (Fricker et al., 1987). In contrast to mammalian liver, only a single 54 kDa bile salt-binding polypeptide was found in skate liver plasma membranes and hepatocytes after photoaffinity labeling. Finally, a group of proteins in the M_r range of 48 to 54,000 has been isolated by Levy and coworkers using bile acid-agarose affinity chromatography (Von Dippe et al., 1986). Functional reconstruction of Na^+-dependent taurocholate uptake was demonstrated by incorporation of this group of proteins into synthetic liposomes. With these studies in mind, it was the aim of our laboratory to isolate the basolateral Na^+-dependent bile acid transport protein using its developmental expression in the perinatal period as a means for identification and isolation. This approach is similar conceptually to that of other investigators who have actively sought to reduce, isolate, or induce mutant cell lines which either lack or overexpress a transport activity of interest. Analysis of basolateral plasma membranes by SDS polyacrylamide gel electrophoresis revealed that a protein of approximate M_r 48,000 was absent from 19 day fetal compared with neonatal and adult rat liver (Suchy et al., 1988). Two-dimensional electrophoresis showed the 48 kDa band to be a single protein with a pI of approximately 9. Basolateral plasma membranes from adult rat liver were then subjected to preparative scale SDS-PAGE. The 48 kDa polypeptide was excised, electroeluted from the gels, and used as an immunogen in guinea pigs. The resulting antisera was tested by Western Blot analysis using horseradish peroxidase conjugated goat anti-guinea pig IgG for detection of specifically bound antibody. The antiserum but not preimmune serum specifically reacted with a polypeptide of approximate M_r 48,000 in basolateral but not canalicular membranes. Furthermore, the guinea pig antiserum was able to specifically immunoprecipitate the 48 kDa protein from ^{125}I labeled membranes solubilized in NP-40. Thus, the antiserum recognizes the 48 kDa band in its native as well as its denatured form.

The initial rate of Na^+-dependent taurocholate uptake by isolated rat hepatocytes was significantly inhibited (by 50%) after preincubation with immune serum or gamma globulin compared to pre-immune or normal guinea pig immunoglobulin. Cell viability, transport of ^{86}Rb (a substrate for the Na^+ pump), sulfate transport, and Na^+-independent taurocholate uptake were not affected by antibody treatment. We conclude from these studies that a developmentally regulated protein of approximate M_r is a component of the basolateral Na^+-dependent transport system for bile acids.

The mechanism for intracellular transport of bile acids from the sinusoidal to the canalicular domain remains poorly understood in developing or mature liver. We recently examined the functional significance of possible age-related differences in intracellular compartmentation (Belknap et al., 1988). In hepatocytes isolated from 14 days old rats, taurocholate efflux was significantly more rapid than that measured in adult rats after preloading of cells with bile acid concentrations from 5 to 100 μM. The rate of efflux was a linear function of the preloading concentration of taurocholate at both ages indicating that the transport process was occurring via non-carrier mediated diffusion. It remains unknown why intracellular compartmentation of bile acids in developing liver is reduced in comparison with the adult. However, recently the activity of a cytosolic bile acid-binding protein was found to be markedly decreased in fetal and neonatal rat liver (Stolz et al., 1986). The development of intracellular bile acid-binding capacity seemed to parallel the maturation of mechanisms for bile acid uptake, biotransformation via conjugation and sulfation and biliary excretion. Quantitatively greater back diffusion of bile acids in the developing liver could effectively short circuit transcellular bile acid transport and decrease the bile acid-dependent fraction of bile flow apart from the developmental maturity of the bile acid carrier system on the canalicular membrane.

Canalicular excretion of bile acids is the rate limiting step in the hepatocellular transport of bile acids, but has been difficult to directly examine, because of the inaccessability and small size of the canalicular lumen. The latter problems are even more significant obstacles in studying this system during development. Canalicular plasma membrane vesicles have been used to define mechanisms of canalicular transport in the adult rat liver and have recently been utilized in studies of developing liver (Novak and Suchy et al., 1987). The initial rate of 50 μM taurocholate influx into canalicular plasma membrane vesicles was 2-fold higher in the adult compared with vesicles from 7 and 14 days old rats. Taurocholate uptake by vesicles derived from 14 days and adult but not 7 days old animals was significantly inhibited by the anion transport inhibitor 4,4'-diisothiocyano 2,2' stilbene disulfonic acid (DIDS). The DIDS-sensitive component of uptake was found to be significantly higher in the adult compared with 14 days old vesicles. Kinetic studies were performed by preloading these predominantly right side out vesicles with taurocholate and measuring the initial rate of efflux in the physiologic direction. Computer analysis of the DIDS-sensitive portion of efflux revealed saturable kinetics with a similar V_{max} but a three-fold higher apparent K_m in 14 days compared with adult canalicular vesicles. In contrast, efflux from 7 days old canalicular membrane vesicles increased linearly with increasing concentrations of taurocholate and was not inhibited by DIDS. These studies indicate that the carrier mechanism for taurocholate is either non-functional or not present in the canalicular membrane of neonatal rats, but by 14 days postnatally this membrane contains a similar number of carriers with a lower affinity for taurocholate than in the adult.

Thus, available studies stress that the development of hepatic transport processes involved in bile formation is extraordinarily complex. Carrier mechanisms for bile acids are independently expressed at both poles of the hepatocyte plasma membrane. It should be stressed that a number of factors including available driving forces for transport, the bile acid pool size and composition, the effectiveness of intracellular compartmentation and transfer, and the functional capacity of the membrane carrier for bile acids and other ions can all contribute to low rates of bile flow and bile acid secretion depending on the particular stage of development. Since the expression and insertion of the bile acid carrier proteins are ontogenically regulated, it is likely that defects such as congenitally absent or defective transport proteins will eventually be discovered in children with undefined familial cholestasis.

References

Bellerman P (1981) Amino acid transport and rubidium-ion uptake in monolayer cultures of hepatocytes from neonatal rats. Biochem J 198:475-483

Belknap WM, Zimmer-Nechemias L, Suchy FJ (1988) Bile acid efflux from suckling rat hepatocytes. Pediat Res 23: 364-367

Blitzer BL, Boyer JL (1982) Cellular mechanisms of bile formation. Gastroenterology 82:346-357

Blitzer BL, Donovan CB (1984) A new method for the rapid isolation of basolateral plasma membrane vesicles from rat liver: characterization, validation, and bile acid transport studies. J Biol Chem 259:9295-9301

Fricker G, Hugentobler G, Meier PJ, Kurz G, Boyer JL (1987) Identification of a single sinusoidal bile salt uptake system in skate liver. Am J Physiol 253:G816-G822

Inoue M, Kinne R, Tran T, Biempica L, Arias IM (1983) Rat liver canalicular membrane vesicles. Isolation and topographical characterization. J Biol Chem 258:5183-5188

Kramer W, Bickel U, Buscher HP, Gerok W, Kurz G (1982) Bile salt polypeptides in plasma membranes of hepatocytes revealed by photoaffinity labeling. Eur J Biochem 129: 12-24

Novak DA, Suchy FJ (1987) Postnatal expression of the canalicular bile acid transport system in rat liver. Hepatology 7:1037 (Abstract)

Piccoli DA, Muller ER, Vanderslice RR (1986) Maturation of biliary steady state uptake and secretion capacity in the perfused rat liver. Hepatology 5:1185 (Abstract)

Ricour C, Rey J (1972) Study of the hydrolysis and micellar solubilization of fats during intestinal perfusion. I. Results in the normal child. Europ J Clin Biol Res 17: 172-178

Shaffer EA, Zahavi I, Gall DG (1985) Postnatal development of hepatic bile formation in the rabbit. Dig Dis Sci 30:558-562

Stolz A, Sugiyama Y, Kuhlenkamp J (1986) Cytosolic bile acid binding protein in rat liver: radioimmunoassay, molecular forms, developmental characteristics and organ distribution. Hepatology 6:433-439

Suchy FJ, Ananthanarayanan M, Bucuvalas JC, Osadchey B, Yamada T, Belknap W, Balistreri W, Kaplowitz N (1988) An antibody to a developmentally regulated 48 kDa liver plasma protein inhibits taurocholate by isolated rat hepatocytes. Gastroenterology 94:A596 (Abstract)

Suchy FJ, Balistreri WF (1982) Uptake of taurocholate by hepatocytes isolated from developing rats. Pediat Res 16:282-285

Suchy FJ, Bucuvalas JC, Goodrich A, Moyer MS, Blitzer BL (1986) Taurocholate transport and Na^+, K^+-ATPase activity in fetal and neonatal rat liver plasma membrane vesicles. Am J Physiol 251: G665-G673

Suchy FJ, Bucuvalas JC, Novak DA (1987) Determinants of bile formation during development: ontogeny of hepatic bile acid metabolism and transport. Seminars Liver Dis 7:77-84

Suchy FJ, Couchene SM, Blitzer BL (1985) Taurocholate transport by basolateral plasma membrane vesicles isolated from developing rat liver. Am J Physiol. 248:G648-G654

Tavolini N, Jones NJT, Berk PD (1985) Postnatal development of bile secretory physiology in the dog. J Pediat Gastroent Nutr 4:256-267

Van Dyke RW, Scharschmidt BF (1983) Na^+,K^+-ATPase mediated cation pumping in cultured rat hepatocytes: rapid modulation by alanine and taurocholate transport and characterization of its relationship to intracellular sodium concentration. J Biol Chem 258:12912-12919

Von Dippe P, Ananthanarayanan M, Levy D (1986) Purification and reconstitution of the bile acid transport system from hepatocyte sinusoidal plasma membranes. Biochim Biophys Acta 862:352-360

Identification and Function of Bile Salt Binding Polypeptides of Hepatocyte Membrane

Gerhart Kurz, Michael Müller, Ursula Schramm and Wolfgang Gerok
Institut für Organische Chemie und Biochemie und Medizinische Klinik, Universität Freiburg, D-7800 Freiburg i. Br., F.R.G.

Introduction

In the course of their enterohepatic circulation bile salts are taken up through the sinusoidal and secreted through the canalicular membrane of the hepatocyte. In these different membrane domains different bile-salt-binding polypeptides are present, as demonstrated unequivocally by photoaffinity-labelling studies (Kramer et al., 1982; Buscher et al., 1987; Fricker et al., 1987). All these binding polypeptides could somehow be involved in hepatobiliary bile salt transport.

Results

In sinusoidal membranes photoaffinity labelling studies with the photolabile bile salt derivative (7,7-azo-3α,12α-dihydroxy-5ß-[3ß-^{3}H]cholan-24-oyl)-2-aminoethanesulfonate (Kramer and Kurz, 1984) resulted in the labelling of polypeptides with the apparent M_r's of 67 000, 54 000, 48 000 and 43 000. The polypeptide with the apparent M_r's of 67 000, identified as albumin, and that with 43 000, being actin, were practically not labelled by photoaffinity labelling of freshly isolated hepatocytes. Therefore only two polypeptides with apparent M_r's of 54 000 and 48 000 seem to be involved in hepatic uptake of bile salts.

Hepatic Transport in Organic Substances
Ed. by E. Petzinger, R. K.-H. Kinne, H. Sies

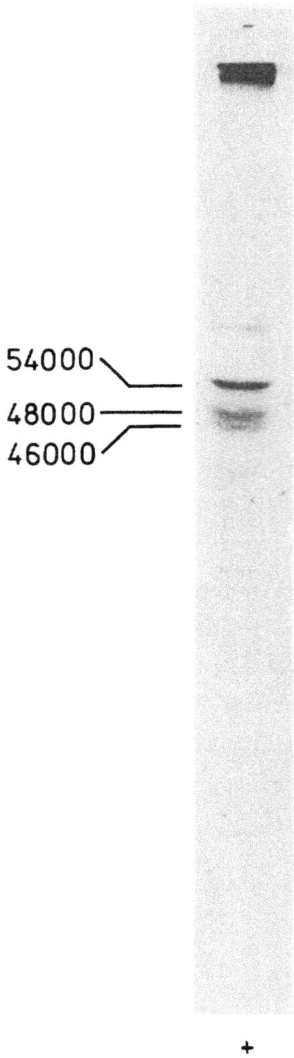

Fig. 1: Identification of bile-salt-binding polypeptides in sinusoidal membranes. 500 µg of protein of plasma membrane subfractions were preincubated with 1.25 µM (10 µCi) (7,7-azo-3α,12α-dihydroxy-5ß-[3ß-^{3}H]-cholan-24-oyl)-2-aminoethanesulfonate for 10 minutes at 30 °C. Photoaffinity labelling was performed at 350 nm for 10 minutes. Sodium dodecylsulfate/polyacrylamide gel electrophoresis was carried out at total acrylamide concentration of 10.5 % at a ratio of acrylamide:bisacrylamide of 97.2 : 2.8. The labelled polypeptides were visualized by fluorography of the slab gels.

Fluorographic analysis of labelled membranes (Fig. 1) revealed after photoaffinity labelling of sinusoidal membrane subfractions and isolated hepatocytes that in addition to the bile-salt-binding polypeptides with the apparent M_r's of 54 000 and 48 000 a third polypeptide with the apparent M_r of 46 000 became labelled. Whether a further splitting up of the bands in the M_r range of about 48 000 and 46 000, as occasionally observed, is of significance will be worked out in future.

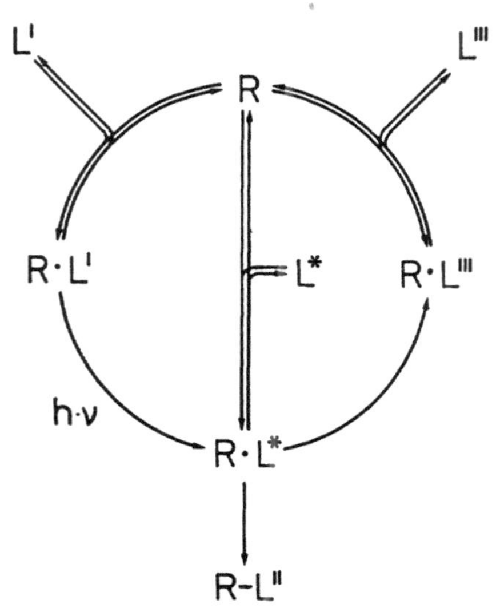

Fig. 2: Reactions following photolysis of a photolabile ligand in solution

The polypeptide with the apparent M_r of 54 000 can easily be removed from the membrane by 2 M NaCl-solution, by sonication, and even by freezing and thawing. Thus, as a peripheral membrane protein, it may not be directly involved in membrane transport of bile salts, but may have an auxiliary function. At present, only the two polypeptides with the apparent M_r's of 48 000 and 46 000 seem to be involved directly in hepatic membrane transport of bile salts.

Photoaffinity labelling in solution occurs by a relative complex sequence of reactions (Fig. 2). A receptor R and a photolabile ligand L' must combine in some way during the course of the biological event forming the corresponding receptor-ligand-complex R·L' in a reversible reaction. Due to irradiation the photolabile compound is converted in an irreversible reaction to a species of very high reactivity R^* , at best a nitrene or carbene. This applies to both the free and the receptor bound ligand. In the scheme only the reactions in which the receptor participates are depicted. The highly reactive species generated by irradiation can undergo different reactions in the receptor-ligand-complex:

The main reaction leads in an intramolecular rearrangment of the reactive species L^* to a much less or inreactive compound L'''. Whilst this reaction is also an irreversible one, the corresponding receptor-ligand-complex R·L''' decomposes reversibly.

The second reaction, the only wanted one, results in the formation of a covalent bond between the receptor and the highly reactive species in an irreversible reaction. As the result, L'', a derivative of the ligand is fixed at the receptor.
The third reaction, the dissociation of the highly reactive species from the complex is in principle reversible and obtaines importance if the half-life time of the reactive species is relatively long. This could be the cause of unspecific labelling.
The time needed for photolysis of the photolabile ligand is extremely long as compared to the rates of all consecutive reactions. During this time the part of the receptor which has not been modified covalently becomes free and susceptible for new interaction with residuary photolabile ligand. With decreasing concentration of the photolabile compound L' and increasing concentration of the unreactive ligand L''' covalent labelling of the receptor will grow slower and finally reaches a constant value. In this way photoaffinity labelling occurs relative effectively until all of the photolabile ligand has been photolyzed. The extent of labelling is dependent upon different rate constants and does not reflect the actual binding state at a defined time.
The described procedure has clear advantages for the identification of binding structures, but does not allow the correlation of the results obtained from kinetic studies with those from photoaffinity labelling experiments. In order to achieve this, two approaches are practicable:

1. Photoaffinity labelling by high intensity flash photolysis
2. Photoaffinity labelling in the frozen state

High intensity flash photolysis reduces the irradiation time to the order of magnitude necessary for time resolved photoaffinity labelling (Frimmer and Ziegler, 1986). This approach has the advantage that intact cells may be recovered after the photoaffinity labelling experiment. However, nonspecific labelling by long-living reactive intermediates can not be excluded.
Photoaffinity labelling in the frozen state makes use of the facts that photolytic reactions occur practically temperature-independently and that solid state conditions favour the yield of insertion reactions (Torres et al., 1986). In a frozen system it may be assumed that all the transient complexes formed by a receptor and a photolabile ligand are fixed at their actual state.

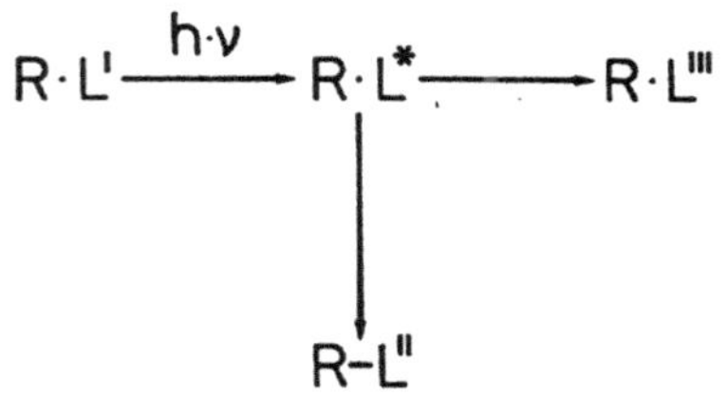

Fig. 3: Reactions following photolysis of a photolabile ligand in the solid state

By irradiation in the solid state the highly reactive intermediate is formed fixed in the active site of the receptor and only the two irreversible reactions, intramolecular rearrangement and insertion, take their course (Fig. 3). Because the solid state conditions prevent diffusion and favour the yield of insertion reactions this approach may be particularly advantageous for photoaffinity labelling with photolabile compounds, which form relatively long-living reactive intermediates. A disadvantage of the method may be the time needed to freeze the specimens: far less than one second but long enough that in rare cases a receptor may change its conformation and loose its ligand.

Photoaffinity labelling of freshly isolated hepatocytes in solution with (7,7-azo-3α,12α-dihydroxy-5ß-[3ß-^{3}H]cholan-24-oyl)-2-aminoethanesulfonate by irradiation for 10 minutes revealed after fluorographic analysis of the particulate cell fraction, that the highest labelling occurred in a polypeptide with the apparent M_r of 33 000 (Fig. 4, Lane A and B). This polypeptide originates from the inner mitochondrial membrane and is not localized in or at the cell membrane. The labelled cell membrane polypeptides have apparent M_r's of 54 000, 48 000, and 46 000. The one with the Mr of 54 000 is the most highly labelled membrane polypeptide. In the solid state under otherwise identical conditions this polypeptide became only very faintly labelled, as shown for incubation times of 5 and 10 minutes, respectively (Fig. 4, Lane C and D). The polypeptides with the apparent M_r's of 48 000 and 46 000 were labelled in the expected intensity. The low labelling of the polypeptide with the M_r of 54 000 would be

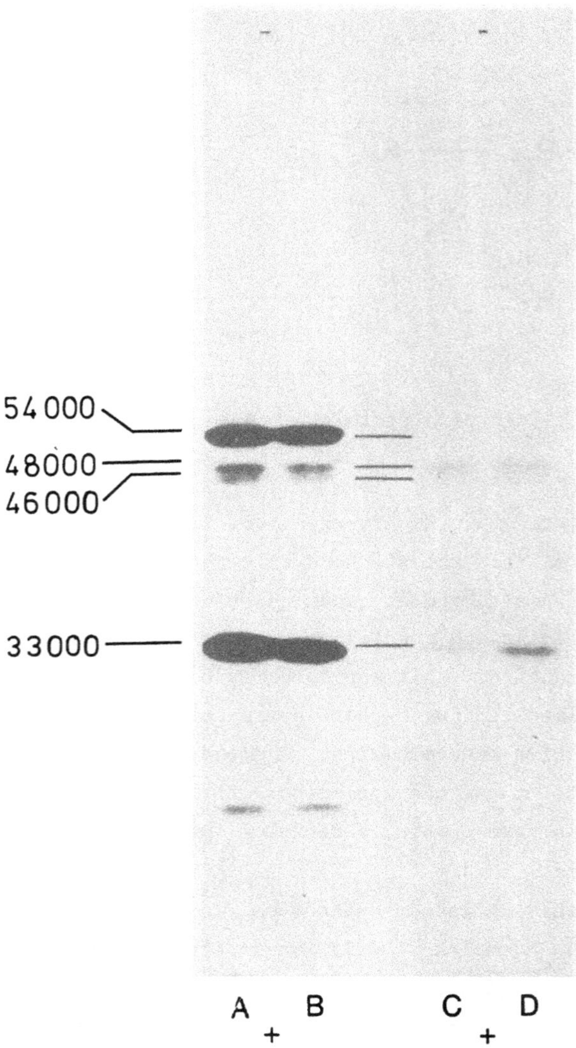

Fig. 4: Identification of bile-salt-binding polypeptides in the particulate fraction after photoaffinity labelling of isolated hepatocytes. 2 x 10^6 cells were photolabelled in 1 ml for 10 minutes with 0.5 µM (10 µCi) of (7,7-azo-3α,12α-dihydroxy-5ß[3ß-^{3}H]cholan-24-oyl)-2-aminoethanesulfonate:
Lane A and B: Photoaffinity labelling in solution at 30 °C
Lane C : Photoaffinity labelling in solid state after 5 minutes of preincubation
Lane D : Photoaffinity labelling in solid state after 10 minutes of preincubation
All other conditions are described in the legend to Fig. 1.

in accordance with its presence in low amounts and with a high turnover number for bile salt binding. This would be consistent with an auxiliary function of this peripheral protein in bile salt transport.

The investigation of bile salt uptake by photoaffinity labelling in the frozen state at low ligand concentrations should present evidence for the membrane polypeptides directly involved in the uptake process. At ligand concentrations one order of magnitude or more lower than the the lowest K_T-value for bile salt uptake into isolated hepatocytes (Buscher et al.,

1987), the steady state assumption demanding that the concentrations of all intermediary receptor-ligand-complexes do not change with time, is not even approximately valid. For this reason, at low concentrations of the ligand, the concentrations of all intermediary complexes should continuously decline.

Photoaffinity labelling in the frozen state of hepatocytes incubated for different lengths of time with low concentrations of the photolabile bile salt derivative should result in the highest labelling of transport proteins at the beginning of the experiment. With increasing time incorporation of radioactivity into polypeptides involved in membrane transport should decrease.

The uptake of (7,7-azo-3α,12α-dihydroxy-5ß-[3ß-^{3}H]cholan-24-oyl)-2-aminoethanesulfonate in freshly isolated hepatocytes and the disappearence of the photolabile bile salt derivative in the medium is demonstrated in Fig. 5. It is obvious that within about 30 seconds half of the substrate was taken up and that after 3 minutes a stationary state was reached. The ratio between bile salts inside and outside of the cells remained constant for about 50 or 60 minutes.

Photoaffinity labelling in the solid state was performed after shock freezing the same incubation assays at the different times shown (Fig. 5). Contrary to all expectations, fluorographic analysis revealed that the incorporation of radioactivity into the polypeptides with the apparent M_r's of 48 000 and 46 000 increased during the first short time interval and remained constant after the uptake had reached its stationary state. The relative density of the fluorographies in the M_r range from 48 000 to 46 000 runs, within the experimental errors, parallel with the amount of bile salts taken up into the hepatocytes. This result may best be explained by the assumption that the polypeptides with the apparent M_r's of 48 000 and 46 000 were labelled from the interior of the cell. When the concentration of the photolabile derivative in the area behind the uptake systems increases, the extent of incorporation of radioactivity into the polypeptides mediating the uptake also increases. The interaction of transport systems with bile salts at the inner surface of the membrane is valid if two transport systems are existing or if only one system exhibiting negative cooperativity is postulated. It may be of regulatory significance and could account for simple trans-inhibition or for a more sophisticated regulation.

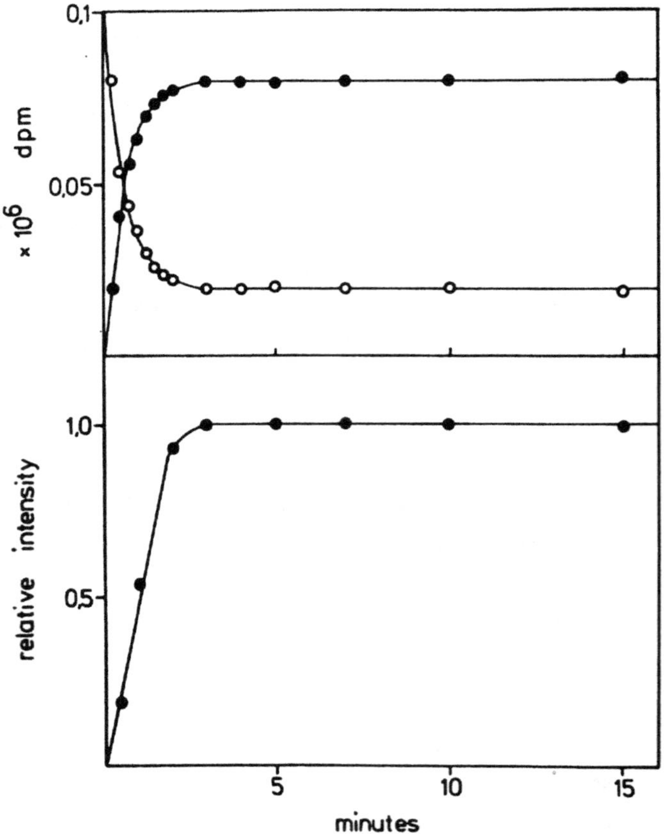

Fig. 5: Correlation of uptake of the photolabile bile salt derivative into freshly isolated hepatocytes with extent of photoaffinity labelling of the sinusoidal integral membrane polypeptides. 1 x 10^6 cells per ml were incubated with 1 µM (20 Ci/mmol) of (7,7-azo-3α,12α-dihydroxy-5ß-[3ß-^{3}H]cholan-24-oyl)-2-aminoethanesulfonate at 30 °C.
Upper panel: o-o-o radioactivity in the incubation medium
•-•-• radioactivity within the hepatocytes
Lower panel: •-•-• relative intensity of incorporation of radioactivity into the integral membrane polypeptides with the apparent M_r's of 48 000 and 46 000

The uptake of chemically quite different compounds into the liver interferes with bile salt transport. In order to identify the membrane polypeptides involved in their transport and to gain insight into the specificity of hepatic transport processes photoaffinity labelling studies with photolabile derivatives of characteristic compounds interfering with bile salt transport were performed (Wieland et al., 1984; Buscher et al., 1986; 1987; Kröncke et al., 1987).

The interference of distinct amphipathic quaternary cations with bile salt uptake into hepatocytes was investigated, using N-(4,4-azo-n-pentyl)-21-deoxy[21-^{3}H]ajmalinium (Fig. 6 A), a photolabile derivative of the anti-arrhythmic compound N-propylajmaline. Its uptake into hepatocytes is competitively inhibited by bile salts and vice versa, the ajmalinium derivative inhibits hepatic bile salt uptake, however at relatively high concentrations (Buscher et al., 1987).

Photoaffinity labelling of sinusoidal membranes with the photolabile ajmalinium derivative N-(4,4-azo-n-pentyl)-21-deoxy[21-^{3}H]ajmalinium revealed by fluorographic analysis that predominantly a polypeptide with the apparent M_r of 48 000 became labelled (Fig. 7, Lane A). The labelling of actin with the M_r of 43 000 is not of relevance. The comparison with the labelling pattern obtained after photoaffinity labelling of the same membrane subfraction with (7,7-azo-3α,12α-dihydroxy-5ß-[3ß-^{3}H]cholan-24-oyl)-2-aminoethanesulfonate (Fig. 7, Lane B), demonstrates that out of all the bile-salt-binding membrane polypeptides only that with the apparent M_r of 48 000 interacts with ajmalinium salts.

In contrast to amphipathic cations having a rigid ring structure like the ajmalinium derivative, small cations do not interfere with bile salt uptake into liver. In order to identify the sinusoidal membrane polypeptides being involved in the uptake of these small cations a photolabile derivative of quinuclidinium, N-(4,4-azo-n-pentyl)-[3-^{3}H]quinuclidinium-3-ol

A

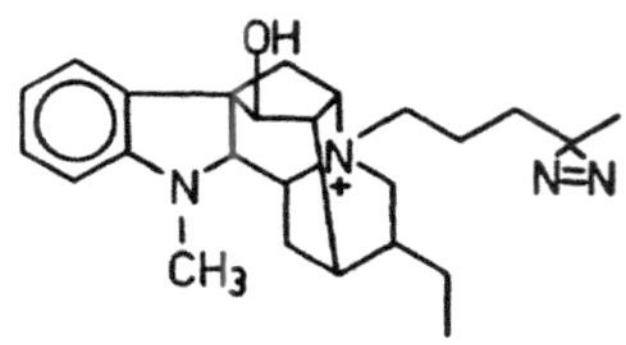

B

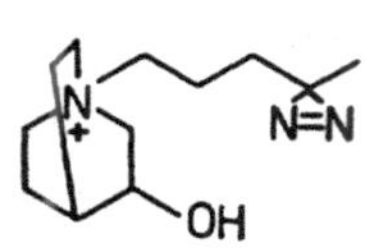

Fig. 6: Photolabile derivatives of ajmaline and quinuclidine
A. N-(4,4-azo-n-pentyl)-21-deoxy-[21-^{3}H]ajmalinium
B. N-(4,4-azo-n-pentyl)-[3-^{3}H]quinuclidinium-3-ol

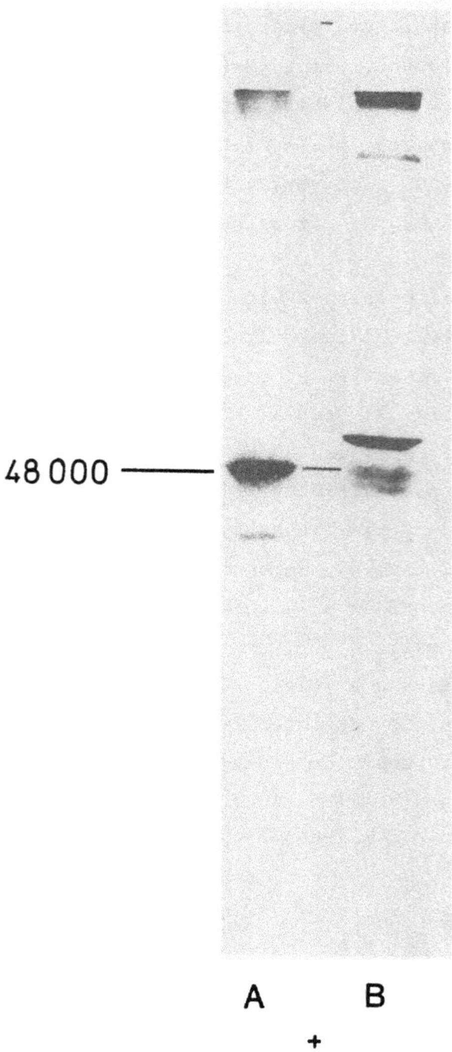

Fig. 7: Identification of ajmaline- and bile-salt-binding polypeptides in sinusoidal membranes.
Lane A. 500 µg protein of plasma membrane subfractions were incubated with 10 µM (10 µCi) of N-(4,4-azo-n-pentyl)-21-deoxy[21-^{3}H]ajmalinium.
Lane B. 500 µg protein of plasma membrane subfractions were incubated with 1.25 µM (10 µCi) (7,7-azo-3α,12α-dihydroxy-5ß-[3ß-^{3}H]cholan-24-oyl)-2-aminoethanesulfonate.
All other conditions are described in the legend to Fig. 1.

(Fig. 6 B) was synthesized. Ajmalinium derivatives inhibit hepatic uptake of quinuclidinium derivatives, but vice versa inhibition could not be demonstrated.

Photoaffinity labelling studies using the N-(4,4-azo-n-pentyl)-[3-^{3}H]quinuclidinium-3-ol resulted in the incorporation of radioactivity in only one polypeptide with the apparent M_r of 48 000 (Fig. 8, Lane B). For comparison the labelling pattern obtained with the ajmalinium derivative with the same membrane subfraction is shown (Fig. 8, Lane A).

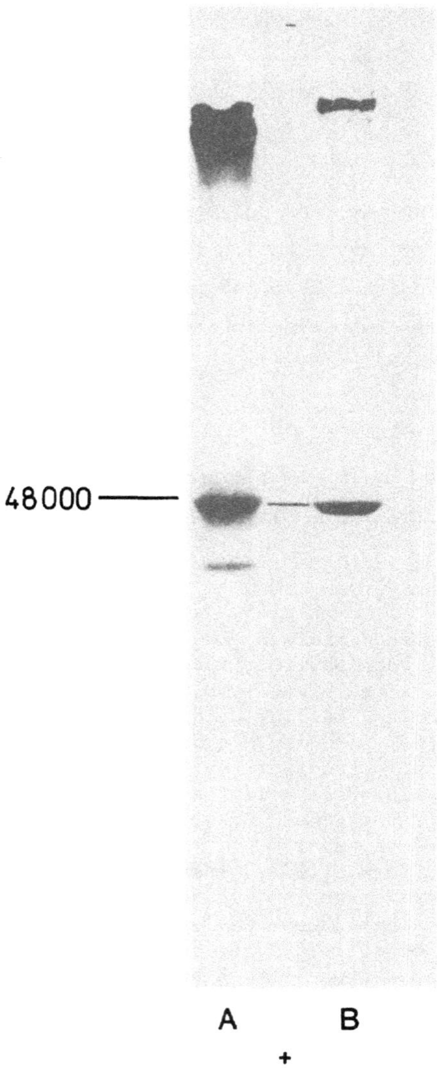

Fig. 8: Identification of ajmaline- and quinuclidine-binding polypeptides in sinusoidal membranes.
Lane A. 500 µg protein of plasma membrane subfractions were incubated with 10 µM (10 µCi) of N-(4,4-azo-n-pentyl)-21-deoxy[21-^{3}H]ajmalinium
Lane B. 500 µg protein of plasma membrane subfractions were incubated with 10 µM (10 µCi) of N-(4,4-azo-n-pentyl)-[3-^{3}H]quinuclidinium-3-ol.
All other conditions are described in the legend to Fig. 1.

As a result of our studies one may assume that a sinusoidal integral membrane polypeptide with the apparent M_r of 48 000 is of central significance for hepatic uptake and additional polypeptides with auxiliary functions may be involved in the transport of substances of chemically different classes of compounds.

Summary

1. Two integral membrane polypeptides with apparent M_r's of 48 000 and 46 000 are involved in hepatic uptake of bile salts through the sinusoidal membrane. In intact hepatocytes both polypeptides were labelled with photolabile bile salts from the inner side of the membrane.
2. The peripheral bile-salt-binding membrane polypeptide with the M_r of 54 000 is assumed to have an auxiliary function in bile salt uptake.
3. The membrane polypeptide with apparent M_r of 48 000 is not only involved in the uptake of bile salts but also in the uptake of distinct cationic compounds.

Acknowledgements

This investigation was supported by the Deutsche Forschungsgemeinschaft, the Boehringer Ingelheim Fonds and the Fritz Thyssen Stiftung.

References

Buscher HP, Fricker G, Gerok W, Kramer W, Kurz G, Müller M and Schneider S (1986) Membrane transport of amphiphilic compounds by hepatocytes. in: Receptor-Mediated Uptake in the Liver (Greten H, Windler G and Beisiegel U, eds.), Springer-Verlag, Berlin, Heidelberg, 189-199

Buscher HP, Fricker G, Gerok W, Kurz G, Müller M and Schneider S, Schramm U and Schreyer A (1987) Hepatic transport systems for bile salts: localization and specificity, in: Bile Acids and the Liver with an Update on Gallstone Disease (Paumgartner G, Stiehl A and Gerok W, eds.) MTP Press Ltd., Lancaster, UK, 95-110

Fricker G, Schneider S, Gerok W and Kurz G (1987) Identification of different transport systems for bile salts in sinusoidal and canalicular membranes of hepatocytes. Biol Chem Hoppe-Seyler 368: 1143-1150

Frimmer M and Ziegler K (1986) Photoaffinity labeling of whole cells by flashed light: a simple apparatus for high-energy ultraviolet flashes. Biochim Biophys Acta 855: 143-146

Kramer W, Bickel U, Buscher H-P, Gerok W and Kurz G (1982) Bile-salt-binding polypeptides in plasma membranes of hepatocytes revealed by photoaffinity labelling. Eur J Biochem 129: 13-24

Kramer W and Kurz G (1983) Photolabile derivatives of bile salts. Synthesis and suitability for photoaffinity labeling. J Lipid Res 24: 910-923

Kröncke KD, Fricker G, Meier PJ, Gerok W, Wieland T and Kurz G (1986) α-Amanitin uptake into hepatocytes. Identification of hepatic membrane transport systems used by amatoxins. J Biol Chem 261: 12562-12567

Torres MJ, Zayas J and Platz MS (1986) A formal CH insertion reaction of an aryl nitrene into an alkyl CH bond. Implications for photoaffinity labelling. Tetrahedron Lett 27: 791-794

Wieland T, Nassal M, Kramer W, Fricker G, Bickel U and Kurz G (1984) Identity of hepatic membrane transport systems for bile salts, phalloidin, and antamanide by photoaffinity labeling. Proc Natl Acad Sci U S A 81: 5232-5236

Characterization of the Bile Acid Transport System in Normal and Transformed Hepatocytes Using Monoclonal Antibodies

Daniel Levy
Department of Biochemistry
University of Southern California
School of Medicine
Los Angeles, California

Introduction

The identification of the bile acid carrier protein in hepatocyte sinusoidal plasma membranes has been the subject of numerous investigations (von Dippe et al. 1983ab; Kramer et al. 1982; Ziegler et al. 1984a). Two proteins (49 and 54 kDa) have been identified as possible candidates for this transport system utilizing the techniques of chemical and photoaffinity labeling with reactive bile acid derivatives. Liposome reconstitution studies using a membrane fraction enriched in components of 49-54 kDa resulted in proteoliposomes which mediated bile acid transport in a Na^+-dependent fashion, indicating that one or more of the isolated membrane components exhibited functional properties consistent with this transport system (von Dippe et al. 1986). Additional evidence to suggest that one of these membrane components was the transporter was obtained using the anion transport inhibitor 4,4'-diisothiocyanostilbene-2,2'-disulfonic acid (DIDS) which also labeled hepatocyte plasma membrane proteins of 49 and 54 kDa (von Dippe et al. 1988; Ziegler et al. 1984b). The above studies however, have not unequivocally established the identity of the Na^+-dependent bile acid transport protein. The characterization of several membrane transport systems has been greatly assisted using monoclonal and polyclonal antibodies. In this report the use of immunological probes to characterize the bile acid transport protein and its expression during development and in transformed hepatocytes is described.

Hepatic Transport in Organic Substances
Ed. by E. Petzinger, R. K.-H. Kinne, H. Sies

Experimental Procedures

Preparation of monoclonal antibodies- SJL/J mice were immunized with a partially purified bile acid transport protein preparation (von Dippe et al. 1986;1988). Hybridoma cultures producing antibodies against the 49 and/or 54 kDa proteins were cloned and used to prepare ascites fluid from which the antibodies were isolated by chromatography on a protein A-Sepharose 4B column. **Cell preparations-** Isolated hepatocytes were obtained from the livers of adult Sprague-Dawley rats (250 g) using a collagenase perfusion technique (von Dippe et al. 1983a). Cells from fetal livers (day 20) were obtained by incubating the livers in a collagenase solution as previously described (Leffert et al. 1979). Hepatoma tissue culture (HTC) cells were grown in suspension cultures as previously described (Mackenzie and Stellwagen 1974). H4-IIEC3 cells (Pitot et al. 1964) were grown in monolayer culture. **Antibody binding procedures-** Hepatocytes and HTC cells were incubated with ^{125}I-antibodies (25D-1)(3 mg/ml) for 30 min at 37°C and the cell associated radioactivity evaluated following centrifugation through dibutylphthalate (von Dippe et al. 1988). **Immunoprecipitation procedures-** Cells or liver homogenates were solubilized and immunoprecipitated with antibody 25D-1 as previously described (von Dippe et al. 1988). The precipitates were analyzed by SDS-PAGE and the resultant Coomassie stained bands quantitated on a Zeineh scanning densitometer. Cell surface proteins were labeled by iodination with NaI and the Enzymobead reagent while total cell proteins were labeled with [^{35}S]methionine. Immunoprecipitates were evaluated by autoradiography. **Transport measurements-** The uptake of taurocholic acid by the different cell systems was evaluated using a dibutlyphthalate rapid centrifugation procedure (von Dippe et al. 1983a).

Results and Discussion

Several antibodies have been shown to react with the Na^{+}-dependent bile acid transport protein. This conclusion was based on studies which demonstrated that DIDS labeling of hepatocyte plasma membranes as well as inhibition of bile acid transport could be significantly blocked by preincubation of membranes or cells with the appropriate antibody. Immunoprecipitation procedures identified this carrier protein as a 49 kDa component located in the sinusoidal plasma membrane domain as shown in

Figure 1. Antibody binding studies indicated that there are approximately $3.3x10^6$ transport proteins/hepatocyte which accounts for 0.8% of the surface membrane protein with a density of 2200 sites/um^2 in the sinusoidal domain. One of the antibodies (25A-3) recognized a common epitope on the 49 kDa and on a 54 kDa protein. Since both proteins were also labeled by DIDS as well as several photoprobes, it is possible that this 54 kDa component represents the Na^+-independent organic anion transporter for substrates such as sulfobromophthalein and bilirubin and that the two transporters are structurally related. This hypothesis is under current investigation.

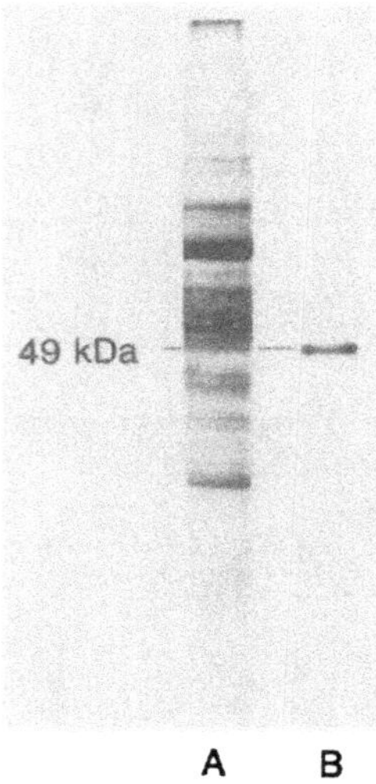

Fig. 1. Immunoprecipitation of the bile acid transport protein. Iodinated sinusoidal plasma membranes were solubilized and incubated with 25D-1 bound to protein A-Sepharose and the resulting immunoprecipitated analyzed by SDS-PAGE. Lane A, Coomassie Blue-stained gel of 100 ug of membrane protein. Lane B, autoradiogram of the 25D-1 immunoprecipitate.

Using hepatoma tissue culture (HTC) cells derived from Morris hepatoma 7288c and H4-IIEC3 cells derived from Reuber hepatoma H-35, studies (von Dippe at al. 1983b) have demonstrated the loss of mediated bile acid transport activity as shown in Fig. 2.

A monoclonal antibody against this transport system was thus used to characterized the molecular basis for the observed loss of transport capacity. Antibody binding to HTC cells demonstrated that only 2% of the 49 kDa protein was on the surface of this cell system as compared to normal hepatocytes as shown in Table 1.

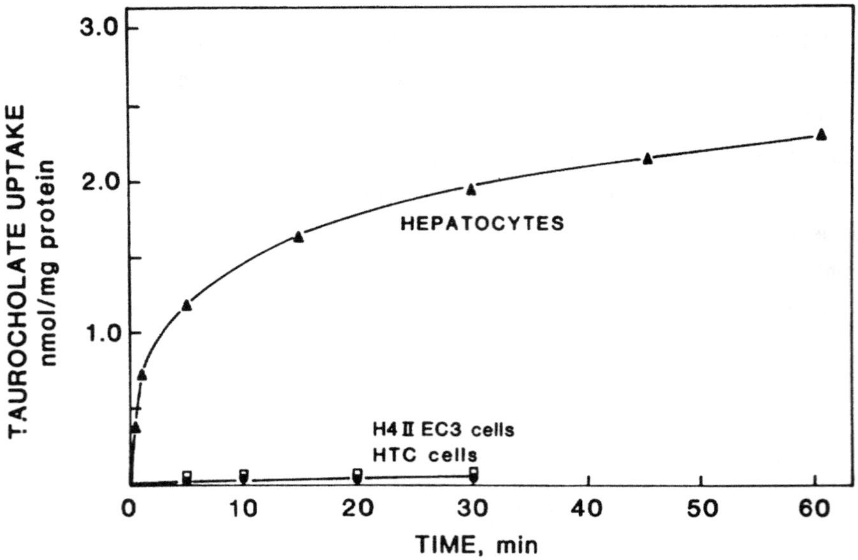

Fig.2 Uptake of taurocholic acid by hepatocytes, HTC and H4-IIEC3 cells was measured at 27°C at the indicated times. Cells were isolated and associated radioactivity determined.

Table 1. Antibody binding to mature, fetal and transformed hepatocytes.

Cell System	25D-1 Binding ug Mab/mg protein	Carriers/ mg protein	Decrease (%)
Hepatocyte	0.472	3.3×10^6	-
HTC cells	0.010	6.9×10^4	97.9
Fetal Hepatocytes (20d)	0.049	3.4×10^5	89.6

The possibility that this carrier protein had been synthesized but not inserted in the surface membrane as a result of a defect in the processing mechanism was investigated using [^{35}S]methionine labeled cells. As shown in Fig. 3, immunoprecipitation of total cell protein from

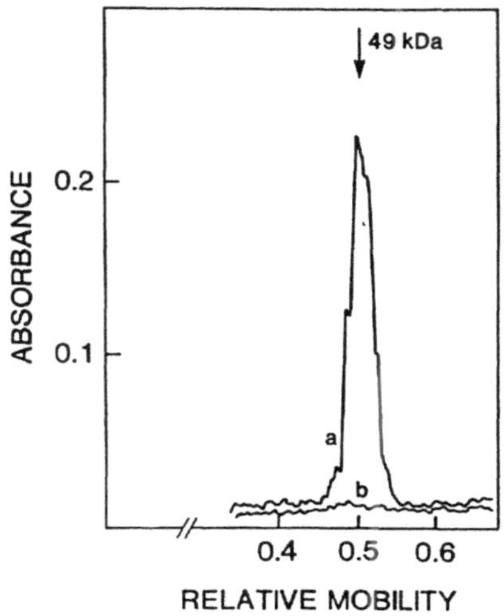

Fig. 3. Quantitation of the bile acid transport protein in normal and transformed hepatocytes. Cells were labeled with [^{35}S]methionine, solubilized and immunoprecipitated with 25D-1. Immunoprecipitates were analyzed by SDS-PAGE . The resultant autoradiogram was analyzed using a densitometer. a, hepatocytes; b, HTC cells.

hepatocytes and HTC cells indicated that the 49 kDa protein was not expressed in the hepatoma cell system or that an altered form of the carrier was synthesized which was not recognized by the antibody.

Several reports have indicated the existance of reduced levels of bile acid transport activity in neonatal hepatocytes and in vesicles derived from fetal hepatocytes (Suchy and Balistreri 1982; Suchy et al. 1986). We have shown that taurocholic acid uptake in 20 day fetal cells was only 5% of the level observed in mature hepatocytes. Monoclonal antibodies against the bile acid transport system have thus been utilized to study the expression of this transport protein during liver development. Binding of 25D-1 to 20 day fetal cells indicated that the surface carrier concentration was reduced by 90% (Table 1), a result consistent with the observed decrease in transport capacity. The expression of this transport protein in liver from 16 and 20 day fetal rats and from 7, 18, 28 and 56 day old rats was also measured using immunoprecipitation procedures. As shown in Fig. 4 the level of expression of the carrier protein increases in approximately a linear fashion, reaching 90% of the maximum adult level in 28 days. The increase in expression of this protein during development is similar to the observed increase in taurocholic acid uptake over this time period (Suchy and Balistreri 1982). The factors which modulate the expression of this protein during development are under investigation.

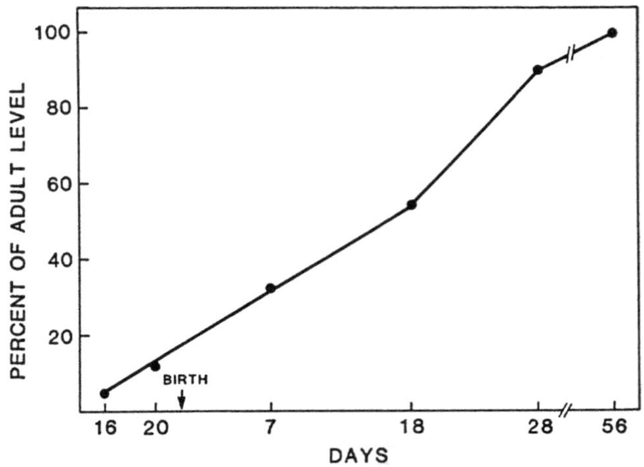

Fig. 4. Expression of the bile acid transport protein during liver development. Liver homogenates were incubated with 25D-1 and the resultant immunoprecipitates analyzed by SDS-PAGE. The levels of the 49 kDa protein present in liver at the indicated time points was evaluated using a scanning densitometer.

In conclusion, a monoclonal antibody (25D-1) which specifically reacts with the 49 kDa Na^{+}-dependent bile acid transport protein in hepatocytes has been used to measure the expression of this protein in hepatoma cells and in hepatocytes during development. Antibody binding, protein immunoprecipitation and transport studies indicate that the transport activity of this cell system is directly related to the level of expression of this transport protein. This antibody has been used to purify this protein to homogeniety. The structure and biosynthesis of this carrier protein in under investigation using cDNA and immunological procedures.

References

Kramer W, Bickel U, Buscher H-P, Gerok W, Kurz G (1982) Bile salt binding polypeptides in plasma membranes of hepatocytes revealed by photoaffinity labeling. Eur J Biochem 129:13-24

Leffert HL, Koch KS, Moran T, Williams M (1979) Liver cells. In: Jakoby WB, Pastan IH (eds) Methods in Enzymology, vol 58. Academic Press, New York, p 536

Mackenzie CW, Stellwagen RH (1974) Differences between liver and hepatoma cells in their complements of adenosine 3';5'-monophosphate-binding proteins and protein kinases. J Biol Chem 249:5755-5762

Pitot HC, Peraino C, Morse PA, Potter VR (1964) Hepatomas in tissue culture compared with adapting liver in vivo. Natl Cancer Inst Monogr 13:229-245

Suchy FJ, Balistreri WF (1982) Uptake of taurocholate by hepatocytes isolated from developing rats. Pediatric Res 16:282-285

Suchy FJ, Bucuvalas JC, Goodrich AL, Moyer SM, Blitzer BL (1986) Taurocholate transport and Na^{+}-K^{+}-ATPase activity in fetal and neonatal rat liver plasma membrane vesicles. Am J Physiol 251:G665-673

von Dippe P, Drain P, Levy D (1983a) Synthesis and transport characteristics of photoaffinity probes for the hepatocyte bile acid transport system. J Biol Chem 258:8890-8895

von Dippe P, Levy D (1983b) Characterization of the bile acid transport system in normal and transformed hepatocytes. Photoaffinity labeling of the taurocholate carrier protein. J Biol Chem 258:8897-8901

von Dippe P, Ananthanarayanan M, Drain P, Levy D (1986) Purification and reconstitution of the bile acid transport system from hepatocyte sinusoidal plasma membranes. Biochim Biophys Acta 862:352-360

von Dippe P, Ananthanarayanan M, Levy D (1988) Identification of the hepatocyte Na^{+}-dependent bile acid transport protein using monoclonal antibodies. J Biol Chem 263:0000

Ziegler K, Frimmer M, Müllner S, Fasold H (1984a) 3'-Isothiocyanatobenzamido[^{3}H]cholate, a new affinity label for hepatocellular membrane proteins responsible for the uptake of both bile acids and phalloidin. Biochim Biophys Acta 773:11-22

Ziegler K, Frimmer M, Fasold H (1984b) Further characterization of membrane proteins involved in the transport of organic anions in hepatocytes. Biochim Biophys Acta 769:117-129

Transport of Bile Salts Across the Canalicular Plasma Membrane

P.J. Meier and S. Ruetz
Division of Clinical Pharmacology
Department of Medicine
University Hospital
8091 Zürich
Switzerland

Introduction

Bile salts are the major organic solutes in bile and undergo efficient enterohepatic circulation. Total bile salt concentrations in bile are approximately 100 to 1000 fold higher than in blood plasma and range between 3 to 45 mM in man and various animal species (Boyer, 1986; Erlinger, 1982). However, the free concentrations of monomeric bile salts are considerably lower, since (a) bile salts selfaggregate into micelles above their "critical micellar concentrations" that range between 2 to 11 mM (in 0.15 M NaCl) for the most important physiologic bile salt species (Roda et al., 1983), and (b) bile salts associate with biliary phospholipids (predominantly lecithin) and cholesterol to form "mixed micelles" (Carey, 1984). Because of this "micellar sink" (Scharschmidt and Schmid, 1978), the intermicellar concentration of free monomeric bile salts has been estimated to amount to 3 - 5 mM or less at the level of the bile canaliculus (Boyer, 1986). Thus, assuming intracellular bile salt concentrations of 100 - 300 μM (Oh and Du Pont, 1975; Okishio and Mair, 1966) it would appear that transcanalicular secretion of bile salts is an active transport process that occurs against an unfavourable bile to cell concentration gradient of at least 10:1. However, the exact energetic conditions of net transcanalicular secretion of bile salts *in vivo* are not known, since it has as yet not been possible to determine the intracellular and intracanalicular activities of free bile salts. Nevertheless, there is

Hepatic Transport in Organic Substances
Ed. by E. Petzinger, R. K.-H. Kinne, H. Sies

kinetic evidence that canalicular secretion represents the rate limiting step in overall transport of bile salts from blood into bile (O'Maille et al., 1966; Reichen and Paumgartner, 1975; 1980). Furthermore, secretion of bile salts into bile is a saturable process that exhibits surprisingly similar maximum secretory rates per unit body weight in different animal species (Heath et al., 1970; Herz et al., 1971; O'Maille et al., 1965).

Basically two mechanisms have been proposed for canalicular bile salt secretion. These include (a) transport by intracellular vesicles followed by canalicular exocytosis, and (b) transcellular transport in cytosol (in solution or bound to protein(s)) and "carrier"-mediated secretion at the canalicular membrane.

Evidence for Vesicle Transport of Bile Salts

The suggestion of bile salt transport by intracellular vesicles stems from the observation that (a) nascent mixed micelles may be formed in the endoplasmic reticulum or Golgi complex and travel intact to the canalicular membrane (Gregory et al., 1975; Reuben et al., 1983), (b) microvesicular structures, vacuoles, and Golgi complex accumulate in the pericanalicular region under conditions of increased bile salt transport (Boyer et al., 1979; Jones et al., 1979), (c) the intracellular distribution of the bile salt analogue ^{125}I-cholylglycyltyrosine exhibits compartmentation within the endoplasmic reticulum and Golgi apparatus (Suchy et al., 1983), and (d) isolated intracellular vesicles distinct from the endoplasmic reticulum (endosomes?) and Golgi-derived membrane vesicles exhibit saturable taurocholate uptake suggesting that certain intracellular organelles might, indeed, possess the "machinery" for active accumulation of bile salts (Simion et al., 1984). While these observations indicate that a portion of the intracellular bile salt pool may preferentially associate with specific subcellular membrane-bound compartments, none of these studies provides quantitative evidence for the major proportion of bile salts being secreted into bile canaliculi by transcytotic and/or exocytotic processes. In fact, overall transport of bile salts from portal plasma into bile requires 4 to 6 minutes to reach peak secretion (Goldman, 1983), whereas vesicular transport of biliary proteins such as IgA_2 or epidermal growth factor lasts considerably longer (20 to 60 min) (Burwen et al., 1984; Fisher et al., 1979; Kloppel et al., 1986; Limet et al., 1985; Renston et al., 1980;

Schiff et al., 1984). In addition, since intracellular movement of vesicles to the canaliculus is mediated by microtubuli (Jones and Burwen, 1985), the microtubular poison colchicine should significantly decrease bile salt output into bile if vesicle exocytosis represents a major mechanism for canalicular secretion of bile salts. Although such an inhibitory effect of colchicine could be demonstrated under conditions of bile salt stimulated choleresis (Dubin et al., 1980; Gregory et al., 1978), most studies agree that under basal bile flow conditions colchicine has no effects on the biliary secretion of bile salts (Barnwell et al., 1984; Erlinger, 1981; Goldman et al., 1983; Kacich et al., 1983). Yet, under similar conditions, colchicine strongly inhibited the biliary output of IgA_2 and horseradish peroxidase (Goldman et al., 1983; Kacich et al., 1983; Lowe et al., 1985). These findings argue against a major role of endocytotic vesicles in overall transport of bile salts from blood into bile, but they do not rule out that a minor portion of the intracellular bile salt pool is associated with a special subpopulation of vesicles in order to sustain ongoing biliary secretion of lipids (Gregory et al., 1978).

Evidence for a Carrier-mediated Canalicular Secretion of Bile Salts

More convincing evidence has recently been obtained for the second proposed mechanism of canalicular bile salt secretion, i.e. the involvement of a bile salt transport system at the canalicular membrane. This possibility was initially supported by kinetic transport studies in isolated canalicular membrane vesicles. These studies have shown that (a) uptake of taurocholate into canalicular vesicles exhibits saturation kinetics (apparent K_m 34 μM; V_{max} 0.7 nmol/mg/min at 37°C) (Inoue et al., 1984), (b) $[^3H]$-taurocholate uptake into as well as efflux from the vesicles can be stimulated by the addition of an excess of unlabelled taurocholate on the opposite (i.e. trans-)side of the membrane (phenomenon of "transstimulation" or "countertransport") (Inoue et al., 1984; Meier et al., 1984), (c) taurocholate uptake into and efflux from canalicular vesicles can be inhibited by the anion transport inhibitors probenecid and 4'4-diisothiocyano-2,2'disulfonic acid stilbene (DIDS) (Inoue et al., 1984; Meier et al., 1987), (d) cholate and conjugated dihydroxy bile salts also cisinhibit and transstimulate vesicle efflux of taurocholate suggesting that the strate specificity of the canalicular bile salt "carrier" is similar to sinusoidal taurocholate uptake system (Hardison et al., 1984; Meier et , 1987; Van Dyke et al., 1982), and (e) high concentrations of glutaone disulfide (GSSG; 1 mM) and the thiol-binding reagent N-ethyleimide inhibit taurocholate efflux from the canalicular vesicles icating a possible involvement of protein-SH groups in the canalicular retion of bile salts (Griffiths et al., 1987). In addition, the vesicle nsport studies also provide evidence that the intracellular negative brane potential of approximately -30 to -40 mV (Fitz and Scharschmidt, 7; Graf et al., 1984) might be a driving force for canalicular bile salt retion. Thus, sodium-independent taurocholate uptake into and efflux n isolated canalicular vesicles could be stimulated by the induction of ravesicular positive or negative potassium diffusion potentials, pectively (Inoue et al., 1984; Meier et al., 1984). This potentialven portion of taurocholate efflux was also sensitive to DIDS, suggest- that it is mediated by an electrogenic "carrier" mechanism rather than eparate conductive leak pathway (Meier et al., 1987). Thus, canalicular retion of anionic bile salts could be primarily mediated by passive ilitated diffusion down a favourable electrochemical gradient. This umption is further supported by recent findings in isolated rat hepatoe couplets (Weinman et al., 1987) as well as by the observation that atively charged bile salts are better secreted by the intact liver pared to uncharged or positively charged bile salt derivatives (Anwer et , 1985).

The concept of "carrier"-mediated canalicular secretion of bile salts further strengthened by recent photoaffinity labeling studies with the tolabile bile salt derivative (7,7-azo-3α,12α-dihydroxy-5ß-[3ß-^{3}H] lan-24-oyl)-2-aminoethanesulfonate. In these studies a canalicular ain-specific bile salt binding protein with an apparent molecular weight approximately 100,000 has been identified in both intact liver tissue icker et al., 1987) as well as isolated canalicular vesicles (Ruetz et , 1987). This protein has been isolated and polyclonal antibodies raised inst it. The antibodies inhibited both taurocholate uptake into as well efflux from canalicular vesicles (Ruetz et al., 1987). Furthermore, onstitution of DIDS-sensitive, transstimulatable and electrogenic rocholate anion transport into proteoliposomes was invariably dependent n the presence of the 100 kDa bile salt binding polypeptide within the osomal membrane (Ruetz et al., 1988). These results virtually prove the bile salt transport function of a canalicular 100 kDa glycoprotein. However, the studies permit no definite conclusions yet with respect to the type of transport system or the gating mechanism(s) involved, since both simple "carriers" as well as complex "channels" can demonstrate the phenomenon of transstimulation or countertransport (Macey, 1986). Furthermore, it remains to be investigated whether the reconstituted 100 kDa bile salt transporting polypeptide is identical with one of the recently characterized 100 to 110 kDa canalicular-specific glycoproteins of unknown functions (Bartels et al., 1985; Becker et al., 1985; Cook et al., 1983; Hixson et al., 1983, 1985; Hong and Doyle, 1987; Kreisel et al., 1980; Odin et al., 1986; Petell et al., 1987; Sadoul et al., 1985). One of these proteins has recently been cloned and its primary amino acid sequence determined (Hong and Doyle, 1987). Reportedly, this so-called gp110-glycoprotein contains only one membrane spanning domain, which makes it a rather unlikely candidate for a membrane transport protein, unless it is organized in vivo as a multimeric membrane protein complex.

Conclusions

Vesicle transport studies as well as photoaffinity labeling and reconstitution experiments strongly indicate that canalicular secretion of bile salts is mediated by a membrane transport protein with an apparent molecular weight of 100,000. Since this canalicular bile salt "carrier" is a heavily glycosylated polypeptide (Ruetz et al., 1987), its final processing must occur within the Golgi complex (Farquhar, 1985), thus providing an explanation for the reported bile salt uptake activity of Golgi membranes both in vivo (Lamri et al., 1987) as well as in vitro (Simion et al., 1984). Further experimentation is required to elucidate whether intracellular direction of the newly synthesized 100 kDa bile salt transport protein to the canalicular membrane occurs via direct Golgi-bile canalicular or rather via indirect Golgi-basolateral-bile canalicular vesicle pathways, as it has recently been suggested for a variety of canalicular membrane constituents (Bartles et al., 1987).

Own work cited in this summary was supported by the Swiss National Science Foundation (Grants 3.983.0.84 and 3.992.0.87).

References

Anwer SM, O'Maille ERL, Hofmann AF, Di Pietro RA, Michelotti E (1985) Influence of side-chain charge on hepatic transport of bile acids and bile acid analogues. Am J Physiol 249:G479-G488

Barnwell SG, Lowe Ph, Coleman R (1984) The effects of colchicine on secretion into bile of bile salts, phospholipids, cholesterol and plasma membrane enzymes: bile salts are secreted unaccompanied by phospholipids and cholesterol Biochem J 220:723-731

Bartles JR, Braiterman LT, Hubbard AL (1985) Biochemical characterization of domain-specific glycoproteins of the rat hepatocyte plasma membrane. J Biol Chem 260:12792-12802

Bartles JR, Feracci HM, Stieger B, Hubbard AL (1987) Biogenesis of the rat hepatocyte plasma membrane in vivo: Comparison of the pathways taken by apical and basolateral proteins using subcellular fractionation. J Cell Biol 105:1241-1251

Becker A, Neumeier R, Park CS (1985) Identification of a transformation-sensitive 110-kDa plasma membrane glycoprotein of rat hepatocytes. Eur J Cell Biol 39:417-423

Boyer JL, Itabashi M, Hruban Z (1979) Formation of pericanalicular vacuoles during sodium dehydrocholate choleresis. A mechanism for bile acid transport. In: Preisig R, Bircher J (eds) The Liver: Quantitative aspects of structure and function. Editio Cantor Aulendorf, Berne, p 163-178

Boyer JL (1986) Mechanisms of bile secretion and hepatic transport. In: Andreoli TE, Hoffman JF, Fanestil DD, Schultz SG (eds) Physiology of membrane disorders. Plenum Publishing Corp., New York, p 609-636

Burwen SJ, Barker ME, Goldman JS, Hradek GT, Raper SE, Jones AL (1984) Transport of epidermal growth factor by rat liver: Evidence for a nonlysosomal pathway. J Cell Biol 99:1259-1265

Carey MC (1984) Aqueous bile salt-lecithin-cholesterol systems: Equilibrium aspects. Hepatology 4:1515-1545

Cook J, Hou E, Hou Y, Cairo A, Doyle D (1983) Establishment of plasma membrane domains in hepatocytes. I. Characterization and localization to the bile canaliculus of three antigens externally oriented in the plasma membrane. J Cell Biol 97:1823-1833

Dubin M, Maurice M, Feldmann G, Erlinger S (1980) Influence of colchicine and phalloidin on bile secretion and hepatic ultrastructure in the rat. Gastroenterology 79:646-654

Erlinger S (1981) Hepatocyte bile secretion: Current views and controversies. Hepatology 1:352-359

Erlinger S (1982) Bile Flow. In: Arias I, Popper H, Schacher D, Shafritz DA (eds) The Liver: Biology and Pathobiology. Raven Press Books,Ltd. New York, p 407-427

Farquhar MJ (1985) Progress in unraveling pathways of Golgi traffic. Ann Rev Cell Biol 1:447-488

Fisher MM, Nagy B, Bazin H, Underdown BJ (1979) Biliary transport of IgA: Role of secretory component. Proc Natl Acad Sci USA 76:2008-2012

Fitz JG, Scharschmidt BF (1987) Regulation of transmembrane electrical potential gradient in rat hepatocytes in situ. Am J Physiol 252:G56-G64

Fricker G, Schneider St, Gerok W, Kurz G (1987) Identification of different transport systems for bile salts in sinusoidal and canalicular domains of hepatocytes. Biol Chem Hoppe-Seyler 368:1143-1150

Goldman JS, Jones AL, Hradek GT, Huling S (1983) Hepatocyte handling of immunoglobulin A in the rat: The role of microtubules. Gastroenterology 85:130-140

Graf J, Gautam A, Boyer JL (1984) Isolated rat hepatocyte couplets: A primary secretory unit for electrophysiologic studies of bile secretory function. Proc Natl Acad Sci USA 81:6516-6520

Gregory DH, Vlahcevic ZR, Schatzki P, Swell D (1975) Mechanism of secretion of biliary lipids. I. Role of bile canalicular and microsomal membranes in the synthesis and transport of biliary lecithin and cholesterol. J Clin Invest 55:105-114

Gregory DH, Vlahcevic ZR, Prugh MF, Swell D (1978) Mechanism of secretion of biliary lipids: role of a microtubular system in hepatocellular transport of lipids in the rat. Gastroenterology 74:93-100

Griffiths JC, Sies H, Meier PJ, Akerboom TPM (1987) Inhibition of taurocholate efflux from rat hepatic canalicular membrane vesicles by glutathione disulfide. FEBS Lett 213:34-38

Hardison WGM, Bellentani S, Heasley V, Shellhamer D (1984) Specificity of a Na^+-dependent taurocholate transport site in isolated rat hepatocytes. Am J Physiol 246:G477-G483

Heath T, Caple IW, Redding DM (1970) Effect of the enterohepatic circulation of bile salts and lipids in sheep. Q J Exp Physiol 55:93-103

Herz R, Cueni B, Bircher J, Paumgartner G (1971) The excretory capacity of the isolated perfused rat liver. Naunyn Schmiedeberg's Arch Pharmacol 277:297-304

Hixson DC, Allison JP, Chesner JE, Leger M, Ridge LL, Walborg EF Jr (1983) Characterization of a family of glycoproteins associated with the bile canalicular membrane of normal hepatocytes but not expressed by two transplantable rat hepatocellular carcinomas. Canc Res 43:3874-3884

Hixson DC, McEntire KD, Obrink B (1985) Alterations in the expression of a hepatocyte cell adhesion molecule by transplantable rat hepatocellular carcinomas. Canc Res 45:3742-3749

Hong W, Doyle D (1987) cDNA cloning for a bile canaliculus domain-specific membrane glycoprotein of rat hepatocytes. Proc Natl Acad Sci USA 84:7962-7966

Inoue M, Kinne R, Tran T, Arias IM (1984) Taurocholate transport by rat liver canalicular membrane vesicles. J Clin Invest 73:659-663

Jones AL, Schmucker DL, Mooney JS, Ockner RK, Adler RD (1979) Alterations in hepatic pericanalicular cytoplasm during enhanced bile secretory activity. Lab Invest 40:512-517

Jones AL, Burwen SJ (1985) Hepatic receptors and their ligands: Problems of intracellular sorting and vectorial movement. Sem Liv Dis 5: 136-146

Kacich RL, Renston RH, Jones AL (1983) Effects of cytochalasin D and colchicine on the uptake, translocation, and biliary secretion of horseradish peroxidase and [^{14}C]sodium taurocholate in the rat. Gastroenterology 85:385-394

Kloppel TM, Brown WR, Reichen J (1986) Mechanism of secretion of proteins into bile: Studies in the perfused rat liver. Hepatology 6:587--594

Kreisel W, Volk BA, Büchsel R, Reutter W (1980) Different half-lives of the carbohydrate and protein moieties of a 110,000-dalton glycoprotein isolated from plasma membranes of rat liver. Proc Natl Acad Sci USA 77:1828-1831

Lamri Y, Roda A, Dumont M, et al. (1987) Immunoperoxidase localization of bile salts in rat liver cells. Hepatology 7:1037

Limet JN, Quintart J, Schneider YJ, Courtoy PJ (1985) Receptor-mediated endocytosis of polymeric IgA and galactosylated serum albumin in rat liver. Eur J Biochem 146:539-548

Lowe PJ, Kan KS, Barnwell SG, Sharma RK, Coleman R (1985) Transcytosis and paracellular movements of horseradish peroxidase across liver parenchymal tissue from blood to bile. Biochem J 229:529-537

Macey RI (1986) Mathematical models of membrane transport processes. In: Andreoli TE, Hoffman JF, Fanestil DD, Schultz SG (eds) Physiology of Membrane Disorders. Plenum Publish. Corp., New York, p 111-131

Meier PJ, Meier-Abt ASt, Barrett C, Boyer JL (1984) Mechanisms of taurocholate transport in canalicular and basolateral rat liver plasma membrane vesicles. Evidence for an electrogenic canalicular organic anion carrier. J Biol Chem 259:10614-10622

Meier PJ, Meier-Abt ASt, Boyer JL (1987) Properties of the canalicular bile acid transport system in rat liver. Biochem J 242:465-469

Odin P, Tingström A, ÖbrinK B (1986) Chemical characterization of cell-CAM 105, a cell-adhesion molecule isolated from rat liver membranes. Biochem J 236:559-568

Oh SY, Du Pont J (1975) Identification and quantitation of cholenoic acids in hepatic and extrahepatic tissues of rat. Lipids 10:340-347

Okishio T, Mair PP (1966) Studies on bile acids: some observations on the intracellular localization of major bile acids in rat liver. Biochemistry 5:3662-3668

O'Maille ERL, Richards TG, Short AH (1965) Acute taurine depletion and maximal rates of hepatic conjugation and secretion of cholic acid in the dog. J Physiol (Lond.) 180:67-79

O'Maille ERL, Richards TG, Short AH (1966) Factors determining the maximal rate of organic anion secretion by the liver and further evidence on the hepatic site of action of the hormone secretion. J Physiol (Lond.) 186:424-438

Petell JK, Diamond M, Hong W, Bujanover Y, Amarri S, Pittschieler K, Doyle D (1987) Isolation and characterization of a M_r=110,000 glycoprotein localized to the hepatocyte bile canaliculus. J Biol Chem 262:14753-14759

Reichen J, Paumgartner G (1975) Kinetics of taurocholate uptake by the perfused rat liver. Gastroenterology 68:132-136

Reichen J, Paumgartner G (1980) Excretory function of the liver. In: Javitt NB (ed) Liver and biliary tract physiology I, vol. XXI. University Park Press, Baltimore, pp 103-150

Renston RH, Jones AL, Christiansen WD, Hradek GT(1980) Evidence for a vesicular transport mechanism in hepatocytes for biliary secretion of immunoglobulin A. Science 208:1276-1278

Reuben A, Allen RM, Boyer JL (1983) Intrahepatic source of "biliary-like" bile acid-phospholipid-cholesterol micelles. In: Paumgartner G, Stiehl A, Gerok W (eds) Bile acids and cholesterol in health and disease. MTP Press, Lancaster, p 61-66

Roda A, Hofmann AF, Mysels KJ (1983) The influence of bile salt structure on self-association in aqueous solutions. J Biol Chem 258: 6362-6370

Ruetz St, Fricker G, Hugentobler G, Winterhalter K, Kurz G, Meier PJ (1987) Isolation and characterization of the putative canalicular bile salt transport system of rat liver. J Biol Chem 262:11324-11330

Ruetz St, Hugentobler G, Meier PJ (1988) Functional reconstitution of the canalicular bile salt transport system of rat liver. Proc Natl Acad Sci USA, in press

Sadoul JL, Peyron JF, Ballotti R, Debant A, Fehlmann M, van Obberghen E (1985) Identification of a cellular 110000-Da protein substrate for the insulin-receptor kinase. Biochem J 227:887-892

Biliary Taurocholate Excretion During the Metabolism of 2-Methyl-1, 4-Naphthoquinone (Menadione) in Perfused Rat Liver

T.P.M. Akerboom, T. Bultmann and H. Sies
Institut für Physiologische Chemie I
der Universität Düsseldorf
Moorenstraße 5
4000 Düsseldorf, FRG

Menadione (2-methyl-1,4-naphthoquinone), a blood coagulant drug, is known to cause hemolysis and impaired liver function. We have investigated the metabolism of the drug and its effect on the hepatic uptake and biliary excretion of the bile acid taurocholate in the perfused rat liver.

Infusion with (^{3}H)-menadione results in the biliary excretion of menadione derivatives, which we analyzed by TLC. The predominant menadione conjugates in the bile are thiodione (menadione-S-glutathione) (60%), and hydrolytic products derived therefrom (17%). Menadione glucuronides amount to about 10%.

Menadione elicits substantial oxidation in both the intracellular NADPH and GSH redox systems. Biliary excretion of GSSG is increased several-fold. However, the excretion of taurocholate is strongly inhibited upon infusion with the drug. The inhibition is small in livers with a low content of Se-GSH-peroxidase, obtained by treatment of the animals with a Se-deficient diet. Likewise, inhibition is small in glutathione-depleted livers, obtained from phorone-pretreated rats. In these livers intracellular GSSG and biliary GSSG release

Hepatic Transport in Organic Substances
Ed. by E. Petzinger, R. K.-H. Kinne, H. Sies

are maintained at low values, although menadione still imposes oxidative stress as indicated by an oxidation of intracellular NADPH.

Under anoxic conditions menadione has little influence on both the NADPH and GSH redox systems and also on biliary taurocholate excretion. The amount of thiodione released into bile is similar to that found under normoxia, whereas the amount of glucuronidated products is almost doubled. We conclude that the inhibition of biliary taurocholate excretion by menadione is related to the increased formation of glutathione disulfide, caused by the intracellular generation of hydrogen peroxide by the quinone, and that menadione derivatives show little, if any, contribution to the inhibition of taurocholate excretion.

References

Akerboom TPM, Bultmann T, Sies H (1988) Inhibition of biliary taurocholate excretion during menadione metabolism in perfused rat liver. Arch Biochem Biophys 263:10-18

Comparison of Bile Salt-Binding Proteins in Basolateral Membranes from Small Intestine, Kidney, and Liver

G. Burckhardt[1], W. Kramer[2] and F.A. Wilson[3]
[1] MPI für Biophysik, Kennedyallee 70, 6000 Frankfurt/Main 70, FRG
[2] Hoechst AG, Postfach 800320, 6230 Frankfurt/Main 80, FRG
[3] Dept. Medicine, The Milton S. Hershey Medical Center, The Pennsylvania State University, Hershey, PA, USA

Previous photoaffinity labeling experiments suggested that proteins with apparent molecular weights of 48,000 and 54,000 are components of the Na^+-dependent and Na^+-independent bile salt transport systems, respectively, in the basolateral (sinusoidal) membrane of rat hepatocytes. In this study we attempted to identify the proteins possibly related to Na^+-independent bile salt transport across the basolateral membranes of rat ileal and renal proximal tubule cells and to compare them with the respective proteins from hepatocytes. Basolateral membrane vesicles were prepared from rat ileum and kidney cortex and subjected to photoaffinity labeling with radioactive 3- and 7-diazirine derivatives of taurocholate. Following labeling, membrane proteins were separated by one- and two-dimensional gel electrophoresis and incorporated radioactivity was detected by fluorography and liquid scintillation counting. In basolateral membranes isolated from ileum several polypeptides were labeled with predominant incorporation of radioactivity into a polypeptide of app. MW 54,000. In basolateral membranes from rat kidney cortex also several membrane proteins were labeled. Taurocholate (1 mM) abolished the labeling of three polypeptides with apparent molecular weights

Hepatic Transport in Organic Substances
Ed. by E. Petzinger, R. K.-H. Kinne, H. Sies

of 67,000 (=albumin), 54,000, and 41,000 indicating a saturable interaction of bile salts with these polypeptides. To compare the bile salt-binding proteins of hepatocytes, enterocytes, and proximal tubule cells, two-dimensional co-electrophoresis of labeled membrane proteins was performed. These experiments show that the bile salt-binding polypeptide of MW 41,000 is present only in renal basolateral membranes. The labeled polypeptides of MW 54,000 from liver and kidney comigrated on the gels, whereas the bile salt-binding polypeptide of MW 54,000 from enterocytes exhibited a more alkaline IP (5-6) as compared to the IP of 3.5 for the hepatic and renal MW 54,000 polypeptide. These data suggest identical Na^+-independent bile salt transport systems in basolateral membranes from hepatocytes and proximal tubule cells. In basolateral membranes of the enterocyte a different sytem may be acting.

Effects of Hydrostatic Pressure and Liver Cell Membrane Potential on Hepatic Bile Acid Transport

J. Graf
Dept. of general and experimental Pathology,
University of Vienna, Waehringerstrasse 13,
A-1090 Vienna, Austria

The isolated perfused rat liver was used to analyze bile acid transport during cholestatic conditions induced either by elevated bile duct pressure (maximum secretory pressure) or by perfusion with media containing a high potassium concentration. Hepatic uptake and release of bile acids were determined by the difference of portal and hepatic venous concentrations of ^{14}C-labelled cholic (C) and taurocholic acid (TC) separated by thin layer chromatography.

During continuous infusion of C, acute obstruction of the bile duct had only minor effects on net uptake of C, but caused a release of the conjugated TC into the venous perfusate. Regurgitation of TC increased with time ultimately commensurate to biliary secretion of TC during the period preceding and following bile duct obstruction. TC release into the medium did not correlate with hepatic bile acid (^{14}C) content. Regurgitation of TC was rapidly and completely reversible when bile duct obstruction was relieved and normal bile flow resumed.

In the absence of infused bile acids an increase of the perfusate potassium concentration to 50 mM (in expense of sodium) caused an instantaneous decrease of bile flow rate.

Hepatic Transport in Organic Substances
Ed. by E. Petzinger, R. K.-H. Kinne, H. Sies

Returning the perfusion to control medium (5 mM K^+) resulted in a transient overshoot of bile flow. In liver slices the same maneuver causes cell membrane depolarization and transient hyperpolarization, respectively. In the presence of TC in the perfusate (10 μM) application of high K^+ medium and return to control resulted in larger changes of bile flow rate and the decrease and increase were associated with a transient reduction and stimulation of biliary secretion of TC to 59% and 180%, respectively. Perfusion with high K^+ medium reduced TC uptake by 14%.

The data are consistent with the concepts that i) bile acid transport across the canalicular cell membrane is driven by the intracellular negative cell membrane potential and that ii) elevated biliary hydrostatic pressure causes paracellular back leak of secreted TC probably by reversible opening of the canalicular tight junction barrier.

Characterization and Purification of Bile Acid Transport Proteins

A. Hoffmann, S. Müllner, K. Ziegler, H. Fasold
Institut für Biochemie der J.W. Goethe Universität
Theodor-Stern Kai 7
6000 Frankfurt 70, FRG

In the rat hepatocellular transport system proteins with a molecular weight of 50 kDa and 54 kDa have been identified as components of the bile acid uptake system (Ziegler et al., 1984; von Dippe et al., 1986; Fricker et al., 1987; Müllner et al., 1988; Ziegler et al., 1988)

When isolated plasma membranes were incubated with the bile acid derivative [^{14}C]-2-[3α,12α-dihydroxy-7-(4'-^{3}N-benzamido)-α 5ß-cholanoyl] -24-aminoethanesulfonate (NH_2-BATC), these proteins were labelled. Additionally, a 67 kDa protein (albumin) and faintly, a 30 kDa protein was detected by autoradiography. In a 2D-gelelectrophoresis the 50 kDa protein can be discerned with an isoelectric point of pI = 5.2.

NH_2-BATC is also used for selective solubilization of the bile acid transport proteins from isolated plasma membranes (5). Further purification is achieved by elution with a phosphate gradient on a hydroxylapatite column, followed by preparative SDS-gelelectrophoresis. These separated 50 kDa and 54 kDa proteins were blotted on Immobilon- or on Polybrene-coated glass-fibre and the amino acid sequences of the amino terminus of both proteins determined.

Hepatic Transport in Organic Substances
Ed. by E. Petzinger, R. K.-H. Kinne, H. Sies

References

Fricker G, Schneider S, Gerok W, Kurz G (1987) Identification of different transport systems for bile salts in sinusoidal and canalicular membranes of hepatocytes. Biol Chem Hoppe Seyler 368:1143-1150

Müllner S, Möller W, Ziegler K, Fasold H (1988) J Chrom, submitted

von Dippe P, Ananthanarayanan M, Drain P, Levy D (1986) Purification and reconstitution of the bile acid transport system from hepatocyte sinusoidal plasma membranes. Biochim Biophys Acta 862:352-360

Ziegler K, Frimmer M, Fasold H (1984) Further characterization of membrane proteins involved in the transport of organic anions in hepatocytes. Biochim Biophys Acta 769:117-129

Ziegler K et al. (1988) Biochim Biophys Acta, in press

Transport of Taurocholate in Isolated Cells from Ileum, Liver and Kidney

M. Schwenk and M. Locher
Abteilung Allgemeine Pharmacology
Medizinische Hochschule Hannover
3000 Hannover 61, FRG

Bile acid carriers are localized in ileum, liver, and kidney. It is unknown whether these carriers are identical or different. This work uses isolated cells of the three organs to compare the uptake kinetics of taurocholate.

Cells were isolated from male guinea pigs (400 g) by collagenase perfusion (liver), collagenase treatment of the tissue mash (kidney), and treatment with EDTA + dithiotreitol (ileum). Morphological examination (electron microscopy) and functional tests (respiration, potassium uptake, amino acid incorporation) indicated good viabilities. The cell suspensions were incubated in Minimum Essential Medium (37°) with ^{3}H-taurocholate and cells were separated at various times by centrifugation through a layer of silicone oil.

All three cell preparations transported taurocholate in a saturable and sodium-dependent manner; ouabain (100 μM) and low sodium concentration (10 mM) inhibited between 65 and 75 %. However, the abilities to accumulate taurocholate in the cellular space varied considerably between cell types (ileal cells 90-fold, liver cells 22-fold, kidney tubule fragments

Hepatic Transport in Organic Substances
Ed. by E. Petzinger, R. K.-H. Kinne, H. Sies

14-fold). Moreover, the apparent K_m values of ileal cells (231 μM), and kidney tubule fragments (276 μM) were much higher than those of liver cells (38 μM).

In conclusion, efficient carriers for taurocholate uptake are localized in all three organs. The carriers resemble each other with regard to the energetic coupling of transport. But the different K_m value in liver, compared to ileum and kidney, may suggest that the hepatic carrier differs from the carriers in the other two organs.

Properties of the Taurocholate Uptake System in Basolateral Rat Liver Plasma Membrane Vesicles: Effects of Albumin and Substrate Specificity

B. Zimmerli, J. Valantinas, P.J. Meier
Div. of Clinical Pharmacology
Dept. of Medicine
University Hospital,
Zürich, Switzerland.

Various recent studies in intact hepatocytes have suggested that the hepatocytic bile salt uptake system(s) might also transport a variety of other amphipathic substances (e.g. PNAS 81:5232, 1984; BBA 947:75, 1988). Therefore, we investigated the substrate specificity of mediated Na^+-dependent taurocholate uptake in isolated blLPM vesicles (J. Cell Biol. 98:991, 1984). Kinetic transport studies were performed by a rapid filtration technique in the presence and absence of bovine serum albumin (BSA). Consistent with previous studies (Am. J. Physiol. 249:G34, 1985) BSA (37 μM) selectively stimulated the Na^+-dependent portion of taurocholate uptake with a decrease of the apparent K_m from 46 ± 6 μM to 17 ± 3 μM and no change in V_{max} (4.2 ± 0.2 $nmol \cdot min^{-1} \cdot mg^{-1}$ protein) (means $\pm$SD; n = 9). Because of this specific effect on Na^+-stimulated taurocholate uptake, BSA was included in all cis-inhibition experiments with the exception of DIDS (4,4'-diisothiocyano-stilbene-2,2'-disulfonic acid) and BSP (Bromosulfophthalein), the inhibitory effects of which were completely abolished by their high degree of BSA binding. Based on complementary analysis by Dixon- and Cornish-Bowden plots the following compounds were indentified as <u>competitive inhibitors</u> of Na^+-dependent

Hepatic Transport in Organic Substances
Ed. by E. Petzinger, R. K.-H. Kinne, H. Sies

taurocholate uptake: Cholate (K_i = 140±30 μM), taurochenodeoxycholate (K_i = 9±3 μM), chenodeoxycholate (K_i = 53±6 μM), progesterone (K_i = 110±30 μM), estradiol-3-sulfate (K_i = 28±4 μM), bumetanide (K_i = 440±85 μM), furosemide (K_i = 460±140 μM), verapamil (K_i = 65±35 μM), and phalloidin (K_i = 850±350 μM). In contrast, non-competitive inhibition was found with BSP (K_i = 12±2 μM), cyclosporin A (K_i = 3±1 μM), and DIDS (K_i = 45±7 μM). These results generally support the concept of multispecificity of the basolateral bile salt uptake system(s). However, the findings also indicate that bile salts and BSP are taken up into rat hepatocytes by different transport systems supporting the assumption of multiple basolateral organic anion "carriers" with distinct, yet partially overlapping substrate specificities (Hepatology 7:165, 1987).

Hepatic Transport of Xenobiotics

Uptake of Foreign Cyclopeptides by Liver Cells

M. Frimmer
Institut für Pharmakologie und Toxikologie
Justus-Liebig-Universität
Frankfurter Straße 107
6300 Giessen (BRD)

Introduction

Chemical modification of naturally occuring peptides is a successful strategy to obtain superagonists or antagonists for peptidergic receptors or inhibitors for proteinases. For therapeutic use of such compounds it is desirable that they cannot be destroyed by peptidases or proteinases. However, the half life of synthetic peptides is not only influenced by enzymatic degradation but also by renal or biliary elimination. The latter phenomenon is the subject of the following report. How are peptides recognized by liver cells? How can they be internalized into hepatocytes and excreted into the bile? The highlight of my findings is the unexpected fact that numerous foreign peptides are taken up into liver cells by a secondary active carrier-mediated mechanism. At present it cannot be excluded that certain peptide hormones are eliminated by the same pathway.

Phalloidin is a Model for Hepatocellular Clearance of Foreign Cyclopeptides

Phalloidin (Fig. 1) is one of the toxic cyclopeptides of _Amanita phalloides_ (reviewed by Frimmer 1987). When injected intravenously this heptapeptide kills animals within 2-3 hours by a dramatic hemorrhagic liver swelling. The toxin is specifically accumulated in liver cells. In the onset of our studies (1975) we suspected that a specific receptor might be responsible for the hepatospecificity of phalloidin. In the following we observed that the phalloidin response of isolated rat liver cells showed some quantitative properties (Table 1) which are due to the specific uptake of the toxin.

Hepatic Transport in Organic Substances
Ed. by E. Petzinger, R. K.-H. Kinne, H. Sies

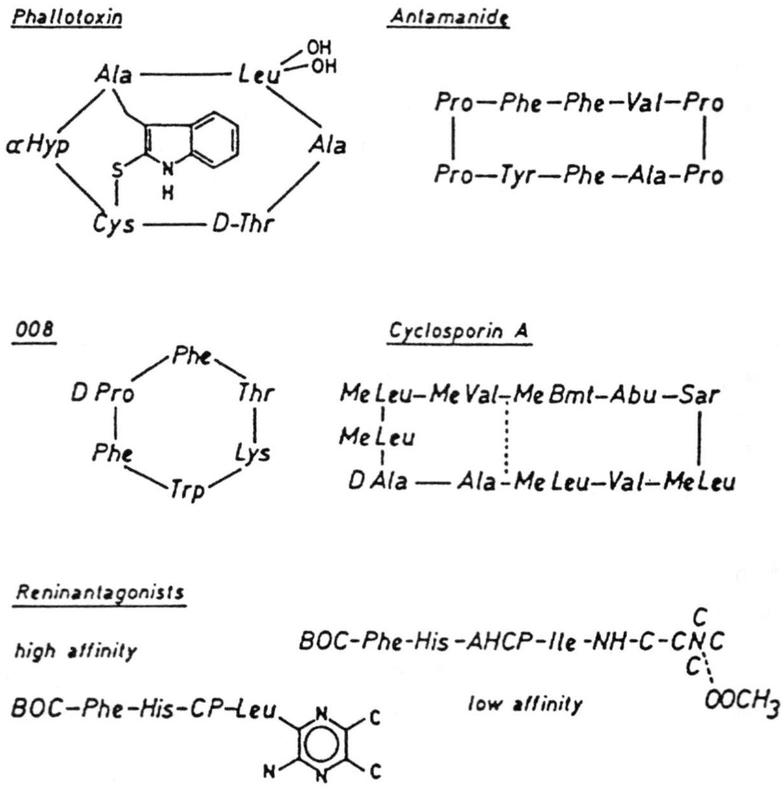

Figure 1: Foreign substrates (cyclic and linear peptides) of the multispecific transporter in the hepatocellular plasma membranes

Table 1.

Properties of phalloidin response on isolated rat hepatocytes

Temperature dependence (no response below 20°C).		
Response is saturable.		(Uptake saturable)
Low response in absense of O_2.		(Low uptake)
Na^+ dependence.		
Valinomycin Monensin Nigericin Gramicidin Cyclosporin A	noncompetitive inhibitors	
Bile acids Fusidic acid Iodipamide Antamanide Cyclosomatostatin	competitive inhibition	
Pretreatment with various hepatoxic compounds (e.g. CCl_4)	decreased response	(drecreased uptake)
Neonatal hepatocytes	less sensitive	(low uptake)
Hepatoma cells	no response	(no uptake)

On the inside of the cell membrane phalloidin binds to microfilamentous F-actin and destroys the mechanical stability of the liver cell membrane. The velocity and quantity of this biological response corresponds to the velocity of uptake. What is the physiological substrate of the phalloidin translocating system? In a series of studies on perfused livers and on isolated hepatocytes we observed that bile acids inhibit both phalloidin response and uptake in a competitive manner. Whenever the transport of bile acids decreased after treatment with drugs or chemicals, metabolic inhibitors, or under anaerobic conditions, the uptake of phalloidin by liver cells decreased to the same degree. The suspicion that one of the bile acid transporters might be responsible for the uptake of phalloidin was confirmed by studies with various affinity labels.

a) (Photo)affinity labels derived from bile acids inhibited both phalloidin uptake and response in a concentration-dependent and irreversible manner (Frimmer 1987, Review).

b) Photoaffinity labels derived either from phalloidin or from antamanide (phalloidin antagonist) bind to the same membrane proteins in liver cells as bile acid labels do (Wieland et al. 1984).

c) Hepatoma cells deficient in both bile acid and phalloidin transport neither bound labels of type a nor of b (Frimmer and Ziegler 1988, Review).

From these and other findings it was concluded that the Na^+-dependent bile acid transporter of the sinusoidal liver cell membrane functions as a phalloidin translocase.

Cyclic Somatostatins are Further Substrates of the Na^+-Dependent Bile Salt Transporter

During our studies reported above we learned that some cyclic modifications (with retro sequence) (Fig. 1) of somatostatin protect liver cells against phalloidin (Rao et al 1982). These latter authors used the in vitro tests developed in our laboratory but did not know that liver cell protection against phalloidin is due to uptake inhibition (Ziegler and Frimmer 1986a). In cooperation with H. Kessler's group (Frankfurt) we tested 114 modified somatostatins for phalloidin protection and inhibition of cholate uptake (in part also for phalloidin uptake). As far as they were studies in detail these compounds inhibited phalloidin response and bile acid uptake in a competitive manner (Ziegler et al 1985). As expected a radioactive photoaffinity label, derived from one of the most potent somatostatin modifications (Cyclo(-Phe(p-NH($1^{14}C$)Ac)-Thr-Lys(CO(pN_3)C_6H_4)-Trp-Phe-D-Pro), bound to the same membrane proteins as bile salt labels (Kessler et al 1987). There is no doubt that cyclosomatostatins are further substrates of the Na^+-dependent bile salt transporter. Some structural regularities were seen to probably be responsible for recognition of those peptides by the transporter (Kessler et al 1986). Cyclosomatostatins were used for affinity electrophoresis of protein components belonging to the carrier system in hepatocellular plasma membranes (Ziegler et al 1988a). Apparently this method purifies the same membrane protein as affinity electrophoresis with bile salt derivatives.

<u>Renin inhibitors</u> (with linear peptide structure)

Recently we were requested by a pharmaceutical company to test about 20 linear peptides with inhibitory effect on the enzymatic activity of renin. Some of them were detected in the bile of treated animals. The question was whether these compounds are recognized by the Na^+-dependent bile salt transporter or not. The above compounds (Fig. 1) possess no plausible structural similarities to phallotoxins, antamanide or cyclosomatostatines. Unfortunately radioactive modifications of those linear peptides are not available as yet, so that we could only study the inhibition of bile acid (and in part of phalloidin) uptake. As expected those new peptides protected liver cells against phalloidin too. Additional studies with radioactive renin inhibitors and affinity labels are projected to determine whether this type of peptide binds to the Na^+-dependent bile salt transporter in liver cell membranes.

<u>Cyclosporines</u>

Searching for new substrates of the bile salt transporter we detected that cyclosporin A (and other cyclosporins) are potent inhibitors of the phalloidin response of liver cells (Ziegler and Frimmer 1984).

Cyclosporin A at 100 nM inhibited the phallotoxin uptake by 50 % and 5 µM inhibited the uptake of cholate (Ziegler and Frimmer 1986b). A large excess of cyclosporin A prevented affinity labeling of membrane proteins with (^3H)-isothiocyanatobenzamidocholate. Detailed investigations of the kinetics showed that cyclosporin A inhibits cholate uptake in a noncompetitive manner. A photolabile diazirine derivative of cyclosporin A inhibited both the uptake of cholate and of phalloidin irreversibly (Ziegler and Frimmer 1986c). Proteins with 200, 85, 54, 50 and 34 kDa were labeled. Comparison of cyclosporin A labeled membrane proteins with those detected after treatment with isothiocyanatobenzamidocholate revealed identity of nearly all binding proteins. Nevertheless cyclosporin A is not transported by the Na^+-dependent bile acid carrier, but as recently shown it penetrates liver cell by diffusion across the lipid barrier (Ziegler et al 1988b). The uptake of cyclosporin does not depend on metabolic energy. Apparently some hydrophobic peptides can interact with carrier components but are not translocated themselves. Photoaffinity labeling of membrane components is not convincing evidence for a carrier-mediated transport.

Nonpeptide substrates of the Na^+-dependent bile salt transporter.
Various curious substrates of the Na^+-dependent bile salt transporter have been detected during the past decade (Frimmer and Ziegler 1988). Besides the peptides discussed above various steroids (fusidic acid, ouabain, ß-estradiol, progesteron), organic anions (DIDS, iopodate, iodipamide, prostaglandin analogs, NAP-taurine) are transported into hepatocytes by the bile salt transporter. The term "multispecific transporter" was recommended in order to characterize its important clearing function in the liver.

Conclusions

The principle of uphill transport of certain polypeptides by mediation of a multispecific carrier system can be postulated, provided that transport kinetics suggest a typical competition of various typical substrates (e.g. bile acids) with these peptides and that binding sites in the sinusoidal membrane of liver cells are identical ones. Excess of any substrate (known to be transported) should prevent coupling of (photo)affinity labels derived from possible substrates. In the case of phalloidin, detailed studies (Petzinger and Frimmer 1988) gave evidence that several characteristics and properties of receptor-mediated pinocytosis or microcytosis are inapplicable for this cyclopeptide. The problem is not the acceptance of an energy-dependent and carrier-mediated transport, but rather the recognition of chemically very different substrates by a single transporter. From a molecular view one can detect some regularities within certain groups of compa-

rable, analogous or similar compounds (e.g. cyclosomatostatins) (Kessler et al 1986). Similarities between antamanide and cyclosomatostatins are obvious, but not between phallotoxins, cyclosomatostatins or the above mentioned renin antagonists. Analogies between cyclopeptides and steroids can hardly be uncovered. Possibly knowledge of three-dimensional distribution of hydrophobic and of charged groups may elucidate the problem in the future. It is, however, difficult to understand that relatively small membrane proteins with 50 kDa (as identified by affinity labels) contain a such variety of different binding sites to explain the extreme broad specificity of possible substrates. We discuss therefore a modular transport system consisting on nonidentical subunits which form a channel. Indeed target-size analysis of the cholate transport system gave a functional molecular weight of 173±13 KDa (Ziegler, unpublished). The native transporter is more than 3 times larger than the subunits identified by affinity labels after solubilization of the membrane (Ziegler unpublished).
Analogous investigations are in progress with the cyclosomatostatin translocating system. Preliminary results indicate that the functional MW is lower than for bile acid transport. We suspect that different combinations of subunits might be involved in the translocation of the various substrates of the multispecific transporter. A further argument for the possible variety of involved components is the degree of glycosylation of the individual subunits. Side by side one detects the same proteins with varied amounts of sugars. Last but not least the concentrative uptake of negatively charged substrates (e.g. bile acids) and of neutral peptides required different driving mechanism (Frimmer and Ziegler 1988). Many questions are open and may be answered after final isolation of all possible protein components of the system. It should be mentioned here that not all components of a modular system must be unconditionally marked by affinity labels.

References

Most of the papers discussed above are comprehensively reviewed in:

Frimmer M, Ziegler K (1988) The transport of bile acids in liver cells. Biochim Biophys Acta 947: 75-99

Frimmer M (1987) What we have learned from phalloidin. Tox Letters 35: 169-182

Further References

Kessler H, Klein M, Müller A, Wagner K, Bats JW, Ziegler K, Frimmer M (1986) Konformative Voraussetzungen für die biologische Aktivität von cytoprotektiv wirksamen Cyclopeptiden. Angew Chemie 25: 927-999

Kessler H, Haupt A, Frimmer M, Ziegler K (1987) Synthesis of a cyclic retro analogue of somatostatin suitable for photoaffinity labeling. Int J Peptide Protein Res 29: 021-628

Petzinger E, Frimmer M (1988) Comparative investigations on the uptake of phallotoxins, bile acids, bovine lactoperoxidase and horseradish peroxidase into rat hepatocytes in suspension and in cell cultures. Biochim Biophys Acta 937: 135-144

Rao GS, Lemoch H, Usadel KH (1982) Behandlung mit Somatostatin schützt Rattenleberzellen gegen Läsionen durch Phalloidin, Äthanol und DMSO. In: Klinische Perspekitiven von Somatostatin. V. Freiburger Kolloquium Attempto Freiburg, p 13-15

Wieland T, Nassal M, Kramer W, Fricker G, Bickel U, Kurz G (1984) Identity of hepatic membrane transport systems for bile salts, phalloidin and antamanide by photoaffinity labeling. Proc Natl Acad Sci USA 81: 5232-5236

Ziegler K, Frimmer M (1984) Cyclosporin protects liver cells against phalloidin. Biochim Biophys Acta 805: 174-189

Ziegler K, Frimmer M, Kessler, H. Damm J, Eiermann V, Koll S, Zarbock J (1985) Modified somatostatins as inhibitors of a multispecific transport system for bile salts and phallotoxins in isolated hepatocytes. Biochim Biophys Acta 845: 86-93

Ziegler K, Frimmer M (1986a) Molecular aspects of cytoprotection by modified somatostatins. Klin Wochenschr 64: 87-89

Ziegler K, Frimmer M (1986b) Cyclosporin A and a diazirine-derivative inhibit the hepatocellular uptake of cholate, phalloidin and rifampicin. Biochim Biophys Acta 855: 136-142

Ziegler K, Frimmer M (1986c) Identification of cyclosporin binding sites in rat liver plasma membranes, isolated hepatocytes, and hepatoma cells by photoaffinity labeling using (^{3}H) cyclosporin-diazirine. Biochim Biophys Acta 855: 141-156

Ziegler K, Polzin G, Frimmer M (1988a) Hepatocellular uptake of cyclosporin A by simple diffusion. Biochim Biophys Acta 938: 44-50

Ziegler K, Frimmer M, Kessler H, Haupt A (1988b) Azidobenzamido c(Phe-Thr-Lys-Trp-Phe-DPro), a new photosensitive substrate for the "multispecific bile acid transporter" of hepatocytes: Evidence for a common transport system for bile acids and cyclosomatostatins in basolateral membranes. Biochim Biophys Acta submitted

Identification of Carrier Proteins in Hepatocytes by (Photo) Affinity Labels Derived from Foreign Cyclopeptides

K. Ziegler
Institut für Pharmakologie und Toxikologie
Justus-Liebig-Universität
Frankfurter Straße 107
6300 Giessen (BRD)

Introduction

Certain cyclopeptides of toxicological (e.g. phallotoxins) or pharmacological interest (e.g. antamanide, cyclosomatostatins) are rapidly eliminated via the biliary pathway (Frimmer 1987, Caldwell 1984). A liver-specific clearance mechanism is responsible for the hepatospecific toxic effects of e.g. phallotoxins and for the short half life of cyclosomatostatins. We did not expect a carrier-mediated uptake to be responsible for this first-pass elimination. The transport of various cyclopeptides is driven by a sodium gradient and the membrane potential (Review by Frimmer and Ziegler 1988). By mutual competition experiments it became evident that bile acids are the physiological substrates of this transport system. Actually, the carrier seems to be unable to discriminate between various structurally different compounds like bile acids and cyclopeptides. It was therefore termed a "multispecific transporter" (Ziegler et al. 1985).

Another class of hydrophobic cyclopeptides, cyclosporins, is cholestatic when applied clinically in high concentrations (Schade et al. 1983). This points to an interaction with the bile acid recirculation. Kinetic data, however, revealed that cyclosporins are non-competitive inhibitors of the above multispecific transporter (Ziegler and Frimmer 1986a) and permeate the membrane by diffusion (Ziegler et al. 1988a).

Hepatic Transport in Organic Substances
Ed. by E. Petzinger, R. K.-H. Kinne, H. Sies

The hypothesis of a common transport system for bile acids, phallotoxins, antamanide and cyclosomatostatins (but not for cyclosporins) is based on kinetic studies. In the following, additional biochemical experiments are described in order to verify the above hypothesis. Reactive analogues of bile acids and cyclopeptides, which show similar kinetic properties, are used as labels to identify hepatocellular binding proteins.

Hepatocellular Uptake of the Cyclosomatostatin c(Phe-Thr-Lys-Trp-Phe-DPro) (in the following termed 008) and Its Photosensitive Analog cyclo(Phe(p-NH-^{14}C-Ac)-Thr-Lys(CO(p-N_3)C_6H_4)-Trp-Phe-D-Pro) (= Azidobenzamido -008)

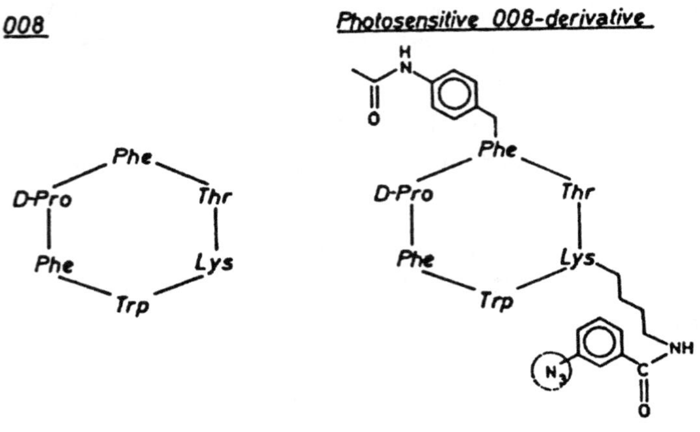

(Fig. 1)

Native somatostatin binds with low affinity to the phallotoxin binding site (Ic_{50} = 45 µM) and with yet lower to the bile acid binding site (Ic_{50} = 220 µM) of the multispecific transporter. Modification in the amino acid sequence however, and introduction of conformational ridgidities led to the development of analogues with about 70-fold more active inhibition of phalloidin and bile acid uptake (no enzymatic degradation; hormonal activity of these compounds decreased). One of the most active analogues synthesized by Kessler and his group c(Phe-Thr-Lys-Trp-Phe-DPro) contains the retro sequence of the amino acids 7-11 of somatostatin bridged by a D-proline residue.

This derivative (008) inhibits the hepatocellular uptake of cholate (K_i = 3 µM) and of phallotoxins (K_i = 2 µM) competitively. The kinetic properties of the hepatocellular uptake of 008, of azidobenzamido-008, of phallotoxins and of taurocholate are very similar (Table 1).

A sodium gradient and the membrane potential are the main driving forces for the uptake of all substrates. Mutual competitive inhibition of hepatocellular transport and a low transport capacity in neonatal hepatocytes support the hypothesis of a common transport system.

For identification of binding proteins an azido derivative of 008 was synthesized by H. Kessler and A. Haupt (Kessler et al. 1988). In order to demonstrate the usefulness of this compound for labeling it must be tested whether the label itself is taken up by hepatocytes. In the dark azidobenzamido-008 is taken up by a sodium-dependent mechanism. It competitively inhibits both the sodium-dependent uptake of cholate (K_i = 1 µM) and of taurocholate (K_i = 5 µM). By flash photolysis of azidobenzamido-008 the bile acid transport inhibition becomes irreversible (Ic_{50} = 2 µM for cholate- and 9 µM for taurocholate uptake inhibition) (Ziegler et al. 1988b).

Cyclosomatostatin Binding Proteins in Isolated Rat Hepatocytes, Basolateral Plasma Membranes from Newborn and Adult Rats (Ontogeny of Binding Proteins) and of Hepatoma Cells

Azidobenzamido-008 binds preferentially to membrane proteins of MW 67, 54, 50 and 37 kDa and to a lesser extent to a 43 kDa protein in isolated hepatocytes (a) as well as in isolated basolateral plasma membranes (b) (Fig. 2).

The predominantly labeled 50 and 54 kDa proteins are heterogenously glycosylated integral membrane proteins, whereas the 67 kDa protein is hydrophilic and non-glycosylated.

Table 1.

Kinetic properties of the hepatocellular uptake of:

	taurocholate	phalloidin	008	azidobenzamido-008
Km (μM)	19 - 57	20	5,8	6
V max (nmol/mg/min)	0,75 - 13	0,03	0,165	0,165
replacement of Na^+ by K^+	▼	▼	▼	▼
Choline	▼	▼	▼	▼
Li^+	▼	▼	▼	▼
replacement of Cl^- by SCN^-	▼	▲	▼	▼
NO_3^-	▼	▲	▼	▼
SO_4^{2-}	▼	▼	▼	▼
temperature dependency	+	+	+	+
activation energy	29 Kcal/mol	30	9	10
+ metabolic inhibitors	▼	▼	▼	▼
N_2atmosphere	▼	▼	▼	▼
mutual competitive inhibition	+	+	+	+
uptake into hepatocytes of 9 day old rats	▼	▼	▼	▼
uptake into AS 30 D ascites hepatoma cells	n.u.	n.u	n.u	n.u.

\- = no effect on uptake
▼ = inhibition of uptake
n.u. = no measurable uptake
▲ = stimulation of uptake

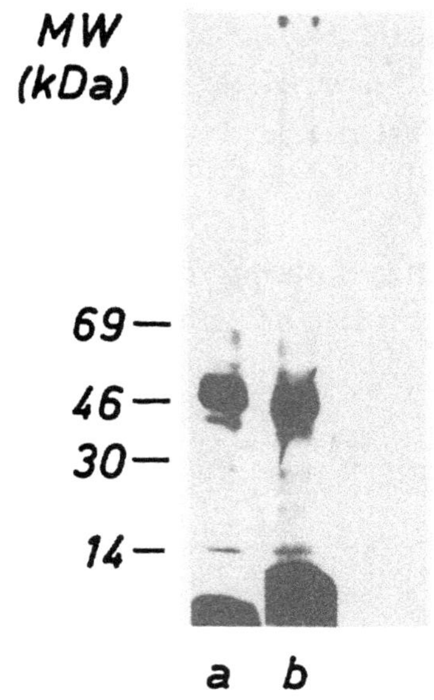

Fig. 2 Isolated hepatocytes (a) or isolated plasma membranes (b) were labeled with (^{14}C)azidobenzamido-008. Shown is a fluorogram of an SDS slab gel.

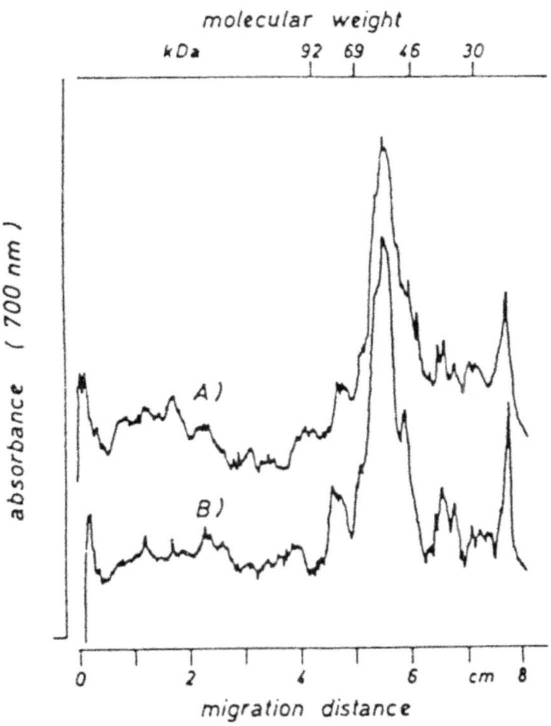

Fig. 3 Isolated plasma membranes were labeled with the above photosensitive analogues. Shown is a scan of a fluorogram of an SDS slab gel. (A) = azidobenzamido-008; (B) = azidobenzamido-taurocholate.

Binding could be prevented by 008 but also by other substrates of the multispecific transporter (cholate, phalloidin) and by photosensitive bile acid analogues e.g. azidobenzamido-taurocholate. The latter has been shown to bind to proteins in the same molecular weight range as cyclosomatostatins (Ziegler et al. 1988c).
(Fig. 3)
The physico-chemical properties of binding proteins for bile acids and cyclosomatostatins and the ontogeny of binding proteins are very similar (Table 2).

Table 2.

Properties of binding proteins for bile acids and for cyclosomatostatin

	Bile acid binding proteins	Cyclosomatostatin binding proteins
MW (kDa)	67, 54, 50, 43, 37	67, 54, 50, 43, 37
glycosylated	54, 50	54, 50
non-glycosylated	67	67
pI	7.5; 7; 6.5; 5.5; 5.25;	7; 5.5; 5.25;
in 9 day old rat liver	67, 54, 37 kDa	67, 54, 37 kDa
in AS 30 D ascites cells	no specific binding	no specific binding
functional molecular weight of the transport system in membrane vesicles	173 ± 16 kDa	100 kDa (p.e.)

(p.e. = pilot experiments)

In rat liver plasma membranes of 9 day old rats neither azidobenzamido-008 nor azidobenzamido-taurocholate bound to a 50 kDa protein. This protein is not fully expressed in membranes of 9 day old rats (Fig. 4). Those cells are immature in bile acid and cyclosomatostatin transport. Transformed cells like AS 30 D ascites hepatoma cells, which are bile acid and cyclosomatostatin transport deficient, are not specifically labeled by either substrate.

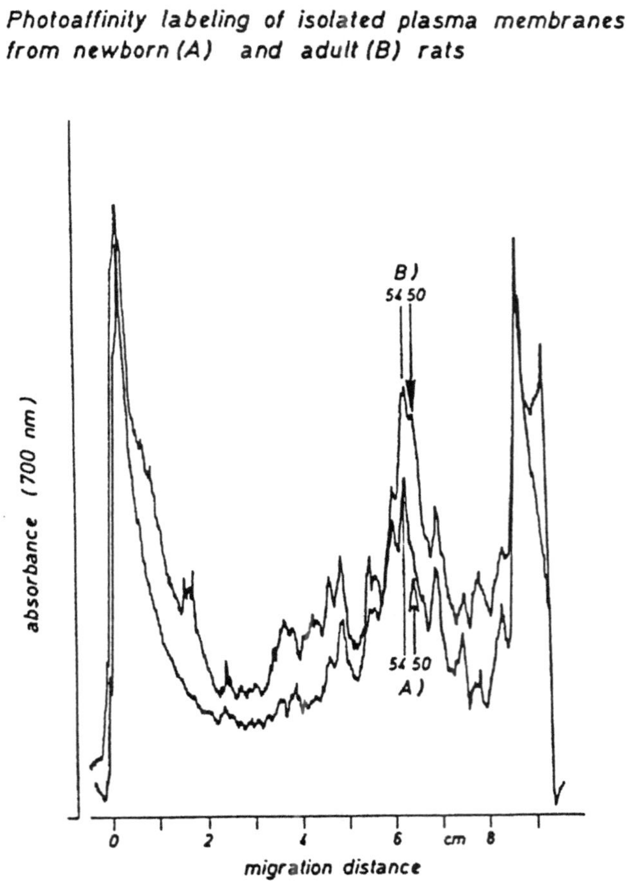

Fig. 4 Isolated plasma membranes from 9 day old adult rats were photolabeled with (^{14}C) azidobenzamido-008. The distribution of radioactively labeled proteins was analysed by scanning a fluorogram of an SDS slab-gel.

Cyclosporin Binding Proteins in Isolated Hepatocytes and Isolated Basolateral Rat Liver Plasma Membranes

In contrast to cyclosomatostatin or bile acids, cyclosporins bind predominantely to a 85 kDa protein in isolated basolateral plasma membranes (Fig. 5) (Ziegler and Frimmer 1986b).

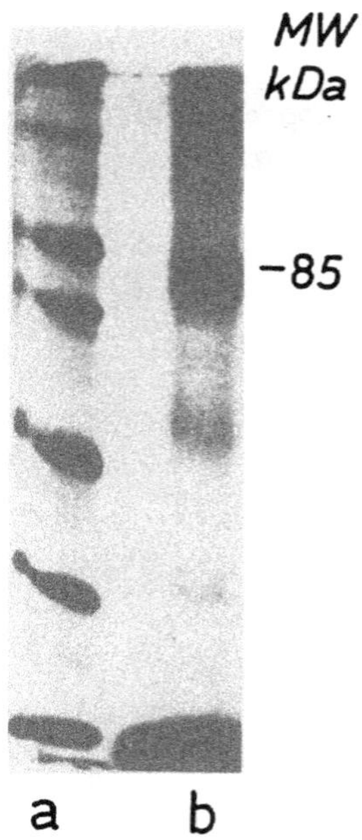

Fig. 5 a = standard proteins; b = labeled plasma membrane proteins

The affinity to 50 and 54 kDa proteins is low. On the other hand, labeling isolated hepatocytes with cyclosporin-diazirine resulted in a time- and temperature-dependent binding to a 50, 54 and 85 kDa protein (Fig. 6).

Conclusion

The hypothesis of a common transport system for bile acids and cyclosomatostatins is supported by photoaffinity labeling studies (as was also reported for phallotoxin and antamanide by Wieland et al. 1984). Nevertheless the identity of the detected proteins can only be proven by amino acid sequencing of purified proteins.

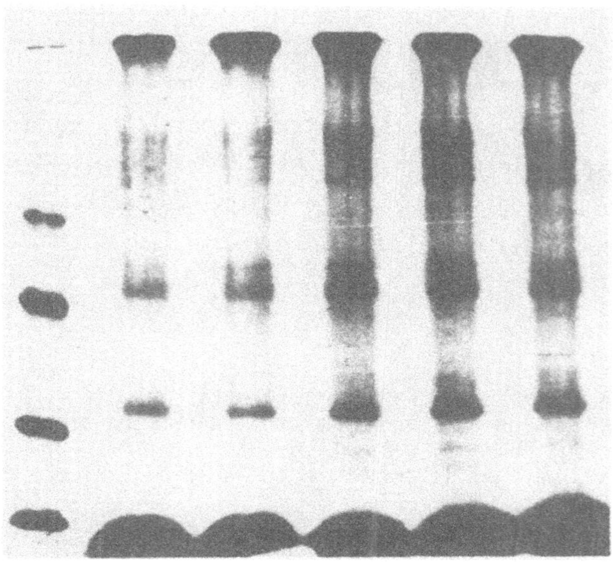

Fig. 6 a = standard proteins. Isolated hepatocytes were preincubated at 25°C for 1 (b); 5 (c); 10 (d); 15 (e), and 30 (f) min in the dark before treatment with a single UV flash

Our studies with cyclosporin may serve as a warning that binding of photoaffinity labels to certain membrane proteins does not provide unconditional proof for a carrier function of these proteins. In the case of cyclosporins a simple diffusion across the lipid barrier is responsible for uptake. This means that caution must be introduced in regard to the specificity and the interpretation of photoaffinity labeling unless it has been shown that the photosensitive probe has uptake kinetics comparable to the starting compound or the system under study (in this case the bile acid transporter).

Furthermore, it is possible that additional proteins, which are not detected by photoaffinity labeling, are additional components of the functional carrier system. Evidence for this assumption comes from measurements of the functional molecular weight of the Na^+-dependent uptake of taurocholate and 008 by radiation inactivation (target size analysis). The molecular weight of the taurocholate carrier was determined to be 173 $\pm$ 13 kDa and that of the 008 uptake system 100 kDa.

It seems that the channel transporting such structurally different substrates is composed of nonidentical subunits and that a different combination of subunits is responsible for the transport of various substrates. On the other hand the broad substrate specificity might be due to the existence of a variety of proteins with nearly identical molecular weight and with overlapping substrate specificity, comparable to that of the cytochrome P 450 system.

References

Caldwell LJ, Parr A, Beihn RM, Agha BJ, Mlodozemiec AR, Jay M, Digenis GA (1985) Drug distribution and biliary excretion pattern of a cyclic somatostatin analog. Pharmaceut Res 2: 55-96

Frimmer M (1987) What we have learned from phalloidin. Toxicol Lett 35: 169-183

Frimmer M, Ziegler K (1988) The transport of bile acids in liver cells. Biochim Biophys Acta 947: 75-99

Kessler H, Haupt A, Frimmer M, Ziegler K (1987) Synthesis of a cyclic retro analogue of somatostatin suitable for photoaffinity labeling. Int J Peptide Protein Res 29: 621-628

Schade RR, Guglielmi A, van Thiel DH, Thompson ME, Warty V, Griffith B, Sanghri A, Bahnson H, Hardesty R (1983) Cholestasis in heart transplant recipients treated with cyclosporine. Transplant Proc 4: 2757-2760

Wieland T, Nassal M, Kramer W, Fricker G, Bickel U, Kurz G (1984) Identity of hepatic membrane transport systems for bile salts, phalloidin and antamanide by photoaffinity labeling. Proc Natl Acad Sci USA 81: 5232-5236

Ziegler K, Frimmer M, Kessler H, Damm J, Eiermann V, Koll S, Zarbock J (1985) Modified somatostatins as inhibitors of a multispecific transport system for bile acids and phallotoxin in isolated hepatocytes. Biochim Biophys Acta 845: 86-93

Ziegler K, Frimmer M (1986a) Cyclosporin A and a diaziridine derivative inhibit the hepatocellular uptake of cholate, phalloidin and rifampicin. Biochim Biophys Acta 855: 136-142

Ziegler K, Frimmer M (1986b) Identification of cyclosporin binding sites in rat liver plasma membranbes, isolated hepatocytes and hepatoma cells by photoaffinity labeling using (^{3}H) cyclosporin-diazirine. Biochim Biophys Acta 855: 147-156

Ziegler K, Polzin G, Frimmer M (1988a) Hepatocellular uptake of cyclosporin A by simple diffusion Biochim Biophys Acta 938: 44-50

Ziegler K, Frimmer M, Kessler H, Haupt A (1988b) Azidobenzamido c(Phe-Thr-Lys-Trp-Phe-DPro), a new photosensitive substrate for the "multispecific bile acid transporter" of hepatocytes: Evidence for a common transport system for bile acids and cyclosomatostatins in basolateral membranes. Biochim Biophys Acta (submitted)

Ziegler K, Frimmer M, Müllner S, Fasold H (1988c) Identification of bile acid binding proteins in hepatocellular membranes of newborn and adult rats. Biochim Biophys Acta (submitted)

Uptake of Xenobiotics by the Bile Acid Transport System in the Liver: Relationship Between the Organic Anion Transport Systems in the Kidney and the Liver

E. Petzinger
Max-Planck-Institut für Systemphysiologie,
4600 Dortmund 1, F.R.G.
present address: Institute of Toxicology, University of Mainz, Obere Zahlbacher Strasse 67, 6500 Mainz, F.R.G.

Introduction

The purpose of this chapter is to describe recent studies concerning the uptake of certain xenobiotics into hepatocytes by the bile salt transport system. These studies indicate that the organic anion secretion system(s) of the kidney, the so-called para-aminohippurate (PAH) transport system(s) is/are equivalent to the bile salt uptake system(s) of the liver in terms of their multispecificity for xenobiotic transport. Studies on the uptake of loop diuretics into hepatocytes also indicate similarities between the comparative protein entities of both transport systems. Thus the hypothesis is put forward that differences in the organo-selectivity of drug clearance via the transport systems for organic anions in both organs is due to a microheterogeneity in otherwise identical carrier proteins.

Historical Background

Drug clearance by liver and kidney is in part mediated by secretory pathways. Transport mechanisms are involved in this process of secretion and there is a long and continuing discussion about the identity and diversity of different secretory transport systems.in liver and kidney.

Hepatic Transport in Organic Substances
Ed. by E. Petzinger, R. K.-H. Kinne, H. Sies

In the early fifties Sperber (1954) suggested the unitary hypothesis. For the excretion of organic anions he proposed separate and distinct pathways; in liver the L-System and in kidney the H-System (H stands for hippurate). Later Kim and Hong (1962) found identical transport systems for the excretion of phenolsulfalein dyes by liver and kidney. The same suggestion was put forward by Despopoulos (1966; 1968; 1971) who described identical transport systems in both organs for the excretion of hippurate, food dyes and phenolsulfalein compounds. Deviations in the organo-selectivity of the clearance data were interpreted in terms of different drug metabolism subsequent to transport. In Sweden Bárány (1971; 1972; 1973 a/b; 1974; 1975) from uptake and competition studies with tissue slices proposed composite transport systems with overlapping substrate specificity. Barany considered whether Sperber's L-System from the liver was also present in the kidney and found in addition that the hepatic bile-salt transport system was indeed present in the kidney. Alpert et al. (1969) and later Jansen et al. (1983) found evidence in the liver for separate pathways for the excretion of organic anions. For sheep and rats with inherited disorders of organic anion excretion and with permanent hyperbilirubinemia two pathways were postulated: the bilirubin-bromosulfophthalein (BSP) transport system and the bile-salt (BS) transport system. These results were confirmed by Scharschmidt et al. (1975) in uptake experiments with bilirubin, bromosulfophthalein and indocyanine green versus uptake of bile acids. Two separate carriers for the uptake or organic anions in hepatocytes were postulated.

An interesting observation by Alpert and Arias 1969 noted that although the hepatic bilirubin-BSP transport system was impaired on the mutant sheep, the renal para-aminohippuric acid excretion was not. These authors concluded that this may indicate a diversity of hepatic and renal excretory pathways for organic anions. The conclusion was based on the assumption that the hepatic bilirubin-BSP transport system is equivalent to the renal para-aminohippurate transport system.

Outline of the Studies Performed

As reviewed in chapter 25 of this book, findings for the organo-selectivity of the toxicity of phalloidin suggested that the hepatic bile-salt transport system is responsible for the selective uptake of this cyclic peptide into hepatocytes. From this initial observation and from subsequent comparative uptake studies with other cholephilic xenobiotics it has become clear over the last decade that the bile-salt uptake system of hepatocytes is a multispecific transport system for xenobiotics (for references see Petzinger et al. 1987). Thus for the correct comparative analysis regarding the identities of secretory transport systems in liver and kidney this finding allows us to switch the focus, away from the bilirubin-BSP transport system, to the bile-salt transport system. From a summary of the latest results obtained for the multispecific bile-salt transport system and a comparison with reported literature data for the renal para-aminohippurate transport system a number of similarities are evident (Table 1).

Tab.1 Common properties of the multispecific transport systems for xenobiotics in liver and kidney

MULTISPECIFIC BILE ACID CARRIER	MULTISPECIFIC p-AMINOHIPPURATE CARRIER
8 TRANSPORTATES	212 TRANSPORTATES (BUMETANIDE)
IDENTIFICATION BY AFFINITYLABELS DERIVED FROM	
BILE ACIDS	
NAP-TAURINE	NAP-TAURINE
DIDS	DIDS
LABELED PROTEINS	
24 kDa	26 kDa
34 - 37 kDa	
48 - 54 kDa	52 kDa
(110) kDa	108 kDa
INHIBITORS	
PROBENECID	PROBENECID
BROMOSULFOPHTHALEIN	BROMPHENOL BLUE
IODIPAMIDE	IODIPAMIDE

Both transport systems are multispecific, and are both inhibited by probenecid, phenolsulfalein dyes and iodipamide. DIDS is also an irreversible inhibitor of substrate uptake (Koschier et al. 1980; Petzinger et al. 1978; Petzinger and Frimmer 1980) in both systems. Finally photoaffinity labelling studies with plasma membranes from liver (Frimmer and Ziegler 1988) and kidney (Goldinger et al. 1984; Kramer and Burckhardt 1988) have revealed very similar putative carrier proteins for organic anion uptake.

As recently reviewed (Frimmer and Ziegler 1988) and discussed in chapters 22 and 23 in this book, the most likely candidates for bile acid carrier proteins are proteins of molecular weights 48-50 kDa and 52-54 kDa. These proteins are also involved in xenobiotic uptake (see chapter 26). The finding that a 52 kDa protein is a putative carrier protein of the renal PAH-System (Goldinger et al. 1984; Kramer and Burckhardt 1988) has prompted the hypothesis that similar or even identical proteins are present for organic anion transport in kidney and liver. To test this hypothesis we initiated experiments on the uptake of loop diuretics into hepatocytes.

The Kinetic Approach

Uptake of loop diuretics into isolated rat hepatocytes

Loop diuretics are classical substrates for the renal PAH-transport system and their clinical value depends on their ability to be secreted into urine (Odlind 1979; Odlind and Beermann 1980). The benzosulfonamide diuretics furosemide and bumetanide (Fig. 1) are secreted at a rate corresponding to that of p-aminohippuric acid (Ostergaard et al. 1972). They are organic anions and subsequent to tubular secretion the compounds inhibit a Na^{+}-K^{+}-Cl^{-} ion cotransport system (Geck et al. 1980). Proteins of this ion cotransport system are not

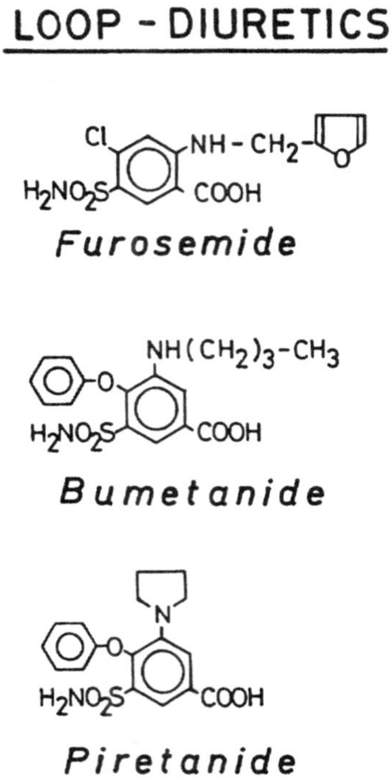

Fig.1 Chemical structure of benzosulfonamide loop diuretics

The pKa value of the COOH-group is 3.8 in furosemide and 3.6 in bumetanide. The pKa of the imino group is 7.5 in furosemide and 7.7 in bumetanide (Orita et al. 1976). Additional information on the pharmacology of bumetanide, furosemide and piretanide can be found by reference to Ostergaard et al. 1972; Burg et al. 1973; Zeuthen et al. 1978 and Patarca et al. 1983.

present in hepatocytes (Bakker-Grunewald 1983) so that an interference of binding with these proteins is ruled out on hepatocytes. We have studied the uptake of (^{3}H)-bumetanide into isolated rat hepatocytes (Petzinger et al. 1988a) and identified two saturable pathways plus an additional diffusion process (Fig. 2). Their kinetic constant analysis revealed a sodium - dependent high affinity but low capacity transport system, and a sodium - independent low affinity but high capacity transport system. In addition a release of hydroxylated and conjugated bumetanide metabolites out of hepatocytes was observed. These metabolites were not taken up again into the cells. Other loop diuretics (furosemide, piretanide, torasemide) were also transported into hepatocytes via the same transport system since all diuretics inhibited bumetanide uptake to an identical extent, as did the competitive inhibitor furosemide (Petzinger et al. 1988a) and since (^{14}C)-furosemide uptake was also inhibited by bumetanide.

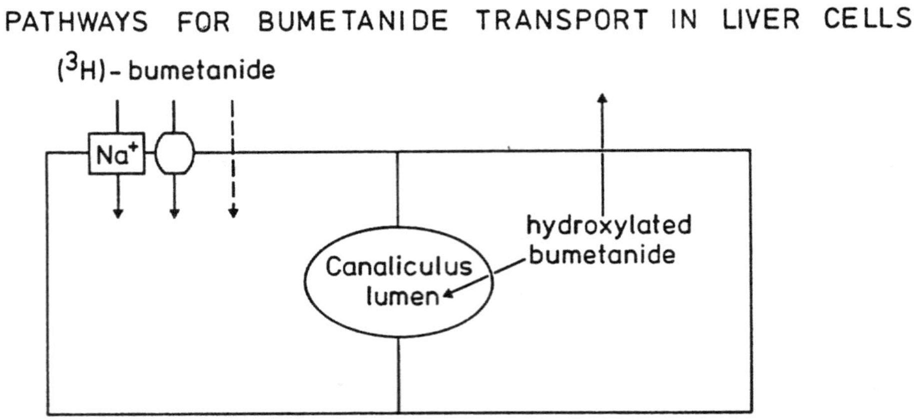

Uptake:	K_m [μM]	V_{max} [pmol/mg × min^{-1}]
Na$^+$ dependent :	21 ± 7	73 ± 25
Na$^+$ independent :	370 ± 109	1006 ± 139
diffusion :	no saturation	P = 1.16 × 10^{-6} cm/sec

FIG.2 Transport systems for the uptake of bumetanide into isolated rat hepatocytes. Metabolites of bumetanide are released by as yet undefined pathways.

Bumetanide has been reported to be a competitive inhibitor of taurocholic acid uptake into isolated rat hepatocytes (Blitzer et al. 1982). Since bile acids are taken up into hepatocytes by both sodium-dependent as well as sodium-independent pathways (Anwer and Hegner, 1978) the effect by taurocholate on bumetanide uptake was studied in the presence and absence of sodium ions. These results demonstrated that taurocholate at 44 μM inhibits only the sodium-dependent uptake of (^{3}H)-bumetanide and that this inhibition is competitive (Fig.6).

The K_i value for taurocholate in these experiments was 24 µM whereas the K_i of bumetanide in the study of Blitzer et al. was 26 µM. In the presence of 7 µM taurocholic acid the uptake of bumetanide was, however stimulated (Petzinger et al. 1988a) indicating a trans-stimulation via a common carrier pathway. So far, the kinetic data supports the assumption of hepatocellular uptake of loop diuretics by the sodium-dependent multispecific bile - salt transport system. This was further supported by the finding that AS-30D ascites hepatoma cells, which lack the multispecific bile-salt transport system (Kroker et al. 1978; Petzinger and Frimmer 1980), are unable to take up (^{3}H)-bumetanide (Petzinger et al. 1988a).

The Biochemical Approach

Affinity labelling of bumetanide carrier proteins in the liver

Bumetanide has been used as a photosensitive label of the Na^{+}- K^{+}-Cl^{-} cotransport system (Amsler and Kinne 1986). Consequently, intact isolated rat hepatocytes, AS 30D hepatoma cells, or basolateral membrane preparations from rat liver were incubated in the presence of 100 µCi (^{3}H)-bumetanide (Petzinger et al. 1987; Petzinger et al. 1988b). Following activation by photoflashes bumetanide covalently binds to proteins from hepatocytes or from liver plasma membranes but not to proteins from transport deficient AS-30D ascites hepatoma cells (Fig.3). When intact hepatocytes were labelled proteins of apparent molecular weight of 52-54, 48, (42), 33, 27, 25, and 23 kDA, whereas in plasma membranes the dominant band was a 52 and a 33 kDa protein. Labelling of these proteins was sodium-dependent in intact hepatocytes but not in plasma membranes. This might indicate that transport through the membrane was necessary for labelling of hepatocyte's proteins and that the position of the label in the most intensively labelled 52 kDa protein is not at the surface side of this protein but either in the membrane spanning or cytoplasmic side.

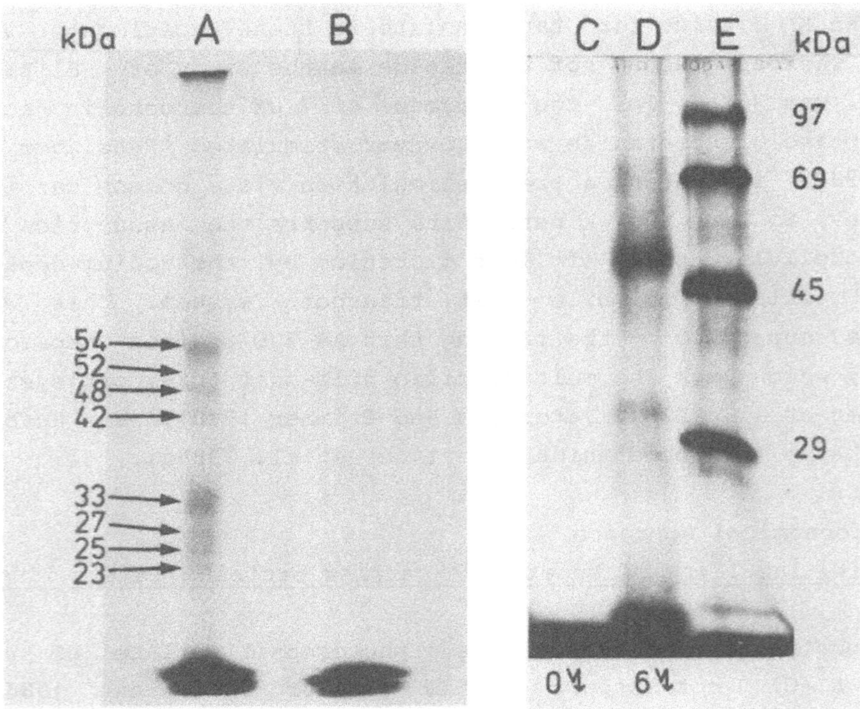

FIG.3 Photoaffinity labelling of carrier proteins for bumetanide in isolated rat hepatocytes, hepatoma cells and liver plasma membranes.
Autoradiography of the radiolabelled proteins from: (A) isolated hepatocytes, (B) AS-30D hepatoma cells, (C) basolateral liver plasma membranes plus (^{3}H)-bumetanide without photoactivation (zero flash) and (D) after photoactivation (six flashes)(according to Petzinger et al. 1988b).

If the position is at the cytoplasmic side in the 52 kDa protein this might explain why purified plasma membranes were labelled even in the absence of sodium ions. The membrane preparation contains not only right side out but also inside out vesicles so that the cytoplasmic side of a carrier protein is exposed at least in a part of the vesicle population.

Finally, we attempted to isolate bumetanide carrier proteins from basolateral liver plasma membranes by affinity chromatography. On a bumetanide affinity column proteins with molecular weight of 93, 66, 52 and 50 kDa could be eluted by ligand competition. We subsequently raised polyclonal antibodies against the 52 kDa protein which was labelled by(^{3}H)-bumetanide. In Western blots the antibodies recognized a 50-52

Fig.4 Affinity purification of putative bumetanide carrier proteins from rat liver plasma membranes on an amino-bumetanide affinity column:
Rat liver plasma membranes were prepared according to Blitzer and Donovan 1984, solubilized in octylglycoside and applied onto the column in a buffer with 20 mM HEPES, 1.2 mM Mg $(NO_3)_2$, pH 7.6 adjusted with NaOH. The elution of bound proteins was carried out with buffers containing either bumetanide or dehydrocholic acid as competing compounds. The protein pattern of an elution fraction is given in (B). (A) is the protein pattern of the applied membrane homogenate. (C) are marker proteins.

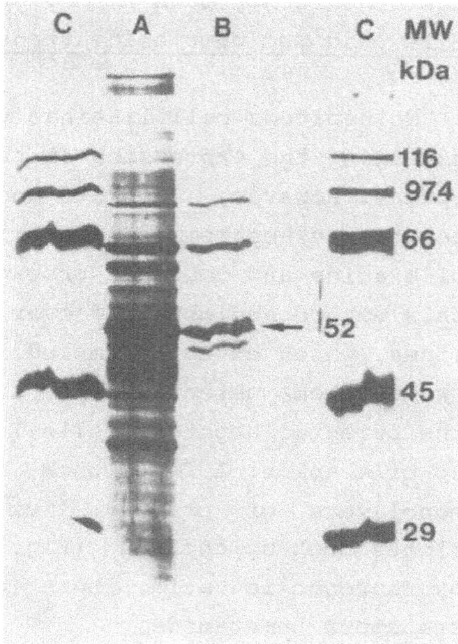

kDa protein from rat liver cells and a 52 kDa protein from kidney cortex basolateral plasma membranes (Schreiber and Petzinger unpublished). Whether this renal 52 kDa protein belongs to the PAH-transport system is yet unproven. Final conclusion on the identity of both proteins presumably involved in bumetanide transport affords their primary amino acid sequence which is unknown. With immunological tests it should, however, be possible to characterize binding epitopes and to determine the degree of antibody crossreactivity. Recent studies with monoclonal antibodies against bile-salt carrier proteins have indicated that a 49 kDa protein is responsible for hepatic sodium-dependent bile-salt uptake while a 54 kDa protein is responsible for non-sodium-dependent bile acid uptake (Ananthanarayanan et al. 1988). The bumetanide carrier proteins are in terms of their molecular weight in between these bile-salt transport proteins but this may also depend on the comparability of molecular weight determination in SDS-PAGE gels.

The Genetic Approach

Bile acid and bumetanide transport in hepatocytoma cells

No hepatoma cell line has yet been found that is capable of maintaing the expression of the carrier proteins for bile acid uptake. However, Polokoff and Everson 1986 have established hepatocyte/hepatoma cell hybrids which were able to synthesize bile acids and could be propagated as a cell line. We adopted this method and obtained over 50 clones of "hepatocytoma" cell lines, which often resembled in their morphological appearance the hepatoma parental cell line. These cells were similar to the parental hepatoma cells in that they were unable to take up bile acids. A few clones, however, resembled the epithelial monolayers of primary liver cell cultures. Three of these clones took up cholate (Fig.5) and this uptake was inhibited by taurocholic acid. These clones, however, were unable to transport bumetanide.

FIG.5 Uptake of 14C-cholate into immortalized hepatocytes obtained by polyethyleneglycol-induced fusion of rat hepatocytes with reuber hepatoma cells. The "hepatocytoma" clones 1B-2F9/2E9 and 1E3 take up cholate but not bumetanide.

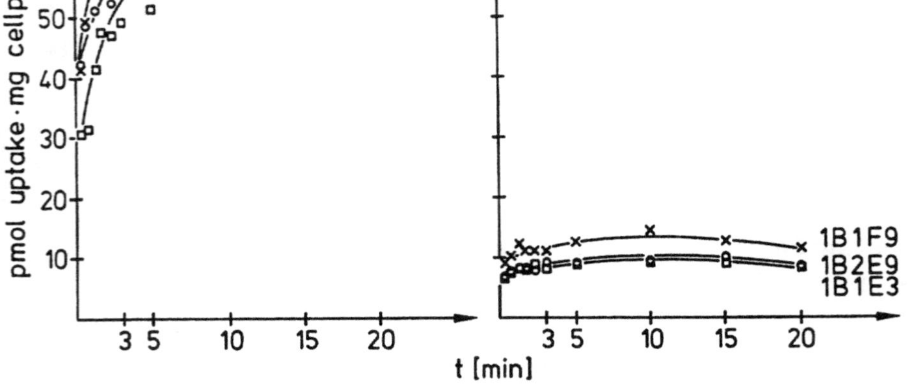

This, to our knowledge, is the first indication of a genetic variance of carrier functions in the multispecific bile acid transport system. It is premature to speculate on the kind of variation which has evoked this unexpected diversity between bile-salt and loop diuretic transport in hepatocytoma cells. However it indicates that for an appropriate analysis of multispecific transport systems different experimental approaches reveal identities as well as diversities of such complex transport functions (Fig.6). Hitherto, the amino acid composition of none of the proteins of either the hepatic or renal multispecific organic anion transport system has to date been elucidated. However, from studies, such as those which are presented here, it can be concluded that diversities in the organoselectivity of organic anion excretion is presumably due to a microheterogeneity in the primary structure of carrier proteins which are present in both kidney and liver.

IDENTITIES → DIVERSITIES

LEVEL	Compared properties	Uptake into hepatocytes of Bumetanide	vs. Taurocholic acid
KINETICS	Na^+	Yes	Yes
	Comp. inhibition	K_i = 24 /uM	K_i = 26 /uM
	K_m/V_{max}	K_m = 21 /uM	K_m = 5-57 /uM
		V_{max} = 0.07-0.1	V_{max} = 0.9-1.5
CARRIER PROTEIN	labeled protein (kDa)	52-53/48/33/25	52-54/48-50
	Binding inhibition	no	?
	Protein purification (kDa)	50/52	48/54
GENETICS	Hepatoma (AS-30 D)	absent	absent
	Hepatocytehybrids	absent	present (Cholate)

FIG.6 Identities and diversities of bumetanide uptake versus bile acid uptake in the liver

Multiple Transport System for the Uptake of Organic Anions in the Liver and in the Kidney

As has already been mentioned the bile-salt transport system is not the only one available for the transport of xenobiotics into hepatocytes. Since this system transports neutral compounds, in addition to anionic ones, its range of substrates is even broader than the range of xenobiotics which are transported by the classic hepatic organic anion transport system, the bilirubin/bromosulfophthalein system. Carrier proteins which have an apparent molecular weight of 55 kDa from this system have been already purified (Stremmel et al. 1983; Wolkoff and Chung 1980), although were immunologically distinct from those of the bile acid carrier proteins (Stremmel and Berk 1985; Wolkoff et al. 1985). Recently Hugentobler and Meier (1986) have found a sulfate/hydroxyl exchange system which could also transport not only dicarboxylic acids, (such as oxalate and succinate) but also cholic acid, although it was unable to transport taurocholic acid (Hugentobler et al. 1987).

A very similar situation for the transport of organic anions exists in the kidney and a comparison indicates a broad homology (Fig.7).

As reviewed by Ullrich and Rumrich 1988 three contraluminal pathways are involved in the secretion of organic anions in the kidney, the (a) monocarboxylate transport system, which is synonymous to the para-aminohippurate transport system, (b) a dicarboxylate transport system and finally, (c) a sulfate/ bicarbonate exchange system which also transports oxalate. The systems have, as has been demonstrated in the liver, a broad overlapping substrate specificity, the most non-specific of which is the PAH transport system. Indeed, in the literature review reported by Möller and Sheikh (1983) 212 substrates for this latter system were listed. These authors found that it is very likely that the PAH system contains iso-carriers for

FIG.7 Carrier-mediated uptake or organic anions in kidney and liver.

Uptake of organic anions is mediated by three transport systems with overlapping substrate specificity. The systems are a monocarboxylate transporter for the uptake of para-aminohippurate, or of bile acids respectively, and a dicarboxylate transporter for the uptake of succinate or of bilirubin. These systems also transport various xenobiotics. In addition, a sulfate/oxalate exchange system is present which has affinity for only a few organic anions (see Ullrich and Rumrich 1988).

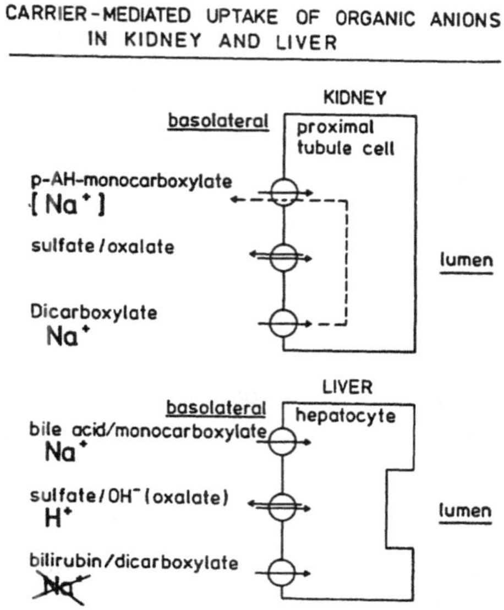

certain subgroups, such as the hippurates, urate, phenolsulfalein dyes, and possibly for other xenobiotics. This concept is substantiated by the present findings for closely related bumetanide carrier proteins in the multispecifc bile acid transport-system and in the renal para-aminohippuric acid transport system. Thus it is concluded as a working hypothesis, that the monocarboxylate transport systems in kidney and liver represent equivalent systems with similar but not identical protein entities.

Conclusions

The long lasting discussion regarding the identity or diversity between the multispecific organic anion transport systems of the liver and kidney has to date been focused on the hepatic bilirubin/BSP transport system and the renal PAH transport system. The PAH transport system, however, transports primarily monocarboxylates, whereas the bilirubin/BSP transport system is primarily a transport system for either dicarboxylates (bilirubin) or disulfonates (BSP). Recent results, obtained for the multispecificity of the bile acid transport system have shown that it is this transport system which should be compared with the renal PAH transport system. In order that correct conclusions regarding possible identical or diverse transport functions in the liver and the kidney can be made, the distinction between mono- and dicarboxylate transport systems is thus essential.

Acknowledgements

This work is dedicated to Professor M. Frimmer the former Director of the Institute of Pharmacology and Toxicology, University Gießen, FRG, on the occasion of his retirement. Professor Frimmer has been the scientific mentor of the author during the past 15 years. The author is also indepted to Professor R.K.H. Kinne, Director of the Max-Planck-Institut für Systemphysiologie, Dortmund, FRG, where he spent with a Heisenberg grant the last two years in a most stimulating scientific atmosphere. The gift of bumetanide by Dr. Feit, Leo Pharmaceuticals, Ballerup, Denmark, and the synthesis of radiolabelled bumetanide by Dr. Deutscher, Max-Planck-Institut für Systemphysiologie, Dortmund, FRG, is thankfully acknowledged. This work was supported by the German Research Foundation, grant PE 250/5-1.

References

Alpert S, Mosher M, Shanske A, Arias IM (1969) Multiplicity of hepatic excretory mechanisms for organic anions. J Gen Physiol 53:238-247

Amsler K, Kinne R (1986) Photoinactivation of sodium-potassium-chloride cotransport in LLC-PK_1/Cl_4 cells by bumetanide. Am J Physiol 250:C799-C806

Ananthanarayanan M, von Dippe P, Levy D (1988) Identification of the hepatocyte Na^+-dependent bile acid transport protein using monoclonal antibodies. J Biol Chem 263:8338-8343

Anwer MS, Hegner D (1978) Effect of Na^+ on bile acid uptake by isolated rat hepatocytes. Evidence for a heterogenous system. Hoppe Seyler's Z Physiol Chem 359:181-192

Bakker-Grunewald T (1983) Potassium permeability and volume control in isolated rat hepatocytes. Biochim Biophys Acta 731:239-242

Bárány EH (1971) Characterization of simple and composite uptake systems in cells and tissues by competition experiments. Acta Physiol Scand 83:220-234

Bárány EH (1972) Inhibition by hippurate and probenecid of in vitro uptake of iodipamide and o-iodohippurate. A composite uptake system for iodipamide in choroid plexus, kidney cortex and anterior uvea of several species. Acta Physiol Scand 86:12-27

Bárány EH (1973a) The liver-like anion transport system in rabbit kidney, uvea and choroid plexus. I. Selectivity of some inhibitors, direction of transport, possible physiological substrates. Acta Physiol Scand 88:412-429

Bárány EH (1973b) The liver-like anion transport system in rabbit kidney, uvea and choroid plexus. II. Efficiency of acidic drugs and other anions as inhibitors. Acta Physiol Scand 88:491-504

Bárány EH (1974) Bile acids as inhibitors of the liver-like anion transport system in the rabbit kidney, uvea and choroid plexus. Acta Physiol Scand 92:195-203

Bárány EH (1975) In vitro uptake of bile acids by choroid plexus, kidney cortex and anterior uvea. I. The iodipamide-sensitive transport systems in the rabbit. Acta Physiol Scand 93:250-268

Blitzer BL, Donovan CB (1984) A new method for the rapid isolation of basolateral membrane vesicles from rat liver. J Biol Chem 259:9295-9301

Blitzer BL, Ratoosh SL, Donovan CB, Boyer JL (1982) Effects of inhibitors of Na^+-coupled ion transport on bile acid uptake by isolated rat hepatocytes. Am J Physiol 243:G48-G53

Burg ML, Stouer J, Cardinal J, Green N (1973) Furosemide effects on isolated perfused tubules. Am J Physiol 225: 119-124

Despopoulos A (1966) Congruence of excretory functions in liver and kidney: hippurates. Am J Physiol 210:760-764

Despopoulos A (1968) Renal and hepatic transport of food dyes. J Pharmacol Exp Ther 163:222-228

Despopoulos A (1971) Congruence of renal and hepatic excretory functions: sulfuric acid dyes. Am J Physiol 220:1755-1758

Frimmer M, Ziegler K (1988) The transport of bile acids in liver cells. Biochim Biophys Acta 947:75-99

Geck P, Pietrzyk G, Burckhardt BC, Pfeiffer B, Heinz E (1980) Electrically silent cotransport of Na^+, K^+ and Cl^- in Ehrlich cells. Biochim Biophys Acta 600:532-447
Goldinger JM, Kahlsa BDS, Hong SK (1984) Photoaffinity labeling of organic anion transport system in proximal tubule. Am J Physiol 247:C217-C227
Hugentobler G, Fricker G, Boyer JL, Meier PJ (1987) Anion transport in basolateral (sinusoidal) liver plasma-membrane vesicles of the little skate (Raja erinacea). Biochem J 247:589-595
Hugentobler G, Meier PJ (1986) Multispecific anion exchange in basolateral (sinusoidal) rat liver plasma membrane vesicles. Am J Physiol 251:G656-G664
Jansen PLM, Peters WH, Lamers WH (1985) Hereditary chronic conjugated hyperbilirubinemia in mutant rats caused by defective hepatic anion transport. Hepatology 5:573-579
Kim JH, Hong SK (1962) Urinary and biliary excretion of various phenol red derivatives in the anaesthetised dog. Am J Physiol 202:174-178
Koschier FJ, Stokols MF, Goldinger JM, Acara M, Hong SK (1980) Effect of DIDS on renal tubular transport. Am J Physiol 238:F99-F106
Kramer W, Burckhardt G (1988) Identification of binding proteins for bile salts, beta-lactam antibiotics and PAH in rat renal basolateral membranes by photoaffinity labeling. Pflügers Arch 411:R103
Kroker R, Anwer MS, Hegner D (1978) Lack of bile acid transport in AS-30D ascites hepatoma cells. Naunyn Schmiedeberg's Arch Pharmacol 303:299-701
Moller JV, Sheikh MI (1983) Renal organic anion transport system: pharmacological, physiological and biochemical aspects. Pharmacol Rev 34:315-358
Odlind B (1979) Relationship between tubular secretion of furosemide and its saluretic effect. Clin Pharmacol Exp Ther 208:515-521
Odlind B, Beermann B (1980) Renal tubular secretion and effects of furosemide. Clin Pharmacol Exp Ther 27:784-790
Orita Y, Ando A, Krakabe S, Abe H (1976) A metal complexing property of furosemide and bumetanide. Determination of pK and stability constant. Arzneim Forsch (Drug Res) 26:11-13
Ostergaard EH, Magnussen MP, Nielsen CK, Eilersten E, Frey HH (1972) Pharmacological properties of 3-n-butylamino-4-phenoxy-5-sulfamoylbenzoic acid (bumetanide), a new potent diuretic. Arzneim Forsch (Drug Res) 22:66-72
Patarca R, Candia OA, Reinach PS (1983) Mode of inhibition of active chloride transport in the frog cornea by furosemide. Am J Physiol 245:F660-F669
Petzinger E, Frimmer M (1980) Comparative studies on the uptake of ^{14}C-bile acids and ^{3}H-demethylphalloin in isolated rat liver cells. Arch Toxicol 44:127-135
Petzinger E, Grundmann E, Veil LB, Frimmer M, Fasold H (1978) Inhibitory effects of 4,4'-diisothio-cyano-stilbene-2,2'disulfuric acid (DIDS) in the response of isolated hepatocytes to phalloidin. Naunyn Schmiedeberg's Arch Pharmacol 304:303-307

Petzinger E, Müller N, Föllmann W, Deutscher J, Kinne RKH (1988a) Uptake of bumetanide into isolated rat hepatocytes and primary liver cell cultures. Am J Physiol, in press

Petzinger E, Müller N, Kinne RKH (1987) Labelling of membrane proteins on isolated rat hepatocytes and liver cell plasma membranes by photoactivated ^{3}H-bumetanide. Biol Chem Hoppe Seyler 368:1266

Petzinger E, Müller N, Zierold K, Kinne RKH (1988b) Photoaffinity labelling by ^{3}H-bumetanide of putative carrier proteins for loop diuretics on isolated rat hepatocytes in the absence of a NaCl/KCl cotransport system. Am J Physiol, submitted

Petzinger E, Ziegler K, Frimmer M (1987) Occurrence of a multispecific transporter for the hepatocellular accumulation of bile acids and various cyclopeptides. In: Paumgartner G, Stiehl A, Gerok W (eds) Bile acids and the liver. MTP-Press, Lancaster, p 111-124

Polokoff MA, Everson GT (1986) Hepatocyte-hepatoma cell hybrids: characterization and demonstration of bile acid synthesis. J Biol Chem 261:4085-4089

Scharschmidt BF, Waggoner JG, Berk PD (1975) Hepatic organic anion uptake in the rat. J Clin Invest 56:1280-1292

Sperber I (1954) Secretion of organic anions in the formation of urine and bile. Pharmacol Rev 11:109-134

Stremmel W, Berber MA, Glezerov V, Thung SN (1983) Physicochemical and immunological studies of a BSP and bilirubin binding protein from rat liver plasma membrane. J Clin Invest 71:1796-1805

Stremmel W, Berk PD (1985) Hepatocellular uptake of sulfobromophthalein and bilirubin is selectively inhibited by an antibody to the liver plasma membrane BSP/bilirubin binding protein. Hepatology 5:1042

Ullrich KJ, Rumrich G (1988) Contraluminal transport systems in the proximal renal tubule involved in secretion of organic anions. Am J Physiol 254:F453-F462

Wolkoff AW, Chung CT (1980) Identification, purification and partial characterization of an organic anion binding from rat liver cell plasma membranes. J Clin Invest 65:1152-1161

Wolkoff AW, Sisiak A, Greenblatt HC, Renswoud JV, Stockert RJ (1985) Immunological studies of an organic anion binding protein from rat liver plasma membranes. J Clin Invest 76:454-459

Zeuthen T, Ramos M, Ellovy JC (1978) Inhibition of active chloride transport by piretanide. Nature 273:678-680

Mechanisms for Hepato-Biliary Transport of Cationic Drugs Studied with the Intact Organ and on the Membrane Level

D.K.F. Meijer, W. Mol, M. Müller*, G. Kurz*
Dept. Pharmacology and Therapeutics
University Center of Pharmacy
Groningen, Netherlands
*Institut für Organische Chemie und Biochemie
Universität Freiburg, FRG

Introduction

About seventy percent of the commonly used drugs belong to the category of (potential) organic cations mainly being quaternary, tertiary, or secondary amines. Quaternary ammonium groups with the nitrogen center linked to four carbon atoms are permanently charged at physiological pH, whereas the tertiary amines become positively charged by acceptance of a proton depending on pH and pK value of the basic group. Cationic drugs can pass membranes by carrier-mediated transport and such processes have been characterized in the kidney tubular cells (Schanker, 1972; Somogyi, 1987; Weiner, 1985) explaining secretion of basic drugs from blood into the urine, in intestine as related to secretion of organic cations from blood into the intestinal lumen (Lauterbach, 1984), in the choroid plexus as a mechanism to remove cationic compounds from the liquor to blood (Schanker, 1972) as well as in the liver mediating hepato-biliary excretion of cationic drugs (Klaassen and Watkins, 1984; Meijer, 1977; 1987; Schanker, 1972).

Hepatic Transport in Organic Substances
Ed. by E. Petzinger, R. K.-H. Kinne, H. Sies

Transport phenomena

The relative contribution of above mentioned organs in the elimination of an organic cation is dependent on the lipophilicity/hydrophilicity balance of the particular molecule. Structure-kinetic relationship has been studied with series of monovalent (Neef and Meijer, 1984; Petzinger and Frimmer, 1984) and bivalent organic cations (Meijer, 1977) in our group and by others (Hirom et al., 1974; Hughes et al., 1973a; 1973b; Wassermann, 1971). Very likely, a two-step carrier-mediated transport is involved: uptake into the cells and excretion from the cells. At both levels saturation and competitive phenomena have been described (Eaton and Klaassen, 1978; Hwang and Schanker, 1973; Meijer et al., 1970; Nakae et al., 1976; 1980; 1981; Schanker, 1972; Solomon and Schanker, 1963). The hepatic transport of monovalent organic cations has traditionally been characterized using the model compound procainamideethobromide (PAEB) or its ^{4}N-acetylated congener (Eaton and Klaassen, 1978b; Hwang and Schanker, 1973; Meijer et al., 1970; 1972; Nakae et al., 1976; 1980; 1981; Schanker, 1972) and more recently with some aliphatic methyl-ammonium derivatives (Neef et al., 1984a; 1984b). Endogenous substrates for this transport system may include choline, and thiamine (Yoshioka, 1984) as well as nicotinamide riboside (Macgregor and Burkhalter, 1973). Although transport of organic cations is sensitive to temperature changes, anoxia, and metabolic inhibitors (Neef et al., 1984a; 1984b), the actual energization of the process remains poorly defined. Secondary active transport through proton-antiport has been described in renal tubular brush border membranes (Hsyu and Giacomini, 1987). Multivalent organic cations only seem to undergo significant hepato-biliary transport, if the presence of the cationic groups is masked by bulky ring structures (Hughes et al., 1973a; Meijer, 1977; 1987). Uptake into hepatocytes was suggested to be mediated by a multispecific transport system (Buscher et al., 1986; Frimmer, 1982; Wieland et al., 1984) differing from that for PAEB and recognizing amphipathic compounds such as bile acids (Buscher et al., 1986; 1987;

Frimmer, 1982; Petzinger et al., 1986; Westra et al., 1981; Wieland et al., 1984) and cardiac glycosides (Eaton and Klaassen, 1978a; Meijer, 1977; 1987) among others. At the canalicular pole of the cell the bulky organic cations and the PAEB-like agents may share one carrier-mediated secretion process as evidenced by mutual interactions at that level (Cohen et al., 1967; Meijer et al., 1970). In some cases net canalicular transport can be promoted by bile salts through choleretic effects, ion pair formation, or binding to biliary micelles (Meijer, 1987; Neef et al., 1984b), but this is certainly not a general rule, especially if uptake into the cells is rate limiting in the hepato-biliary transport (Vonk et al., 1978).

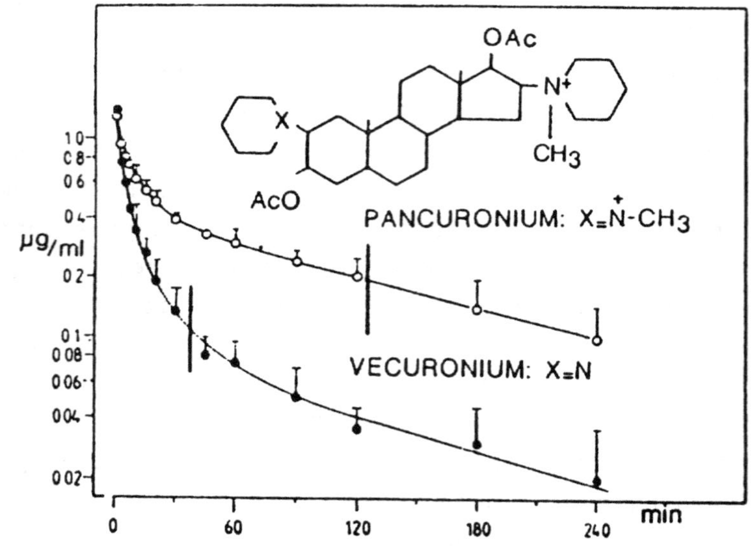

Figure 1: Plasma concentrations of vecuronium and pancuronium after 0.1 mg/kg intravenously, in man. Vertical bars bisecting the plasma curves indicate plasma concentrations at the timepoint of 50% recovery of neuromuscular block. Modified from Sohn et al. (1986).

Hepatic transport of steroidal cations in the intact organ, pharmacokinetic analysis

The liver is capable of extensive accumulation and elimination of drugs including organic cations, and may thus contribute to a short duration of action of certain muscle relaxants (Agoston et al., 1973; 1977; Bencini et al., 1986; Sohn et al., 1986; Waser et al., 1987). We, therefore, decided to study the relation between chemical structure and hepatic disposition of this type of agents in more detail.

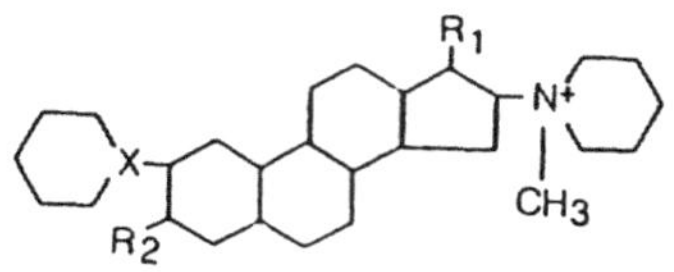

Figure 2: *Structural formulas and physicochemical data of the steroidal muscle relaxants involved in this study.*

	X	R_1	R_2	MW_{cation}	Oct./Krebs part. coeff.
Vecuronium	N	$OCOCH_3$	$OCOCH_3$	557	2.56
3-Hydroxyvecuronium	N	$OCOCH_3$	OH	515	0.032
Org 6368	N^+-CH_3	H	$OCOCH_3$	514	0.0145
Pancuronium	N^+-CH_3	$OCOCH_3$	$OCOCH_3$	572	0.0033

The hepato-biliary transport of several structurally related, but physicochemically distinct steroidal muscle relaxants (Fig. 2) was first studied in isolated perfused rat liver. In this model the liver is the only organ of elimination, thus allowing adequate analysis of the transport phenomena under study. Marked differences were observed in the overall hepato-biliary transport of the four muscle relaxants under study. In brief, the most hydrophilic compound, pancuronium, showed the lowest net transport from perfusate to bile. Reversely, the most lipophilic relaxant, vecuronium (the monoquaternary analogue of pancuronium), showed the most efficient hepato-biliary transport. Vecuronium was very effectively taken up in the liver and within 2 hours more than 60% of the dose was excreted in bile, partly as its 3-hydroxy metabolite. This metabolite also appeared in the perfusate, but only in very low concentrations. The bisquaternary relaxant Org 6368 and 3-hydroxyvecuronium showed an intermediate behaviour with effective uptake in the liver, but only modest excretion in bile. Similar to pancuronium, the amount in the liver versus time curves indicated that hepatic storage in a deep compartment might be involved in the hepatic disposition of steroidal muscle relaxants.

Cardiac glycosides are potent inhibitors of the hepato-biliary transport of bivalent organic cations, whereas they do not affect the transport of monovalent cations (Meijer et al., 1970; 1971a). The effect of cardiac glycosides on the hepatic transport might indicate whether the transport of the muscle relaxants fully depends on the transport system for bivalent organic cations. It appeared that the transport of the three hydrophilic compounds was strongly reduced by K-strophanthoside, whereas the uptake of the more lipophilic vecuronium was only partly affected at the dose ratio used. The rather absolute effect of K-strophanthoside on 3-hydroxyvecuronium might indicate that uptake via the carrier system for monovalent organic cations (Eaton and Klaassen, 1978; Hwang and Schanker, 1973; Schanker, 1972) is not involved. However, this alternative pathway cannot be fully excluded. Similar to vecuronium, a modest effect of K-strophanthoside (Meijer et al., 1971a) was also observed in studies with the rather lipophilic bivalent cation hexafluronium (Meijer et al., 1971b). This may imply that passive diffusion is partly involved in the hepatic uptake of these lipophilic muscle relaxants, but alternatively that the cardiac glycoside fails to compete successfully for the supposed common carrier site.

The data of the perfusion experiments were analyzed with the DIFFIT computer program. This program enabled simultaneous fitting of independently measured plasma disappearance and biliary excretion rate curves by a numerical approach. The kinetic analysis revealed that the hepato-biliary transport of the steroidal muscle relaxants was best described by a three-compartment model with elimination from the peripheral compartment V_2 and storage in a deep compartment V_3, connected to V_2.

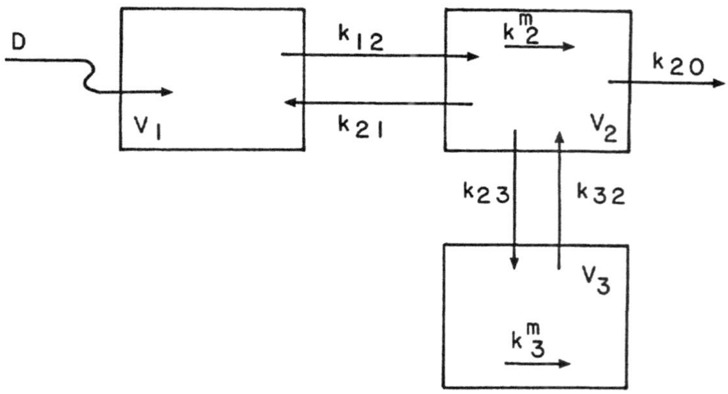

Figure 3
Compartment model for the hepato-biliary transport of steroidal muscle relaxants

Rather than molecular weight (Hughes et al., 1973a), the lipophilicity of the muscle relaxants influenced the kinetic parameters of the hepato-biliary transport. The biliary clearance (Cl_{20}) and the initial hepatic uptake (Cl_{12}) showed a positive relationship with the lipophilicity, confirming that hepato-biliary transport of these organic cations is highly dependent on the hydrophobic character of the compounds (Hirom et al., 1974; Meijer et al., 1972; Petzinger and Frimmer, 1984). In addition, net hepatic uptake of the muscle relaxants (rate of distribution to the liver) is markedly influenced by the liver to plasma transport (hepatic efflux). The tendency for hepatic efflux, expressed as the k_{21} value, is increasing with decreasing lipophilicity. This phenomenon can be envisioned, if the dissociation of the relatively lipophilic compounds from the carrier sites at the outer side of the membrane would be the rate limiting step in the hepato-biliary efflux process.

	$P_{Oct/Krebs}$	V_2 (ml)	Cl_{12} (ml/min)	Cl_{20} (ml/min)	k_{12} (min^{-1})	k_{21} (min^{-1})	k_{23}/k_{32}	k_{20} (min^{-1})	cytosol/ perfusate conc. ratio	bile/ cytosol conc. ratio
Vecuronium	2.56	976	20.2	23.2	0.212	1.35×10^{-7}	6.9	0.0238	15.8	27.7
3-Hydroxyvecuronium	0.032	576	7.10	4.24	0.0716	0.00540	11.6	0.00736	N.D.	N.D.
Org 6368	0.0145	104	7.31	3.31	0.0783	0.0503	42.0	0.0318	1.57	31.2
Pancuronium	0.0033	14.7	1.40	0.138	0.0140	0.0895	4.3×10^{9}	0.00940	0.52	5.7

Table 1

In analogy with the classical muscle relaxant d-tubocurarine (Blom et al., 1982; Weitering, 1975; 1977) the kinetic analysis revealed an intracellular deep compartment. The affinity of the muscle relaxants for this intracellular distribution compartment, expressed as the k_{23}/k_{32} ratio, is inversely related with lipophilicity. However, it should be realized that the absolute amount of drug that can accumulate in the deep compartment is also determined by the fraction of the dose that has penetrated into the hepatocyte. The size of this fraction is determined by the k_{12}/k_{21} ratio, a factor that increases with increasing lipophilicity. Although the combined data indicated that lipophilicity has a major influence on several stages of the hepato-biliary transport of these organic cations, it is important to note that the net uptake of the muscle relaxants (indicated by the k_{12}/k_{21} ratio) is of special importance, inasmuch as the rate of the various subsequent transport steps is dependent on the concentration of the drug in the hepatocyte cytoplasm as the driving force for these processes.

Subcellular distribution

To investigate the differences in hepatic accumulation and hepato-biliary transport between vecuronium, pancuronium, and Org 6368 on a subcellular level, studies were performed to determine the subcellular distribution (De Duve et al., 1955) and hepato-biliary concentration gradients (Meijer et al., 1972) of these steroidal muscle relaxants (Mol and Meijer, 1988). The data from the distribution studies suggested that the marked differences in hepato-biliary transport are not caused by qualitative differences in the association with various intracellular organelles, since the three muscle relaxants exhibited essentially the same distribution pattern within the particulate fraction of liver homogenate at an equal liver load. The steroidal muscle relaxants were predominantly accumulated in the mitochondrial fraction. Accumulation in mitochondria has been earlier demonstrated for various other mono- and bivalent cations (Bakeeva et al., 1970; Gallagher, 1968; Ramsay et al., 1986) and might be explained by passive equilibration according to the mitochondrial membrane potential (Bakeeva et al., 1970). Alternatively, active carrier-mediated uptake of organic cations in mitochondria might be involved (Ramsay et al., 1986). In this context a potential interaction with the endogenous substrate NAD^+ has been inferred (Graf and Petersen, 1978; Ramsay et al., 1986). In comparison with the classical muscle relaxants d-tubocurarine and metocurarine (Meijer et al., 1976; 1979), for which subfractionation studies (Meijer et al., 1972; Weitering et al., 1975; 1977) and electron microscopy (Weitering et al., 1975) of d-tubocurarine molybdate precipitates indicated predominant intracellular association with lysosomes (Echigoya et al., 1972), accumulation in the lysosomal fraction seemed to be of less importance in the case of the steroidal muscle relaxants and decreased relatively to the association with the mitochondrial fraction, if the liver load was increased.

Instead of the relative distribution over the different intracellular organelles, the extent of binding to the potential storage compartments appears to be an important factor. The particle/cytosol concentration ratio of Org 6368 was significantly higher than the values for vecuronium and pancuronium and might explain the effective intracellular storage of Org 6368, as observed in the isolated perfused rat liver experiments (Westra et al., 1981) as well as in cat in vivo (Agoston et al., 1977). From the data of the kinetic analysis, the highest particulate/cytosol concentration ratio would be inferred for pancuronium, because this compound exhibits the highest tendency (k_{23}/k_{32}) for intracellular accumulation. It should be noted, however, that in contrast to vecuronium and Org 6368, particle/cytosol concentration ratios for pancuronium could not be determined under steady-state conditions in the liver, which may explain the unexpectedly low particle/cytosol ratio of pancuronium in relation to the k_{23}/k_{32} value from the kinetic analysis.

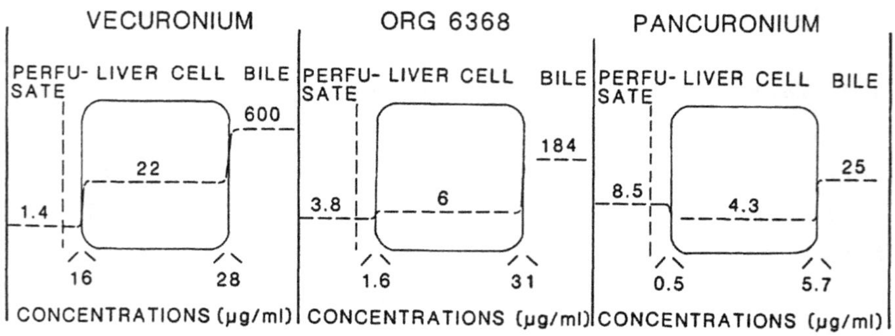

Figure 4: Hepato-biliary concentration gradients of vecuronium, Org 6368, and pancuronium in rat liver. The unbound concentrations in perfusate, hepatic cytosol, and bile are indicated together with the presumable concentration gradients across the sinusoidal and canalicular membranes (Mol and Meijer, 1988).

Determination of the unbound fractions of the muscle relaxants in Krebs-albumin solution, in cytosolic fraction of liver homogenate, and in bile, enabled a rough estimation of the concentration gradients across the sinusoidal and the canalicular membrane.

The data indicated that major differences exist in the net sinusoidal uptake of the three relaxants, which increased in the order pancuronium $<$ Org 6368 $<$ vecuronium. Yet, the cytosol/plasma concentration ratios of all the agents did not exceed the values that would be attained by passive equilibration according to the membrane potential. At the canalicular site "uphill" transport of the muscle relaxants into bile against an electrochemical gradient occurs. Thus, some kind of active transport is inferred, which appears to be more efficient for Org 6368 and vecuronium than for pancuronium. The nearly equal bile/cytosol concentration ratios for Org 6368 and vecuronium suggest that the differences in biliary excretion rate between these two substances are not caused by differences in membrane transport at the canalicular level, but rather by an unequal extent of binding within the liver cells. Since vecuronium and Org 6368 strongly differ in lipophilicity, these results suggest that lipophilicity is not the only factor determining the efficacy of canalicular transport of organic cations, which is in line with data of Neef et al. (1978). Integration of the kinetic and distribution data indicates that the extensive binding of Org 6368 within the cell is a major determinant in the relative efficient hepatic uptake as well as in the modest biliary excretion rate of this muscle relaxant. The limited hepato-biliary transport of pancuronium, however, appears to be due to inefficient net transport both at the sinusoidal and at the canalicular membrane.

Hepatic uptake of muscle relaxants: mechanisms of membrane transport

The mechanism of uptake of steroidal muscle relaxants has been investigated in isolated rat hepatocytes (Schwenk, 1980) using vecuronium as a model compound (Mol et al., 1988). Determination of initial uptake velocity at different vecuronium concentrations demonstrated that vecuronium uptake into the hepatocyte occurs by both a saturable (K_m = 15 μM, V_{max} = 181 pmol/min x 10^6 cells) and a nonsaturable (k = 1.10 pmol/ min x 10^6 cells/μM) process. This implies that in the in vivo situation mainly the saturable component is responsible for the hepatic uptake, since plasma concentrations in the rat usually do not exceed 5 μM.

The uptake of vecuronium satisfies the other criteria for carrier-mediated transport. The uptake is inhibited by structurally related compounds, is temperature-dependent and decreases by various metabolic inhibitors. The inhibitory effect of SH-reagents indicates that sulfhydryl groups may be located at the active site of the transport system.

With respect to potential driving forces, the uptake of vecuronium appears to be independent on the Na^+ concentration in the extracellular medium. On the contrary, the uptake of vecuronium exhibits anion dependency. Both cotransport of the cationic substrate with inorganic anions to maintain electroneutrality (Joppen et al., 1985; Petzinger et al., 1986) as well as carrier-mediated uptake of (electroneutral) ion pairs (Meijer, 1987; Neef et al., 1984b; Ruifrok, 1982) with vecuronium, formed at the fluid/lipid interface, might explain the observed anion dependency. In sucrose medium, that increases the negative transmembrane potential (Graf and Petersen, 1978), the uptake of vecuronium was decreased instead of increased, supporting the concept of electroneutral uptake of the muscle relaxant.

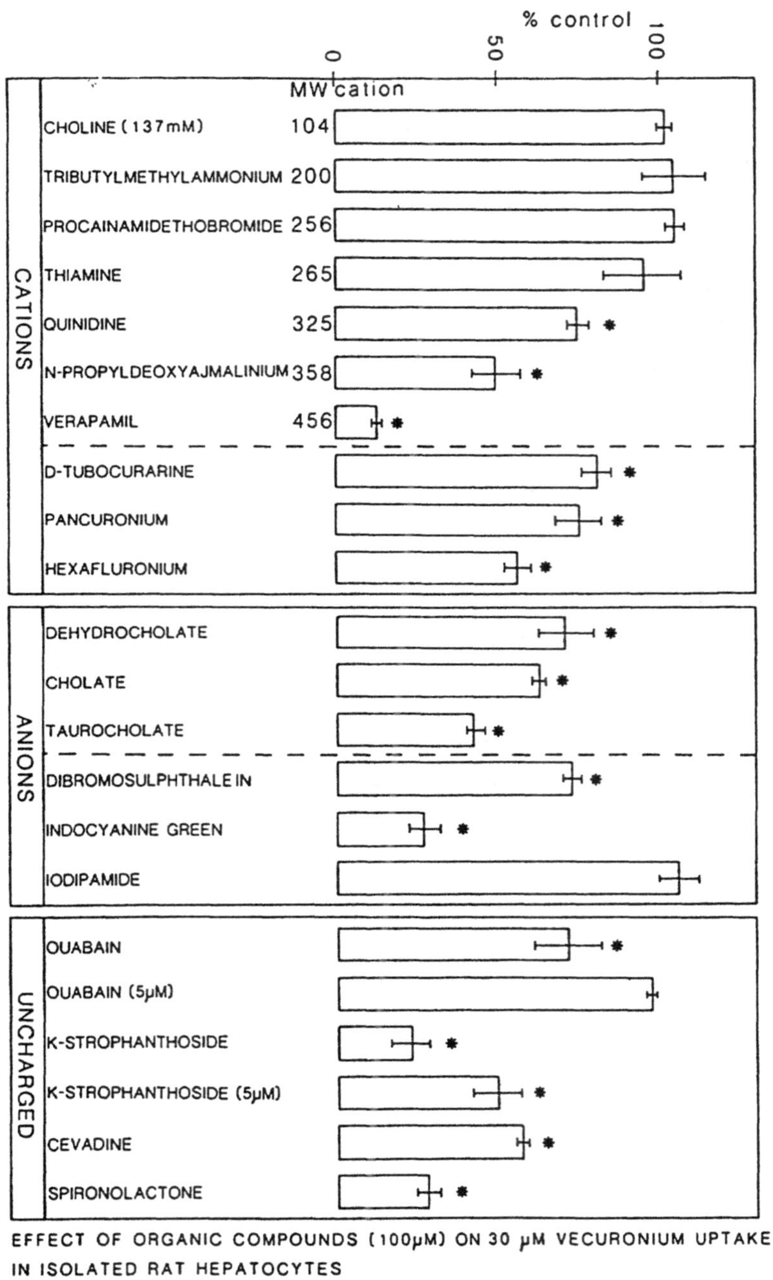

Figure 5: Effect of substrates of different charge on the initial uptake velocity of vecuronium. Vecuronium concentration was 30 μM. Substrate concentration was 100 μM, unless stated otherwise. Results are the mean and S.E.M. of three or four experiments.

The hepatic uptake of drugs and xenobiotics is generally thought to occur via three distinct pathways, depending on the charge of the compound. However, both kinetic and photoaffinity labeling studies provided evidence that the liver is equipped with transport systems with broad overlapping substrate specificity (Buscher et al.,1986; Frimmer, 1982; Wieland, 1984). In contrast to monovalent organic cations like procainamideethobromide (PAEB), uptake of bivalent organic cations in the liver is effectively inhibited by cardiac glycosides (Meijer et al., 1970; 1971a), a phenomenon that might be related to a common uptake for uncharged compounds and the bivalent organic cations studied. Therefore, the structural specificity of the carrier-mediated uptake system for vecuronium was studied in more detail.

The results from the interaction experiments with substances from different classes of model compounds indicated that the uptake system for vecuronium may be shared with bile salts, uncharged compounds, and relatively high molecular weight cations (Mol et al., 1988). A general transport system for "bulky" organic compounds with multiple ring structures, irrespective of charge, presents an attractive explanation for the observed interactions (Buscher et al., 1986; Frimmer, 1982; Meijer, 1987; Meijer and Groothuis, 1986). The detoxification function of the liver may benefit from such a system with broad substrate specificity. Furthermore, the lack of effect of low molecular weight cations on vecuronium uptake suggests that these cations are taken up by a separate uptake system, which is in line with previous studies from our laboratory (Meijer et al., 1970; 1971a).

Characterization of potential carrier proteins

Photoaffinity labeling with a photolabile derivative of vecuronium might be potentially helpful in the identification of binding polypeptides involved in vecuronium uptake. However, no photolabile derivative of vecuronium is available up to now. Yet of N-propyldeoxyajmalinium (NPDA), which was shown to inhibit the uptake of vecuronium in isolated hepatocytes, a photolabile derivative has been synthesized by one of us (M. Müller).

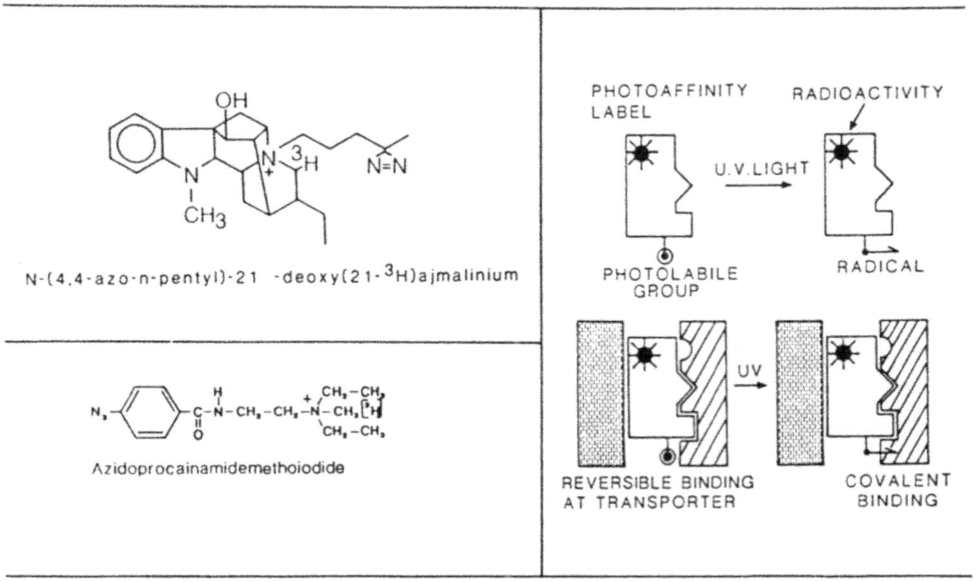

Figure 6: Principle of photoaffinity labeling and photoaffinity probes to label potential carrier proteins for organic cations (Frimmer and Ziegler, 1986).

Therefore, we studied the hepatic uptake of NPDA and vecuronium to determine whether the photolabile NPDA-derivative might serve as a photoprobe for vecuronium-like compounds. Physicochemical parameters such as albumin binding and lipophilicity (expressed as octanol/Krebs partition coefficient) of NPDA and vecuronium appeared to be highly comparable. Studies in isolated perfused rat liver demonstrated mutual uptake inhibition for both compounds. Furthermore, the hepatic uptake of the monovalent cation NPDA was inhibited by taurocholate and by K-strophanthoside, thus showing a striking parallel with vecuronium-like compounds (Westra et al., 1981).

Kinetic analysis of the uptake of NPDA in freshly isolated hepatocytes indicated that two saturable transport systems are involved, which are both competitively inhibited in presence of vecuronium. Reversely, NPDA competitively inhibited the saturable part of the vecuronium uptake in isolated hepatocytes. These data demonstrated that vecuronium and NPDA share at least partly the same uptake system in the liver. This was further supported by the fact that the pattern of the competitive inhibition of NPDA uptake by various model compounds was highly comparable with the pattern of vecuronium uptake inhibition in isolated hepatocytes.

Subsequent photoaffinity labeling with the photolabile NPDA derivative of isolated cells revealed two plasma membrane binding polypeptides with apparent M_r's of 48000 and 50000. Differential photoaffinity labeling studies demonstrated decreased incorporation of radioactivity in the plasma membrane polypeptides in presence of amphipathic cations (vecuronium, pancuronium, d-tubocurarine, quinidine) and in presence of uncharged compounds (k-stophanthoside, ouabain, digitoxin), which all inhibit the uptake of NPDA in cells. These data suggest that the identified binding polypeptides may be involved in the hepatic uptake of the amphipathic cation. Because vecuronium appears to be taken up by one saturable transport system, it remains to be established if the M_r 50000 or the M_r 48000 is involved in the hepatic uptake of vecuronium. In the differential photoaffinity labeling experiments no difference between the two polypeptides could be observed, because the incorporation in both polypeptides was reduced. This might be explained by competitive binding of vecuronium to both transport systems, but net transport only via one system. Additional studies, e.g. by reconstitution experiments with both polypeptides and determination of their transport characteristics, are necessary to establish transport function and specificity of the labeled polypeptides.

It should be noted that the data of the present study fit well within the concept of a multispecific uptake system with apparent M_r of 48000, that has been postulated on the basis of both kinetic and photoaffinity labeling evidence (Buscher et al., 1987; Frimmer, 1982; Wieland, et al., 1984). Photolabile derivatives of compounds that have been shown to interfere kinetically with NPDA and vecuronium uptake, also labeled plasma membrane polypéptides in the same apparent molecular weight range (ouabain: M_r 50000 (Petzinger et al., 1986); bile acids: M_r 48000 (Buscher et al., 1986; Wieland et al., 1984)).

Hepatic uptake of low molecular weight cations: kinetic and photoaffinity labeling studies

A major observation in the experiments with NPDA and vecuronium was the lack of effect of low molecular weight organic cations such as PAEB in both kinetic and differential photoaffinity labeling experiments. This suggests that these low molecular weight cations are taken up by a transport system distinct from the system for NPDA and vecuronium. In order to identify potential transport polypeptides for these low molecular weight cations, a photolabile derivative of the classical model compound PAEB was synthesized: azidoprocainamidemethoiodide (APM). Determination of the octanol/Krebs partition coefficient and binding to albumin revealed that the introduction of the photolabile azido group did not markedly change the physicochemical parameters, as compared to its parent compound PAEB. Experiments in isolated perfused rat liver demonstrated that APM is efficiently taken up in the liver and excreted into bile, partly in the form of metabolites.

The inhibitory effect of other cations such as tributylmethylammonium on the hepato-biliary transport of APM suggests that carrier-mediated mechanisms are involved, as was demonstrated for PAEB (Eaton and Klaassen, 1978; Hwang and Schanker, 1973; Meijer et al., 1970; Nakae et al., 1981; Solomon and Schanker,

1963). Kinetics of the uptake of APM in isolated rat hepatocytes revealed that both saturable and nonsaturable processes are involved. The saturable part of the uptake was best described by two uptake sytems (V_{max1} = 80 pmol/min x 10^6 cells, K_{m1} = 3 μM and V_{max2} = 130 pmol/min x 10^6 cells, K_{m2} = 100 μM). APM uptake was Na^+-independent, but replacement of bicarbonate in the incubation medium by chloride decreased the hepatic uptake via both systems. The mechanism of this bicarbonate effect, that occurs at an unchanged extracellular pH, remains to be clarified. Besides ion-pair formation mechanisms and/or cotransport of anions to maintain electroneutrality, an effect of bicarbonate on secondary mechanisms involved in the driving of the uptake may underly the observed phenomenon. Since both K_m and V_{max} are affected, also a combination of such factors may be involved.

In order to investigate the specificity of the hepatic uptake systems for APM, the influence of several classes of hepatic model compounds on the uptake of APM was studied. The results show that all the mono- and bivalent organic cations studied substantially decreased the uptake of APM into the liver cell. In presence of anionic and uncharged compounds the uptake of APM was not inhibited. These data indicate that APM is taken up by transport systems differing from the systems for anions and uncharged compounds. Integration of the data on the uptake of vecuronium and NPDA furthermore indicates that APM is taken up by a system distinct from the uptake mechanisms for vecuronium and NPDA, but that the latter substrates exhibit considerable affinity for the supposed uptake system for low molecular weight organic cations. It remains to be established whether binding to this carrier for the lipophilic actions also results in efficient membrane transport by this system or that only carrier occupation without net transport occurs.

Photoaffinity labeling with APM was performed on isolated hepatocytes and sinusoidal membrane fractions. The experiments revealed two membrane polypeptides with apparent molecular weight of 48000 and 72000 that are involved in binding with

APM. Differential photoaffinity labeling in presence of substrates that inhibit the uptake of APM suggested that these polypeptides might be involved in the hepatic uptake of APM. Interestingly, with several photolabile derivatives of substances that do not interfere with APM uptake in the present study, polypeptides in the range of M_r 48000 were labeled (Buscher et al., 1986; 1987; Wieland et al., 1984). Yet, in preliminary experiments to localize the M_r 48000 bands in relation to one another a difference in labeling in the M_r 48000 region was hardly distinguishable. Nevertheless, in combination with kinetic data from interaction experiments the observed labeling with APM of a M_r 48000 polypeptide suggests the existence of at least two transport proteins for organic compounds in the molecular weight range of 48000 in the sinusoidal membrane. Autoradiographic data obtained after photoaffinity labeling with several other photoprobes pointed to protein heterogeneity in the range of M_r 48000 (Buscher et al., 1987). Transport protein heterogeneity has also been postulated in the molecular weight range of 55000, where bile acid, BSP, and bilirubin binding polypeptides have been identified (Berk et al., 1987). With respect to organic cations, additional experiments are required to substantiate this concept.

Studies with AS 30 D hepatoma cells, lacking transport function might substantiate the transporting function of the identified M_2 72000 and 48000 polypeptides. Isolation of the membrane polypeptides, the preparation of antibodies and/or reconstitution of transport in liposomes are necessary to demonstrate the role of the identified binding polypeptides in the hepatic uptake of organic cations. The recent introduction of flash-photolysis (Frimmer and Ziegler, 1986) potentially enables investigation of sequential transport phenomena. In combination with the development of highly photolabile cations that are not metabolized, transcellular transport of organic cations could be studied using this technique.

In conclusion, based on the results of this and other studies, the hepatic disposition of organic cations can be schematically pictured, as depicted in Fig. 7.

HEPATIC DISPOSITION OF ORGANIC CATIONS

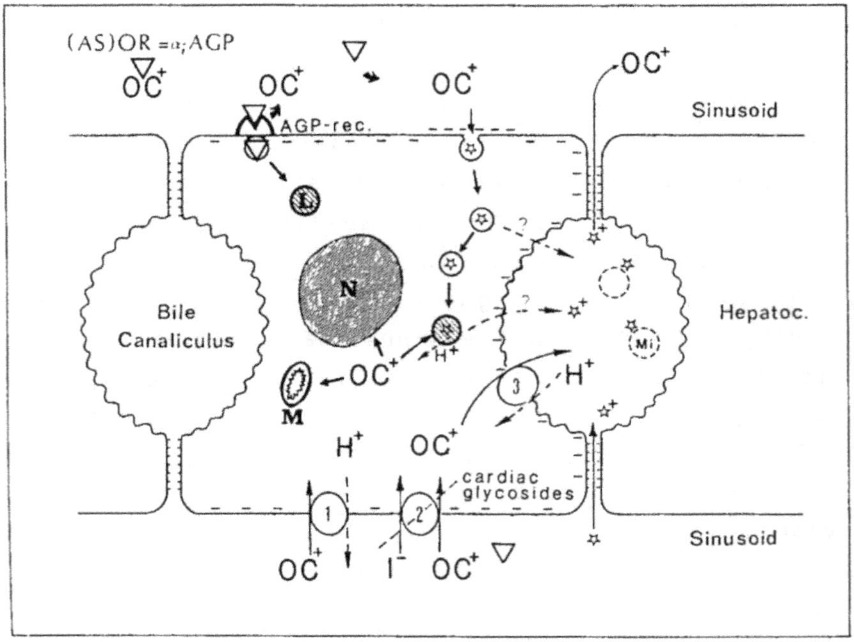

Figure 7: Transport mechanisms for hepato-biliary transport of organic cations (OC^+) (within organelles and canaliculi indicated by ☆). Uptake can occur by at least two systems. System 2 is inhibitable by cardiac glycosides that may share a carrier process transporting relatively high molecular weight organic cations, possibly as ion-pairs with inorganic counter anions. System 1 may preferentially serve low molecular weight monovalent cations and could operate by proton antiport. Lipophilic organic cations bind in plasma to α1-acid glycoprotein, but are not coendocytosed with asialo-forms of this glycoprotein (Sluijs et al., 1987). Accumulation in lysosomes may occur via a specific fluid-phase endocytosis at the plasma membrane or via antiport with protons in the cytoplasma. Direct transport of drug from lysosomes to bile remains to be demonstrated. Apart from association with lysosomes extensive binding can occur to mitochondria and nuclei. Biliary excretion involved carrier-mediated transport, possibly by antiport with inorganic ions. Binding to mixed biliary micelles may facilitate net transport into the bile canaliculus. Adapted from Meijer (1987).

References

Agoston S, Vermeer GA, Kersten UW, Meijer DKF (1973) The fate of pancuronium bromide in man. Acta Anaesth Scand 17:267-275

Agoston S, Crul EJ, Kersten UW, Houwertjes MS, Scaf AHJ (1977) The relationship between dispostion and duration of action of a congeneric series of steroidal neuromuscular blocking agents. Acta Anaesth Scand 21:24-30

Bakeeva LE, Grinius LL, Jasaitis AA, Kuliene VV Levitsky DO, Liberman EA, Severina II, Skulachev VP (1970) Conversion of biomembrane-produced energy into electric form. II. Intact mitochondria. Biochim Biophys Acta 216:13-21

Bencini AF, Scaf AHJ, Sohn YJ, Kersten-Kleef UW, Agoston S (1986) Hepatobiliary disposition of vecuronium bromide in man. Br J Anaesth 58:988-995

Berk PD, Potter BJ, Stremmel W (1987) Role of plasma membrane ligand-binding proteins in the hepatocellular uptake of albumin-bound organic anions. Hepatology 7:165-176

Blom A, Scaf AHJ, Meijer DKF (1982) Hepatic drug transport in the rat. A comparison between isolated hepatocytes, the isolated perfused liver and the liver in vivo. Biochem Pharmacol 30:1809-1816

Buscher HP, Fricker G, Gerok W, Kramer W, Kurz G, Müller M, Schneider S (1986) Membrane transport of amphiphilic compounds by hepatocytes. In: Greten H, Windler E, Beisiegel U (eds.) Receptor-mediated uptake in the liver. Springer, Berlin Heidelberg, p 189-199

Buscher HP, Fricker G, Gerok W, Kurz G, Müller M, Schneider S, Schramm U, Schreyer A (1987) Hepatic transport systems for bile salts: localization and specificity. In: Paumgartner G, Stiehl A, Gerok W (eds) Bile acids and the liver. MTP Press, Lancaster, p 95-110

Cohen EN, Brewer HW, Smith D (1967) The metabolism and elimination of d-tubocurarine-^{3}H. Anesthesiology 28:309-317

De Duve C, Pressman BC, Gianetto R, Wattiaux R, Appelmans F (1955) Tissue fractionation studies. 6. Intracellular distribution patterns of enzymes in rat liver tissue. Biochem J 60:604-617

Eaton DL, Klaassen CD (1978a) Carrier-mediated transport of ouabain in isolated hepatocytes. J Pharmacol Exp Ther 205:480-488

Eaton DL, Klaassen CD (1978b) Carrier-mediated transport of the organic cation procaine amide ethobromide by isolated rat liver parenchymal cells. J Pharmacol Exp Ther 206:595-606

Echigoya Y, Matsumoto Y, Nakagawa Y, Suga T, Niinobe S (1972) Metabolism of quaternary ammonium compounds. I. Binding of tropane alkaloids to rat liver lysosomes. Biochem Pharmacol 21:477-484

Frimmer M (1982) Organotropism by carrier-mediated transport. Trends Pharmacol Sci 2:395-397

Frimmer M, Ziegler K (1986) Photoaffinity labeling of whole cells by flashed light: a simple apparatus for high-energy ultraviolet flashes. Biochim Biophys Acta 855:143-146

Gallagher CH (1968) The effects of neuromuscular blocking agents on mitochondria. IV. Effects of d-tubocurarine, pyrrolizidine alkaloids and magnesium on oxidative phosphorylation. Biochem Pharmacol 17:533-538

Graf J, Petersen OH (1978) Cell membrane potential and resistance in liver. J Physiol Lond 284:105-126

Hirom PC, Hughes RD, Millburn P (1974) The physicochemical factor required for the biliary excretion of organic cations and anions. Biochem Soc Trans 2:327-330

Hsyu PH, Giacomini KT (1987) The pH gradient transport of organic cations in the renal brush border membrane. Studies with acridine orange. J Biol Chem 262:3964-3968

Hughes RD, Millburn P, Williams RT (1973a) Biliary excretion of some diquaternary ammonium cations in the rat, guinea pig and rabbit. Biochem J 136:979-984

Hughes RD, Millburn P, Williams RT (1973b) Molecular weight as a factor in the excretion of monoquaternary ammonium cations in the bile of the rat, rabbit and guinea pig. Biochem J 136:967-978

Hwang SW, Schanker LS (1973) Hepatic uptake and biliary excretion of N-acetyl procainamide ethobromide in the rat. Am J Physiol 225:1437-1443

Joppen C, Petzinger E, Frimmer M (1985) Properties of iodipamide uptake by isolated hepatocytes. Naunyn-Schmiedeberg's Arch Pharmacol 331:393-397

Klaassen CD, Watkins JB (1984) Mechanisms of bile formation, hepatic uptake and biliary excretion. Pharmacol Rev 36:1-67

Lauterbach F (1984) Intestinal permeation of organic bases and quaternary ammonium compounds. In: Csaky TZ (ed) Handbook of experimental pharmacology, Vol. 70/II. Pharmacology of intestinal permeation. Springer, Berlin Heidelberg New York, p 271-299

Macgregor JT, Burkhalter A (1973) Biliary excretion of nicotinamide riboside. A possible role in the regulation of hepatic pyridine nucleotide dynamics. Biochem Pharmacol 22:2645-2658

Meijer DKF (1977) The mechanisms for hepatic uptake and biliary excretion of organic cations. In: Kramer M (ed) Intestinal permeation. Excerpta Medica, Amsterdam, p 196-208

Meijer DKF (1987) Current concepts on hepatic transport of drugs. J Hepatol 4:259-268

Meijer DKF, Bos ES, Van der Laan KJ (1970) Hepatic transport of mono- and bisquaternary ammonium compounds. Eur J Pharmacol 11:371-377

Meijer DKF, Arends JW, Weitering JG (1971a) The cardiac glycoside sensitive step in the hepatic transport of the bisquaternary ammonium compound, hexafluorenium. Eur J Pharmacol 15:245-251

Meijer DKF, Vermeer GA, Kwant G (1971b) The excretion of hexafluorenium in man and rat. Eur J Pharmacol 14:280-285

Meijer DKF, Wester J, Gunnik M (1972) Distribution of quaternary ammonium compounds between particulate and soluble constituents of rat liver, in relation to their transport from plasma into bile. Naunyn-Schmiedeberg's Arch Pharmacol 273:179-192

Meijer DKF, Weitering JG, Vonk RJ (1976) Hepatic uptake and biliary excretion of d-tubocurarine and trimethylcurarine in the rat in vivo and in isolated perfused rat livers. J Pharmacol Exp Ther 198:229-239

Meijer DKF, Weitering JG, Vermeer GA, Scaf AHJ (1979) Comparative pharmacokinetics of d-tubocurarine and metocurarine in man. Anesthesiology 51:402-407

Meijer DKF, Groothuis GMM (1986) Mechanisms for hepatic uptake of xenobiotics. In: Fillastre JP (ed.) Hepatotoxicity of drugs. Editions INSERM, Publications de l'Université de Rouen, p 157-190

Mol WEM, Fokkema GN, Weert B, Meijer DKF (1988) Mechanisms for the hepatic uptake of organic cations. Studies with the muscle relaxant vecuronium in isolated rat hepatocytes. J Pharmacol Exp Ther 244,1:268-275.

Mol WEM, Meijer DKF (1988) Hepatic transport mechanisms for bivalent organic cations. Subcellular distribution and hepato-biliary concentration gradients of some steroidal muscle relaxants. Biochem Pharmacol (in press)

Nakae H, Sakata R, Muranishi S (1976) Biopharmaceutical study of the hepatobiliary transport of drugs. V. Hepatic uptake and biliary excretion of organic cations. Chem Pharm Bull 24:886-893

Nakae H, Takada K, Asada S, Muranishi S (1980) Transport rates of hepatic uptake and biliary excretion of an organic cation, acetyl procainamide ethobromide. Biochem Pharmacol 29:2573-2576

Nakae H, Muranishi S, Asada S and Takada K (1981) Pharmacokinetic study on saturated hepatobiliary transport of acetyl procainamide ethobromide. J Pharm Dyn 4:584-589

Neef K, Jonkman JHG, Meijer DKF (1978) Hepatic disposition and biliary excretion of the organic cations thiazinamium and thiazinamium sulfoxide in rats. J Pharm Sci 67:1147-1150

Neef C, Keulemans KTP, Meijer DKF (1984a) Hepatic uptake and biliary excretion of organic cations I. Characterization of three new model compounds. Biochem Pharmacol 33:3977-3990

Neef C, Oosting R, Meijer DFK (1984b) Structure-pharmacokinetics relationship of quaternary ammonium compounds. Elimination and distribution characteristics. Naunyn-Schmiedeberg's Arch Pharmacol 328:103-110

Neef C, Meijer DFK (1984) Structure-pharmacokinetics relationship of quaternary ammonium compounds. Correlation of physico-chemical and pharmacokinetic parameters. Naunyn-Schmiedeberg's Arch Pharmacol 328:111-118

Petzinger E, Frimmer M (1984) Driving forces in hepatocellular uptake of phalloidin and cholate. Biochim Biophys Acta 778: 539-548

Petzinger E, Fischer K, Fasold H (1986) Role of bile acid transport system in hepatocellular ouabain uptake. In: Erdmann E, Greef K, Skou JC (eds) Cardiac Glycosides 1785 - 1985. Steinkopff, Darmstadt, p 297-304

Ramsay RR, Salach JI, Singer TP (1986) Uptake of the neurotoxin 1- methyl-4-phenylpyridine (MPP^+) by mitochondria and its relation to the inhibition of the mitochondrial oxidation of NAD^+-linked substrates by MPP^+. Biochem Biophys Res Commun 134:743-748

Rennick BR (1981) Renal tubule transport of organic cations. Am J Physiol 240:F83-F89
Ruifrok PG (1982) Uptake of quaternary ammonium compounds into rat liver plasma membrane vesicles. Biochem Pharmacol 31: 1431-1435
Schanker LS (1972) Transport of drugs. In: Hokin LE (ed) Metabolic pathways. Metabolic transport, Vol IV. Academic Press, London, p 556-567
Schwenk M (1980) Transport systems of isolated hepatocytes. Arch Toxicol 44:113-126
Sluijs Van der P, Spanjer HH, Meijer DKF (1987) Hepatic disposition of cationic drugs bound to asialo-orosomucoid: lack of co-endocytosis and evidence for intrahepatic dissociation. J Pharmacol Exp Ther 240:668-673
Sohn YJ, Bencini AF, Scaf AHJ, Kersten UW, Agoston S (1986) Comparative pharmacokinetics and dynamics of vecuronium and pancuronium in anesthetized patients. Anesth Anal 65:233-239
Solomon HM, Schanker LS (1963) Hepatic transport of organic cations: active uptake of a quaternary ammonium compound, procainamide ethobromide, by rat liver slices. Biochem Pharmacol 12:621-626
Somogyi A (1987) New insights into the renal secretion of drugs. Trends Pharmacol Sci 8:354-357
Vonk RJ, Scholtens E, Keulemans GTP, Meijer DKF (1978) Choleresis and hepatic transport mechanisms. IV. Influence of bile salts on the hepatic transport of the organic cations, d-tubocurarine and N^4-acetylprocainamide ethobromide. Naunyn-Schmiedeberg's Arch Pharmacol 302:1-9
Vonk RJ, Jekel PA, Meijer DKF, Hardonk MJ (1978) Transport of drugs in isolated hepatocytes, the influence of bile salts. Biochem Pharmacol 27:397-405
Waser PG, Wiederkehr H, Sin-Ren AC, Kaiser-Schönenberger E (1987) Distribution and kinetics of ^{14}C-vecuronium in rats and mice. Br J Anaesth 59:1044-1051
Wassermann O (1971) Influence of substituents of pharmacokinetics of bisquaternary ammonium compounds. Naunyn-Schmiedeberg's Arch Exp Pharmacol 270:R154
Weiner IM (1985) Organic acids and bases and uric acid. In: Seldin DW, Giebisch GL (eds) The Kidney, Physiology and Pathophysiology, Raven Press, New York, p 1703-1724
Weitering JG, Mulder GJ, Meijer DKF, Lammers W, Veenhuis M, Wendelaar Bonga SE (1975) On the localization of d-tubocurarine in rat liver lysosomes in vivo by electron microscopy and subcellular fractionation. Naunyn-Schmiedeberg's Arch Pharmacol 289:251-256
Weitering JG, Lammers W, Meijer DKF, Mulder GJ (1977) Localization of d-tubocurarine in rat liver lysosomes. Lysosomal uptake, biliary excretion and displacement by quinacrine in vivo. Naunyn-Schmiedeberg's Arch Pharmacol 299:277-281
Westra P, Keulemans GTP, Houwertjes MC, Hardonk MJ, Meijer DKF (1981) Mechanism underlying the prolonged duration of action of muscle relaxants caused by extrahepatic cholestasis. Br J Anaesth 58:217-227

Wieland T, Nassal M, Kramer W, Fricker G, Bickel U, Kurz G (1984) Identity of hepatic membrane transport systems for bile salts, phalloidin, and antamanide by photoaffinity labeling. Proc Natl Acad Sci USA 81:5232-5236

Yoshioka K (1984) Some properties of the thiamine uptake system in isolated rat hepatocytes. Biochim Biophys Acta 778:201-209

The Role of Liver and Extrahepatic Organs in the Metabolism of N-Nitrosodibutylamine (NDBA): Comparison of in Vitro with in Vivo Data

E. Richter, X.-C. Feng and J. Schulze
Walther-Straub-Institute of Pharmacol. & Toxicol.
Nußbaumstr. 26
D-8000 München 2, FRG

The organotropism of NDBA, a powerful carcinogen towards liver and urinary bladder, has been shown to depend upon dose and route of administration. With decreasing oral doses a shift from liver to bladder tumors was observed. On the other hand, after subcutaneous injection of much higher doses, only bladder tumors were obtained. A major urinary metabolite, N-nitrosobutyl-(3-carboxypropyl)amine, NB3CPA, arising from ω-hydroxylation of NDBA to N-nitrosobutyl-(4-hydroxybutyl)-amine, NB4HBA, has been shown to be responsible for its tumorigenic action on the urinary bladder. However, rat liver preparations in vitro either activated NDBA by α-hydroxylation or formed the nearly nontoxic ω-1-hydroxide N-nitrosobutyl-(3-hydroxybutyl)amine, NB3HBA, not explaining the formation of large amounts of ω-hydroxylation products, NB4HBA and NB3CPA, in vivo.

Female Sprague-Dawley rats were used throughout the study. In microsomal preparations of lung, kidney and small intestinal mucosa a twofold excess of NB4HBA over NB3HBA was formed, whereas liver yielded 6-8 times more NB3HBA as compared to NB4HBA. Overall hydroxylation activity was in the order lung> liver >> intestine = kidney. The microsomal reaction was

Hepatic Transport in Organic Substances
Ed. by E. Petzinger, R. K.-H. Kinne, H. Sies

clearly dependent upon cytochrome P450. The differences between organs could be ascribed to different isozyme patterns. No further oxidation of NB4HBA to NB3CPA could be observed in microsomes. In isolated perfused jejunal segments a concentration-dependent first-pass metabolism of NDBA was observed, being as high as 95 % at concentrations < 10 μM and decreasing to about 30 % at 200 μM. The ratio of ω-/ω-1-hydroxylation products was in the range of 10-20. This high ratio was mainly due to the formation of NB3CPA representing 80-90 % of total metabolites formed. These results could be confirmed by studies in autoperfused intestinal segments in situ. Therefore, the much higher absorption rate in situ did not compete with the first-pass metabolism of NDBA. For in vivo studies awake rats supplied with permanent catheters were infused with 50 mg/kg NDBA during 10 min by the arterial (i.a.), venous (i.v.), portal (i.p.) or duodenal (i.d.) route, respectively. For NDBA concentrations in plasma during the first 2 h the area under the curve decreased from 2934 min x ng after i.a. infusion to 1280, 1013, and 664 min x ng after i.v., i.p., and i.d. infusion, respectively. Hence, the fraction eliminated by first pass metabolism was highest in lung (0.56) followed by small intestine (0.34) and liver (0.21). After subcutaneous NDBA injection of 200 mg/kg, the maximal NDBA concentration in plasma was tenfold lower than after duodenal infusion of 50 mg/kg (1.3 vs 14.4 ng/ml) suggesting a high first-pass metabolism in lung.

In conclusion, the portals of entry of NDBA, small intestine or lung, are very effective in metabolizing NDBA in vivo. Because of the prevalence of ω-hydroxylation in these organs leading to the proximate bladder carcinogen NB3CPA, the observation of urinary bladder tumors after low oral as well as high subcutaneous NDBA doses may be explained by first pass metabolism in small intestine and lung, respectively.

Supported by a grant of the DFG and the Dr. Robert Pfleger Stiftung.

The Role of Biliary Excretion and Intestinal Metabolism for Toxokinetics and Toxicity of T-2 Toxin, a Major Trichothecene Mycotoxin

S. Conrady-Lorck, S.G. Schäfer and B. Fichtl
Walther-Straub-Institut für Pharmakologie und Toxikologie, Nußbaumstr. 26
8000 München 2, FRG

The trichothecene mycotoxins are highly toxic fungal metabolites produced by various Fusarium species. Outbreaks of sublethal and lethal toxicoses in man and animals following ingestion of food contaminated with trichothecenes have been observed worldwide. T-2 toxin (3 α-hydroxy-4ß,15-diacetoxy-8 α-(3-methyl-butyryloxy)-12,13-epoxytrichothec-9-ene) is one of the most toxic compounds among the trichothecenes.

The intestinal metabolism of T-2 toxin was investigated using the method of the vascularly autoperfused jejunal loop _in situ_. Tritium labelled T-2 toxin was injected into the tied off intestinal segments from rats. T-2 toxin and its metabolites in the blood draining from the jejunal loops were determined by HPLC and GCMS. During the experimental period of 50 min, up to 50% of the total radioactivity appeared in the effluent blood. There was an extensive metabolic degradation of T-2 toxin. The fraction of total radioactivity attributable to the parent compound amounted to only some 5%. Main metabolite was the hydrolysis product HT-2 toxin accounting for 58 - 74% of the total radioactivity. Furthermore, 3'-OH-T-2 toxin (5%), 3'-OH-HT-2 toxin (11 - 18%), 4-deacetylneosolaniol (4 - 8%), and T-2 tetraol (2 - 5%) were present. No glucoronide or sulfate conjugates could be detected.

Hepatic Transport in Organic Substances
Ed. by E. Petzinger, R. K.-H. Kinne, H. Sies

The biliary excretion of T-2 toxin was studied in urethane-anesthetized rats. The bile was collected from a cannula inserted into the bile duct. Following i.v. injection of labelled T-2 toxin, high concentrations of radioactivity were found in the bile. The concentration ratio between bile and plasma was about 900 after 15 min and decreased to about 200 after 120 min. Most of the radioactivity appearing in the bile was due to 3'-OH-HT-2 toxin and the glucuronide of HT-2 toxin. Furthermore, there were small amounts (about 10%) of HT-2 toxin and more polar metabolites. Unchanged T-2 toxin was not detectable. At the end of the experimental period, i.e. after 120 min, up to 40% of the total dose had already appeared in the bile. In intact rats a total of 40 - 50% of an i.v. dose was excreted with the feces during a five day period with only minute amounts of glucuronides present. This strongly suggests a breakdown of glucuronides in the gut, possibly by intestinal bacteria, and marked enterohepatic recycling. This view is supported by experiments demonstrating that the lethality following intravenous injection of T-2 toxin could markedly be reduced by oral ingestion of activated charcoal.

From the present experiments it is concluded that T-2 toxin is subject to a marked first pass effect due to metabolism in both the intestine and the liver. This raises the question as to the metabolite(s) responsible for toxicity in field cases of T-2 toxicosis.

Binding Proteins for Dipeptides, β-Lactam Antibiotics, and p-Aminohippurate (PAH) in Plasma Membranes from Small Intestine and Kidney

W. Kramer[1] and G. Burckhardt[2]
1 Hoechst AG, Postfach 800320, 6230 Frankfurt/Main 80, FRG
2 MPI für Biophysik, Kennedyallee 70, 6000 Frankfurt/Main 70 FRG

It has been suggested that orally active amino-ß-lactam antibiotics and dipeptides are absorbed from the small intestine by a common H^+-dependent transport system in the brush border membrane of enterocytes. Excretion of absorbed amino-ß-lactam antibiotics occurs mainly in renal proximal tubules. Again, amino-ß-lactam antibiotics and dipeptides share a transport system in the brush border membrane of proximal tubule cells. In the basolateral membrane the antibiotics share the transport system with p-aminohippurate (PAH). In order to identify the binding proteins possibly related to the transport of lactam antibiotics, dipeptides, and PAH, brush border and basolateral membrane vesicles were prepared and subjected to photoaffinity labelling with (^{3}H)benzylpenicillin and the N-(4-azidobenzoyl)derivatives of cephalexin and glycyl-L-proline. With all derivatives a polypeptide of app. MW 127,000 was predominantly labelled in brush border membranes from the small intestine of rats, rabbits, and pigs. The labelling of this polypeptide was abolished by ß-lactam antibiotics (penicillins and cephalosporins), dipeptides, and PAH, but not by amino acids or bile acids. In the brush border membrane from rat renal cortex two membrane proteins with app. MW of 130,000 and 95,000 were labelled. ß-lactam antibiotics,

Hepatic Transport in Organic Substances
Ed. by E. Petzinger, R. K.-H. Kinne, H. Sies

dipeptides, and PAH prevented the labelling of these polypeptides. Since uptake of aminocephalosporins into intestinal vesicles, but not into renal brush border vesicles, was stimulated by an inward H^+ gradient, the respective transport systems are related, but not identical. In the basolateral membrane from rat kidney cortex, benzylpenicillin and N-(4-azidobenzoyl)-cephalexin predominantly labelled a membrane polypeptide of MW 50,000-52,000. The labelling of this polypeptide was decreased by the presence of ß-lactam antibiotics and PAH, whereas bile salts had no effect. In conclusion, membrane polypeptides of MW 127,000 in the small intestine and of MW 130,000 and 95,000 in the kidney are components of the ß-lactam antibiotics/dipeptide transport system in the brush border membrane, whereas a polypeptide of MW 52,000 represents a component of the ß-lactam antibiotics/PAH transporter in the basolateral membrane from rat kidney cortex.

Hepatic Transport of Hormones

Transport Properties of Steroid Glucuronides

Mary Vore, Arunee Changchit, Sherrie Durham and Steve Teo
Department of Pharmacology
University of Kentucky
College of Medicine
Lexington, Kentucky 40536 USA

Introduction

The D-ring glucuronide conjugate of estradiol, estradiol-17β-(β-D-glucuronide) ($E_2$17G) induces a rapid, dose-dependent and reversible cholestasis in the rat when given intravenously (Meyers et al., 1981). This organic anion is taken up into the hepatocyte by a saturable process which is at least partially dependent on metabolic energy and an intact sodium gradient and is inhibited by other anions such as bromosulfophthalein (BSP) (Brock and Vore, 1984; Brock et al., 1985). These data suggested that $E_2$17G shares an uptake mechanism with other organic anion dyes. Such carrier-mediated uptake involves two processes: 1) binding of ligand to the carrier and 2) translocation of ligand across the membrane.

Results

We have developed a receptor binding assay using ^{3}H-$E_2$17G of high specific activity (50 Ci/mmol) and purity (>98% by HPLC) as a ligand to determine if we could identify binding sites whose properties were consistent with those of an

Hepatic Transport in Organic Substances
Ed. by E. Petzinger, R. K.-H. Kinne, H. Sies

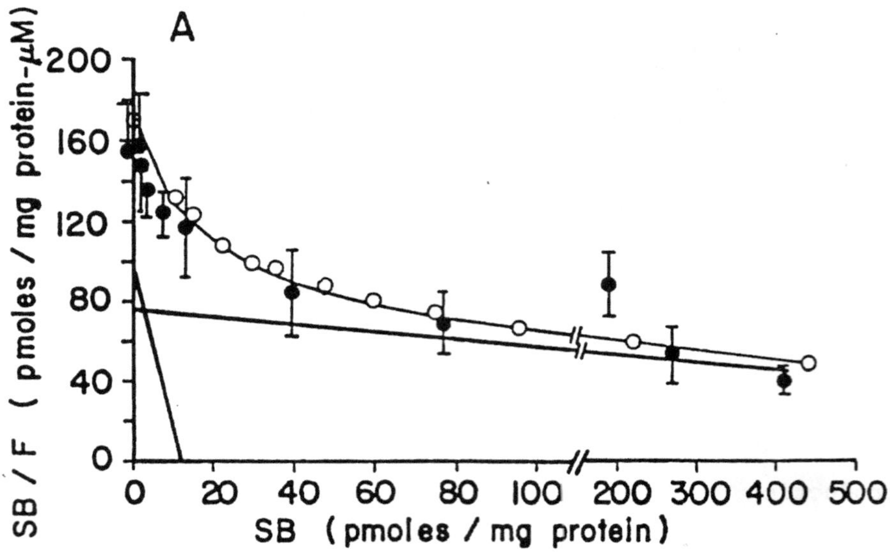

FIG. 1: Scatchard analysis of the saturation experiment. The data are represented graphically by specific bound/free (SB/F) versus specific bound. The binding parameters were determined by LIGAND analysis and the solid lines are those calculated by LIGAND with slopes = $1/Kd_1$ and $1/Kd_2$. ●, observed data points (Mean ± SEM); O, predicted values determined by performing geometric construction of the curved line.

organic anion carrier. Alternatively, such a binding site could represent a site that mediated cholestasis. Liver plasma membranes were incubated with 2-3 nM ^3H-$E_2$17G for 2 min and bound ligand separated from free ligand by microcentrifugation (Takacs and Vore, 1987). Specific binding was shown to reach equilibrium within 1 min and remain stable for up to 60 min and to be maximal at 4°C. Scatchard analysis of a saturation experiment (Figure 1) showed a hyperbola, upwardly concave, suggesting the presence of multiple binding sites. Analysis of the data by the computer program LIGAND (Munson and Rodbard, 1980) showed that a two-binding site model fit the data best with the following parameters: site 1, K_{D_1}= 0.12 μM, B_{max_1} = 11.1 pmol/mg of protein; site 2,

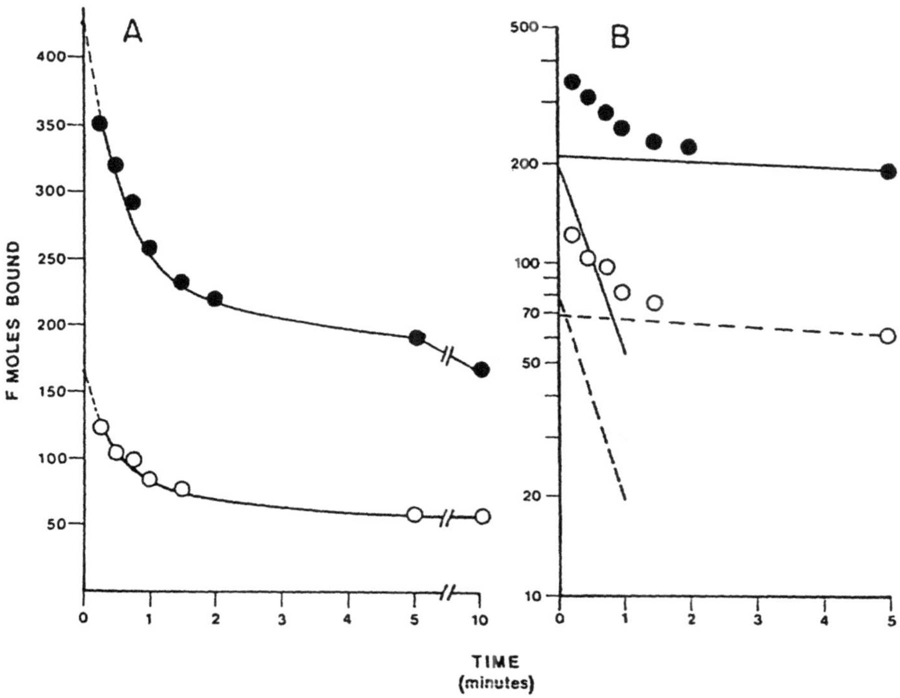

FIG. 2. Linear (A) and log (B) plots of dissociation rate experiments. The data are represented graphically by amount bound versus time after dissociation of bound ligand. ● and ○ , excess $E_2 17G$ dilution and infinite volume dilution, respectively. Each point represents the average of two determinations performed in triplicate. The lines drawn are those calculated by NONLIN in which their slopes are equal to K_{-1} (site 1) and K_{-1} (site 2).

K_{D_2} = 11.1 μM, B_{max_2} = 881 pmol/mg of protein. Dissociation experiments showed that dilution of bound radioligand either by infinite volume or excess unlabelled $E_2 17G$ displaced all specific binding in a biphasic manner, again denoting heterogenous binding sites (Figure 2). The dissociation rate constants determined by NONLIN for the excess $E_2 17G$ experiment were 1.4 and 0.024 min^{-1} and for the infinite volume dilution experiment were 1.28 and 0.025 min^{-1}. The dissociation rate constants do not differ between these two types of experiments, thus ruling out cooperative interactions at either site.

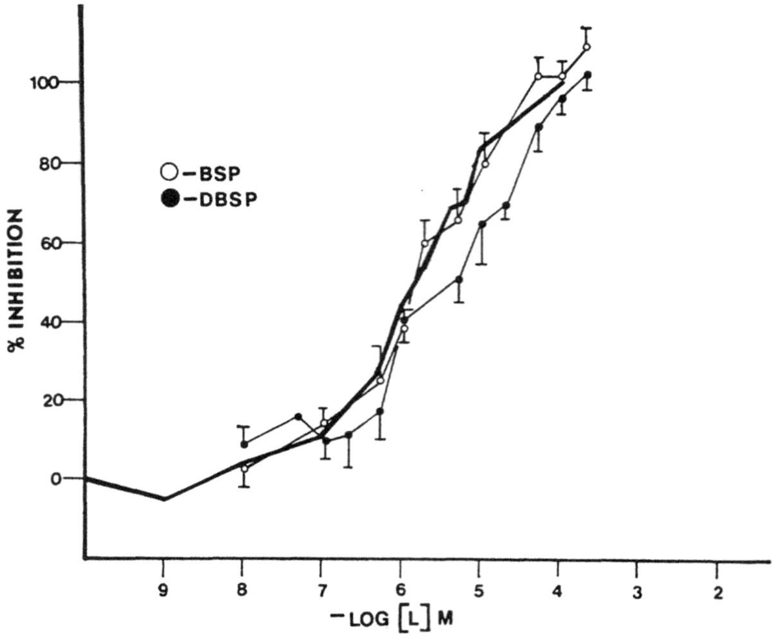

FIG. 3. Inhibition of $^3H-E_217G$ binding by organic anion dyes, BSP and DBSP. The ordinate is the percentage inhibition of the maximal displacement achieved by 100 µM E_217G, and the abscissa is the negative logarithm of concentration of unlabeled ligand employed. Unlabeled BSP and DBSP (10^{-9} - 10^{-4} M) were incubated with 5×10^{-9} M $^3H-E_217G$ and the binding experiment was performed as described earlier.

Further experiments were carried out to determine the specificity of the $^3H-E_217G$ binding sites by examining the ability of a series of ligands to inhibit $^3H-E_217G$ binding. As shown in Figure 3, BSP and dibromosulfophthalein (DBSP) inhibited binding at both E_217G sites, whereas taurocholate (TC), cholate (CHO) and taurodehydrocholate (TDHC) were able to displace E_217G only from the high affinity site (Figure 4). The choleretic estradiol-3-glucuronide (E_23G) and the cholestatic estriol-16- and estriol-17-glucuronide were also able to displace E_217G from both sites, whereas testosterone glucuronide was selective for the high affinity site (Takacs and Vore, 1987). The selectivity of the bile acids and testosterone glucuronide for one site provided a third line of evidence for heterogenous binding sites.

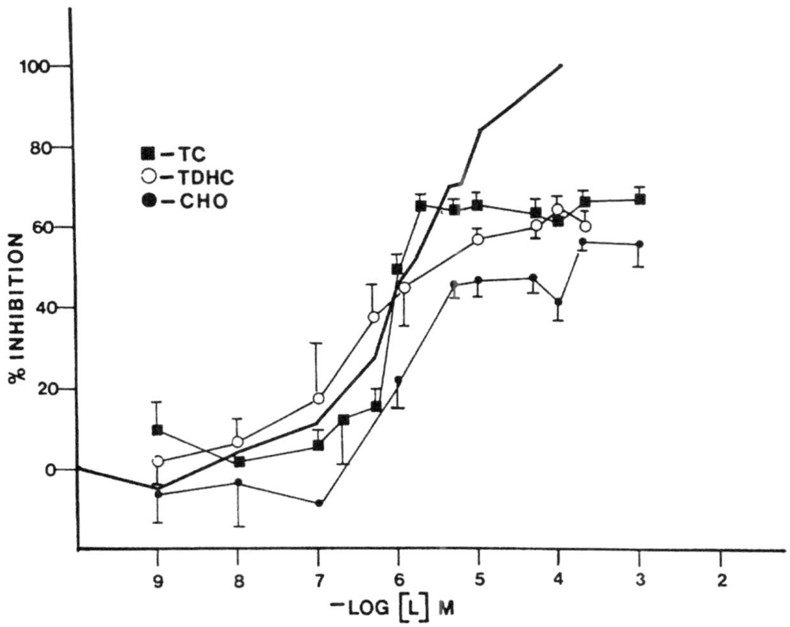

FIG. 4. Inhibition of ^3H-E_217G binding by bile acids. Unlabeled taurocholate (TC), cholate (CHO), and taurodehydrocholate (TDHC) (10^{-9}-10^{-3} M) were incubated in the presence of $5x10^{-9}$ M E_217G and the binding experiment was performed as described earlier (See the legend for Figure 3).

Finally, the ability of both choleretic and cholestatic steroid glucuronides to inhibit E_217G binding argued against these binding sites representing a site mediating cholestasis. The ability of BSP and DBSP, organic anions also taken up into isolated hepatocytes, to inhibit E_217G binding suggested that these binding sites could represent organic anion carriers.

In order to test this hypothesis, the uptake of ^3H-E_217G into isolated hepatocytes from female rats was examined over a broad substrate range (0.1-100 µM) (Figure 5) as described by Brouwer et al. (1987). Michaelis-Menten parameters were estimated using the least squares regression program NONLIN (Metzler et al., 1978). The data were fit best to a two-site model where K_{m_1} = 4.5 µM, V_{max_1} =

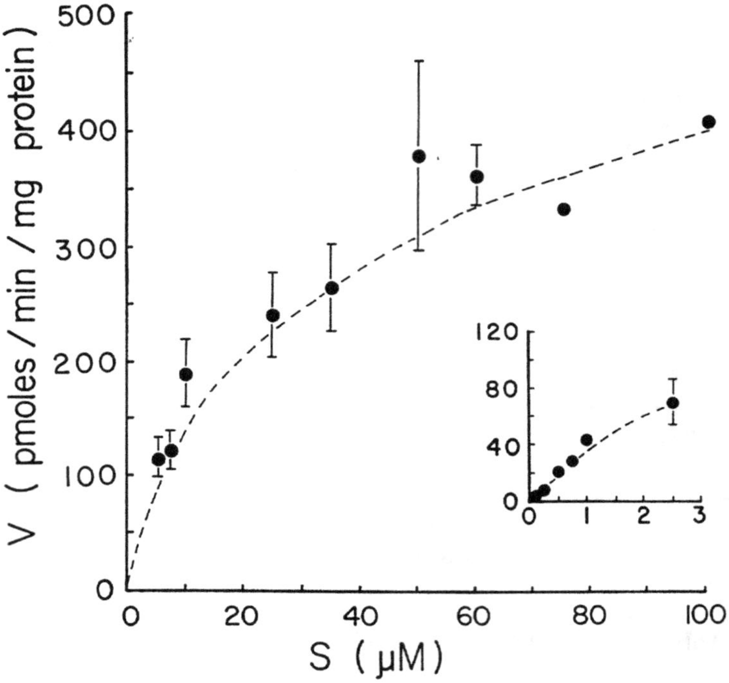

FIG. 5. Initial $E_2$17G uptake velocity (pmol/min/mg protein) by isolated female rat hepatocytes versus extracellular $E_2$17G concentrations (0.1, 0.25, 0.5, 0.75, 1.0, 2.5, 5.0, 10, 25, 35, 50, 60, 75, and 100 μM). The Michaelis-Menten equation for two sites was fitted to the data using NONLIN ($1/y^2$ weighting factor). Each point represents Mean ± SEM of two to four experiments. The ---, computer-generated best fit uptake curve when the two-site model was fit to all points simultaneously. The inset shows initial uptake at low substrate concentrations.

0.149 nmol/min/mg protein, and K_{m_2} = 149.4 μM and V_{max_2} = 0.641 nmol/min/mg protein. Our earlier work had shown that at high substrate concentrations (100 μM $E_2$17G), taurocholate (200 μM) did not inhibit the uptake of $E_2$17G (Brock et al., 1984). If the two $E_2$17G binding sites observed in liver plasma membranes represented uptake carriers for $E_2$17G, then taurocholate should inhibit $E_2$17G uptake at the high affinity site only. Dixon plot analysis (Figure 6) showed that taurocholate competitively inhibited the uptake of low, but not high, concentrations of $E_2$17G with a K_i

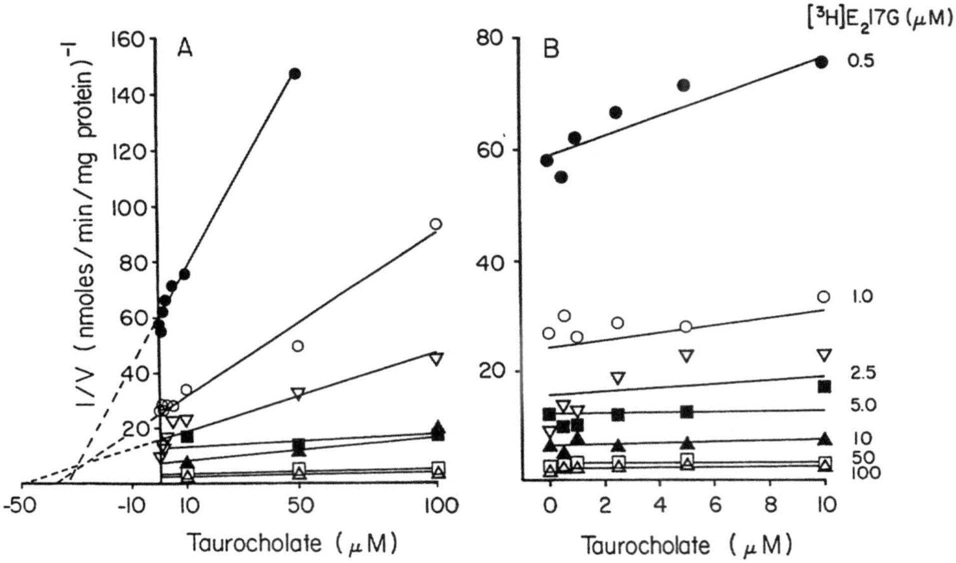

FIG. 6. Dixon plot of the effect of taurocholate (0, 0.5, 1.0, 2.5, 5.0, 10, 50, and 100 µM) on ^{3}H-$E_2$17G (0.5, 1.0, 2.5, 5.0, 10, 50, and 100 µM) uptake into isolated female rat hepatocytes. B: Inset of A showing ^{3}H-$E_2$17G uptake at taurocholate concentrations ranging from 0 to 10 µM. Each point represents the Mean of two to four experiments.

value of 43 µM. Inhibition studies using testosterone glucuronide gave similar results (Brouwer et al., 1987). These data provide strong evidence that the two $E_2$17G binding sites identified in liver plasma membranes represent two organic anion carriers one of which is shared with bile acids. Laperche et al. (1981) identified two uptake sites for BSP, one of which was inhibited by taurocholate, and postulated that this taurocholate-sensitive site for BSP was the same as the Na^+-independent carrier for taurocholate, identified by Anwer and Hegner (1978) and shown to be competitively inhibited by BSP. Our data support the hypothesis that this carrier is also the high affinity carrier for $E_2$17G.

TABLE 1
EFFECT OF ISOSMOTIC ION REPLACEMENT AND INHIBITORS ON THE UPTAKE OF 0.5 μM $E_2$17G

MEDIA	N	UPTAKE (MEAN ± SEM) PMOL/MIN/MG PROTEIN
EXPERIMENT 1		
NaCl	10	29.9 ± 5.0
SUCROSE	10	17.9 ± 2.8*
LiCl	10	24.2 ± 3.7
EXPERIMENT 2		
NaCl	5	28.8 ± 4.8
$NaNO_3$	5	13.1 ± 1.7*
NaGLUCONATE	5	33.5 ± 4.6
KCl	5	18.5 ± 4.7
EXPERIMENT 3		
NaCl	3	42.8 ± 2.9
CHOLINE Cl	3	31.2 ± 7.7
+ROTENONE (10 μM)	3	14.3 ± 5.9*
$+SO_4^=$ (50 mM)	3	30.8 ± 6.0

* P < .05, DIFFERENT FROM NaCl CONTROL: PAIRED T-TEST.

We have investigated the driving forces for uptake of $E_2$17G at this high affinity site, which is probably the more physiologically relevant site (Table 1). Although the uptake of 0.5 μM $E_2$17G in female rat hepatocytes was significantly inhibited by isosmotic replacement of NaCl in the media with sucrose, or by preincubation with the metabolic inhibitor rotenone (10 μM), neither replacement of Na^+ with Li^+, K^+ or choline nor the replacement of Cl^- with NO_3^- or gluconate had any consistent significant effects. These data argue against a sodium co-transport or Cl^- anion exchange system. Likewise, addition of a high concentration of $SO_4^=$ (50 mM)

TABLE 2

DISSOCIATION CONSTANTS (K_D) OF INHIBITING LIGANDS

LIGAND	bLPM			cLPM		
	SITES	K_{D1}	K_{D2}	SITES	K_{D1}	K_{D2}
$E_2$17G	2	2.36 ±0.71nM	1.65±0.5 μM	2	84.1±11.2nM	14.1±1.87μM
BSP	2	59.0 ±1.28nM	8.02±0.17μM	1	-----	2.23±0.62μM
TC	2	0.58 ±0.32μM	0.20±0.11mM	2	0.23±0.04μM	0.10±0.02mM
$E_2$3G	2	1.03 ±0.15μM	0.20±0.05mM	1	-----	0.17±0.02mM

K_D VALUES WERE DETERMINED BY LIGAND. EACH VALUE REPRESENTS THE MEAN ± SEM OF 3-4 DETERMINATIONS, EACH PERFORMED IN QUADRUPLICATE.

had no significant effect, suggesting that the sulfate-hydroxyl exchange system (Hugentobler and Meier, 1986) is not involved. Finally, we have compared the binding characteristics of ^{3}H-$E_2$17G in basolateral (bLPM) vs canalicular (cLPM) liver plasma membranes. Membranes were prepared from male rats as described by Meier et al. (1984) and binding of ^{3}H-$E_2$17G examined as described by Takacs and Vore (1987). $E_2$17G bound to two sites in both bLPM and cLPM (Table 2). BSP was able to inhibit $E_2$17G binding to both sites with high affinity in bLPM but bound only to the low affinity site in cLPM. $E_2$3G behaved similarly to BSP, but had a much lower affinity than did $E_2$17G or BSP. Taurocholate, like $E_2$17G, bound to both sites in bLPM and cLPM, but with much lower affinity than $E_2$17G. These data differ from that in mixed plasma membranes in which taurocholate displaced $E_2$17G binding from only the high affinity site (Figure 4). This difference may be due to the different method of membrane preparation or the fact that these latter studies were done in membranes from male rats. Further studies will be needed correlating the substrate specificities of binding with those of transport in vesicles to determine if these sites represent organic anion carriers. The high affinity sites in cLPM and bLPM clearly differ since neither BSP nor E_23G bind to this site in cLPM, but do in bLPM. The fact that only the cholestatic E_217G and taurocholate bind to the high affinity site in cLPM makes it tempting to postulate that this site represents a site which mediates cholestasis. Again, the substrate specificities of this site must be examined further to test this hypothesis.

In summary we have shown that E_217G, an endogenous organic anion, is taken up into hepatocytes by two systems. The high affinity uptake system, which is probably the more physiologically relevant site, is not Na^+-dependent, and appears to be shared by other organic anions such as BSP and taurocholate.

References

Anwer MS, Hegner D (1978) Effect of Na^+ on bile acid uptake by isolated rat hepatocytes. Evidence for a heterogenous system. Hoppe-Seyler's Z Physiol Chem 359:181-192

Brock WJ, Durham S, Vore M (1984) Characterization of the interaction between estrogen metabolites and taurocholate for uptake into isolated hepatocytes. Lack of correlation between cholestasis and inhibition of taurocholate uptake. J Steroid Biochem 20:1181-1185

Brock WJ, Vore M (1984) Characterization of uptake of steroid glucuronides into isolated male and female rat hepatocytes. J Pharmacol Exp Ther 229:175-181

Brouwer KLR, Durham S, Vore M (1987) Multiple carriers for uptake of $[^3H]$estradiol-17β(β-D-glucuronide) in isolated rat hepatocytes. Mol Pharmacol 32:519-523

Hugentobler G, Meier PJ (1986) Multispecific anion exchange in basolateral (sinusoidal) rat liver plasma membrane vesicles. Am J Physiol (Gastrointest Liver Physiol) 251:G656-G664

Laperche YA, Preaux AM, Berthelot P (1981) Two systems are involved in the sulfobromophthalein uptake by rat liver cells: one is shared with bile salts. Biochem Pharmacol 30:1333-1336.

Meier PJ, Sztul ES, Reuben A, Boyer JL (1984) Structural and functional polarity of canalicular and basolateral plasma membrane vesicles isolated in high yield from rat liver. J Cell Biol 98:991-1000.

Metzler CM, Elfring GL, McEwen AJ (1974) A package of computer programs for pharmacokinetic modeling. Biometrics 30:562-563

Meyers M, Slikker W, Vore M (1981) Steroid D-ring glucuronides: characterization of a new class of cholestatic agents in the rat. J Pharmacol Exp Ther 218:63-73

Munson PJ, Rodbard D (1980) LIGAND: a versatile computerized approach for characterization of ligand binding systems. Anal Biochem 107:220-239

Takacs AL, Vore M (1987) Binding of ^{3}H-estradiol-17β-(β-D-glucuronide), a cholestatic organic anion, to rat liver plasma membranes. Evidence consonant with identification of organic anion carriers. Mol Pharmacol 32:511-518

ABBREVIATIONS. $E_2$17G estradiol-17β(β-D-glucuronide); BSP bromosulfophthalein; DBSP dibromosulfophthalein; $E_2$3G estradiol-3-glucuronide; TC taurocholate; CHO cholate; TDHC taurodehydrocholate.

ACKNOWLEDGEMENTS. This work was supported by the National Institute of Health Grant HD 13250.

Figures 1-4 were taken from Takacs and Vore (1987) and Figures 5 and 6 from Brouwer *et al*. (1987) with permission.

Intestinal Transport of Glucuronides and Sulpho-Conjugates

F. Lauterbach, R.-P. Czekay, A. Fries and M. Schorn
Institute of Pharmacology and Toxicology
Dept. of Biochemical Pharmacology
Ruhr-University
4630 Bochum, FRG

Two anion transport systems have been inferred from their selective inhibition by phenoxybenzamine and the differences in the transport rate of various anions in the jejunum and the colon (1). Transport of acidic conjugates of phenolic compounds was now studied in preparations of the isolated mucosa of guinea pig jejunum and colon mounted in flux chambers (2). 1-Naphthol formed a glucuronide and a sulphoconjugate (3), estradiol mainly glucuronides, and p-nitrophenol (PNP) mainly a sulphoconjugate (PNP-S). In jejunal mucosae, with luminal phenol administration glucuronides and sulphoconjugates accumulated in the luminal solution. With phenol administration to the blood side of jejunal mucosae and to the luminal or blood side of colonic mucosae the main part of glucuronides appeared in the blood side solution, whereas sulphoconjugates were released at both faces of the isolated tissue in comparable amounts. - PNP-S was a much better substrate for transepithelial secretion into the luminal solution than p-nitrophenyl glucuronide (PNP-G), when added to the blood side compartment. The sulphonic acid ß-naphthol orange inhibited the secretion both of extracellular and intracellular PNP-S. On the other hand, PNP-G inhibited the influx of p-aminobenzoic acid at the basolateral border of the colonic mucosa, where PNP-S was without effect. It is concluded from these

Hepatic Transport in Organic Substances
Ed. by E. Petzinger, R. K.-H. Kinne, H. Sies

results that the excretion of intracellularly formed glucuronides and sulphoconjugates is mediated by different transport systems.

The dependence of metabolite pattern and metabolite distribution on the side of administration indicated a compartmentalization of intestinal metabolism of xenobiotics (3). This hypothesis was substantiated by pretreating the isolated jejunal mucosa with a hypertonic, 1 M sodium cyclohexanesulphamate solution. Luminal hypertonicity blocked luminal 1-naphthol and estradiol conjugation almost completely, but influenced metabolism of blood side phenols only slightly. Blood side hypertonicity had the opposite effect. Different compartments became also apparent when the availability of extracellular sulphate for sulphoconjugation was studied. In the jejunum luminal 1-naphthol incorporated luminal ^{35}S-sulphate to a tenfold higher specific activity into 1-naphthol sulphate than blood side ^{35}S-sulphate. Blood side 1-naphthol was conjugated preferentially, but not as exclusively as with blood side ^{35}S-sulphate. In the colon blood side ^{35}S-sulphate was used both for the cojugation of luminal and blood side 1-naphthol.

Acknowledgement

The financial support by the Minister für Wissenschaft und Forschung des Landes Nordrhein-Westfalen is thankfully acknowledged.

References

(1) Lauterbach F (1983) In: Gilles-Baillien M and Gilles R (eds) Intestinal Transport. Springer-Verlag, Berlin Heidelberg New York Tokyo, p 76-86
(2) Lauterbach F (1977) Naunyn-Schmiedeberg's Arch Pharmacol 297:201-212
(3) Sund RB and Lauterbach F (1986) Acta Pharmacol et Toxicol 58:74-83

Transport of Thyroid Hormone (TH) into Hepatocytes

E.P. Krenning[1,2], R. Docter[2], T.J. Visser[2], G. Hennemann[2]
Departments of Nuclear Medicine[1] and Internal Medicine III[2],
Erasmus University,
Rotterdam
The Netherlands

Introduction

Thyroxine (T_4), which is predominantly a prohormone, is produced by the thyroid gland. The biological activity of T_4 is generated after enzymatic conversion of T_4 into 3,3',5-triiodothyronine (T_3). In the euthyroid condition only about 20% of the total production of T_3 is accomplished by the thyroid gland itself; the remainder is mainly produced by the liver from T_4. Once formed, T_3 is transported via the circulation to various tissues for biological action. T_3 has at least two receptor sites: (1) a site of action in the nucleus with a resultant effect on gene expression, and (2) a direct effect on membrane processes, leading to an enhanced transport of amino acids and sugars. A receptor site on mitochondria has been claimed to exist, but sufficient evidence for this is lacking so far. It should be emphasized that several tissues do not only depend on circulating T_3 for biological action, but also to a varying extent on T_3 locally derived from T_4. These differences in various tissues can be explained among other things by the presence of different types of deiodinating enzymes (for a review, see Leonard and Visser, 1986).

Hepatic Transport in Organic Substances
Ed. by E. Petzinger, R. K.-H. Kinne, H. Sies

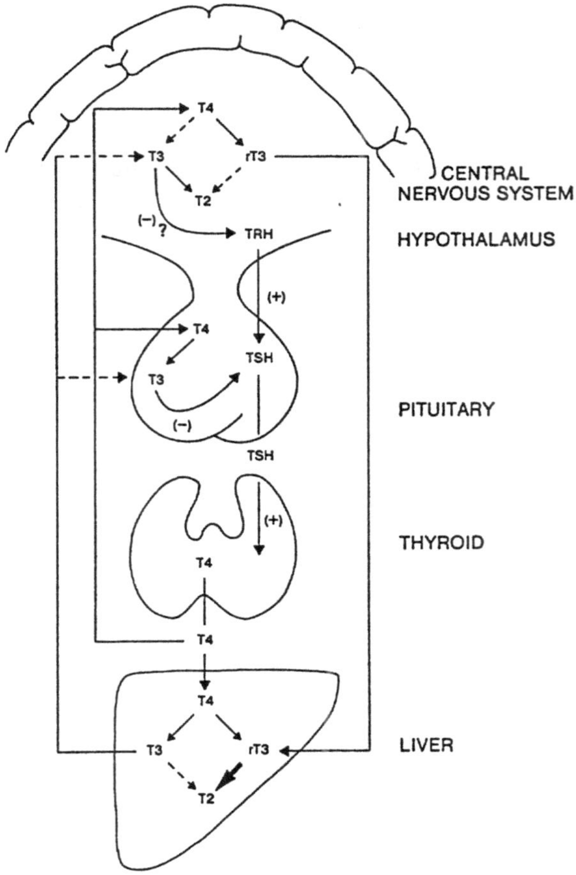

Figure 1. Diagram of thyroid hormone physiology. Indication of the efficiency of deiodination is based on determinations of deiodinase activities in tissues from euthyroid rats.

Another metabolite of T_4 is 3,3',5'-triiodothyronine (rT_3). The metabolism of this compound in liver is so rapid that the rT_3 produced in this organ does not enter the general circulation. The circulating rT_3 is derived from other tissues, for instance the brain. The liver plays also an important role in the clearance of circulating rT_3. In physiological concentrations rT_3 is not biologically active.

The synthesis and secretion of T_4 by the thyroid gland is influenced by thyroid-stimulating hormone (TSH), which is produced by the pituitary. The synthesis and secretion of TSH in its turn is inhibited by both serum T_4 and T_3 concentration (thus a feed-back mechanism) and stimulated by the hypothalamic TSH releasing hormone (TRH) (Fig. 1).

Iodothyronines in the circulation are predominantly bound to serum proteins. Generally, the biological activity is mediated by the free hormone moiety. Because large differences in serum TH binding may occur, either due to changes in the amount of protein or due to interaction of drugs with carrier proteins, it should be clear that the total hormone concentration is not a proper parameter for evaluation of thyroid function. Via the before-mentioned feed-back mechanism the function of the thyroid gland is regulated in such a way that a normal thyroid hormone effect is obtained. If the free hormone concentration were the overriding factor in the determination of the hormone effect, then one would expect that the free hormone concentration always reflects thyroid function, also in case of changes in the capacity of thyroid hormone binding proteins. Since the determination of the free T_4 concentration is a cumbersome procedure, the free T_4 index (FT_4I) has been introduced as a measure of the free T_4 concentration. The FT_4I is the product of the total T_4 concentration and the T_3 resin uptake. However, the FT_4I is not always a true reflection of the free T_4 concentration, if changes in serum proteins occur. Nowadays, the free T_4 concentration can be determined in a relatively simple way by commercially available kits.

It appears that increased free T_4 values (also by the generally accepted reference technique, i.e. equilibrium dialysis) are encountered in many circumstances in the absence of hyperthyroidism. In these circumstances, the increased free T_4 is apparently without biological consequences. A decreased free T_4 concentration without hypothyroidism is only rarely encountered, e.g. thyroxine binding globulin deficiency. The apparent discrepancy between free T_4 concentration and the thyroid status of the patient demonstrates that this parameter not always determines overall thyroid hormone tissue effects and that other factors may be involved, e.g. changes in transport of thyroid hormone over the cell membrane or in thyroid hormone deiodination.

Increased free T_4 concentrations without hyperthyroidism may be seen in patients with "non-thyroidal illness" (NTI). An explanation for the increased free T_4 concentration is a decrease in metabolism of T_4 that is more pronounced than a sometimes co-occuring diminution in T_4 production. It depends on the type of illness or used drug(s) whether the diminished metabolism of T_4 can be explained by a diminished uptake of T_4 by the liver or by a diminished deiodination of T_4 in the liver itself. In starvation and in NTI energy-mediated uptake of T_4 by the liver is impaired in humans (Van der Heijden et al., 1986). Several possibilities can be hypothesized, by which uptake is affected: (1) changes in microcirculation of the liver; (2) the presence of a plasma inhibitor, which not only interferes with the binding of thyroid hormone to serum proteins, but also with the entry of iodothyronines into hepatocytes (Table 1); (3) competition for transmembranal transport by structurally related compounds, e.g. amiodarone; (4) changes in composition of the plasma membrane and/or fluidity; (5) decreased intracellular ATP concentrations; and (6) decreased Na^+,K^+-ATPase activity, possibly related to changes in membrane fluidity.

Table 1.

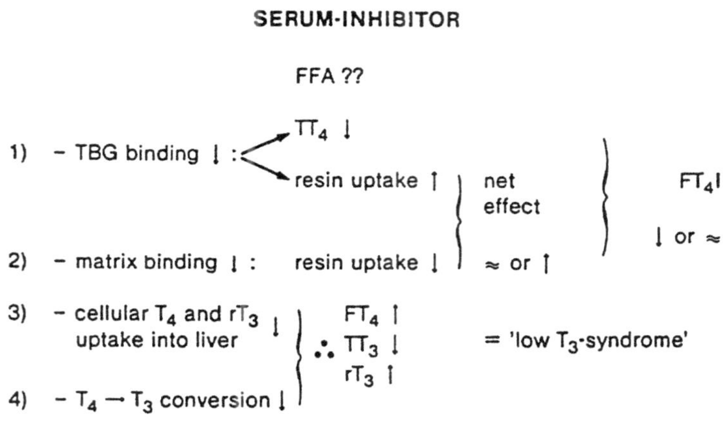

In NTI and starvation, except in the presence of renal insufficiency, serum rT_3 concentration is generally increased. This can be explained by a decreased uptake and/or intracellular metabolism of rT_3 by the liver. In NTI and starvation the serum T_3 level is decreased due to a decreased T_3 production from T_4. This is a logical consequence of (1) a diminished availability of T_4 in the liver due to a decreased uptake, by which indirectly T_4 to T_3 conversion is decreased and (2) a diminished T_4 to T_3 conversion per se. The diminution in T_3 production can be seen as an useful adaptation in illness in order to minimize the catabolic state, in other words, to conserve energy.

In this chapter aspects of TH passage through the plasma membrane, especially of hepatocytes, have been summarized (for review, see Krenning and Docter, 1986).

Thyroid Hormone Transport into Hepatocytes

Largely in consideration of the lipophylic nature of iodothyronines it was generally believed that penetration of the (fatty) plasma membrane by thyroid hormones occurs by simple diffusion. This assumption was supported by findings of high phospholipid-water partition coefficients for T_3 and T_4 as well as apparent nonsaturability of the entry of these substances into tissues. However, in spite of the high tendency of iodothyronines to partition in lipid vesicles, the zwitterionic character of the alanine side chain makes it unlikely that these molecules are translocated across phospholipid bilayers. Indeed, studies using T_3 and T_4 analogs with a spin label on the amino group have demonstrated free rotational and lateral diffusion in phospholipid vesicles, whereas flip flop of the analogs in the membranes was not observed. These observations suggest that diffusion is not an important mechanism for transport of thyroid hormone across cell membranes.

As early as 1954, it was suggested by Christensen et al. that T_3 and T_4 enter Ehrlich ascites cells by transport mechanism common to other, neutral amino acids. The main argument in support of this view was the inhibitory effect of KCN on the cellular accumulation of T_3. Later studies using the same cell line have refuted the idea of the involvement of neutral amino acid pathways, largely because of the lack of effects of Na^+ on T_4 uptake. Instead, facilitated diffusion was suggested as a possible mechanism of T_4 uptake, although no distinction was made between T_4 entry into the cells or adsorption onto the cell surface.

Overwhelming evidence has been reported in the last decade, however, indicating that the transport of thyroid hormone into target cells is an active process mediated by specific carrier systems located in the cell membranes. Thus, it has been demonstrated that T_3 and T_4 bind to plasma membranes isolated from rat liver and kidneys. Biochemical characterization suggested that each T_3 and T_4 interact with two sets of binding sites, of which those with the higher affinities specifically bind either T_3 or T_4 with apparent K_d values in the nM range. The involvement of these sites with the transport of T_3 and T_4 into the cells is suggested by the comparison with studies of the uptake of these compounds by isolated rat hepatocytes. Studying monolayer cultures of freshly isolated rat hepatocytes, we have found that the kinetics of uptake of T_3 and T_4 by these cells consists of three components. Although termed diffusion, the non-saturable process probably does not reflect entry of hormone, but could represent the partition of hormone in the lipid bilayer of plasma membrane. Interaction of T_3 and T_4 with low-affinity, high-capacity sites on the hepatocytes is neither associated with transport into cells, since it lacks all the characteristics necessary for such function, e.g. it is not influenced by incubation temperature or metabolic inhibitors. Therefore, this process probably represents interaction with binding sites on the external cell surface. The third component is characterized by high affinity and low capacity, and it is likely that only by

Table 2

CHARACTERISTICS OF MEMBRANE TRANSPORT OF THYROID HORMONE INTO RAT HEPATOCYTES

- Saturation	:	K_m T4 1 nM K_m T3 60 nM K_m rT3 6 nM
- Stimulation	:	temperature ATP albumin
- Inhibition	:	T3, T4 ouabain amiodarone radiographic agents monoclonal antibody ER-22 NTI-serum factors

this pathway thyroid hormone is able to penetrate the cell membrane. The properties of this transport system are summarized in Table 2.

Entry of T_3, T_4, and rT_3 into hepytocates is saturated at relatively low hormone levels with apparent K_m values of about 60 nM, 1 nM, and 6 nM, respectively. Although T_3 is a competitive inhibitor of the transport of T_4 and vice versa, it is not certain, if there is only one pathway common to both compounds, since K_m and K_i values are quite different. That the entry of hormone into hepatocytes is an energy-dependent phenomenon, is indicated by the decreased uptake of T_3 and T_4 observed at lower incubation temperature or resulting from decrements in cellular ATP content. The latter were induced by pretreatment of cells with T_3 through an as yet unknown mechanism, by addition of metabolic inhibitors such as KCN and oligomycin, or by a decrease in medium glucose levels. Extensive ATP depletion was obtained, if glucose in the medium was replaced by fructose. Uptake of T_4 and rT_3 was more readily affected by ATP decrease than uptake of T_3. This was taken in support of the view that different systems are used for transport of T_4 and rT_3 and that of T_3 across the cell membrane. The way these systems operate may be the same, however, since uptake of all iodothyronines is inhibited by ouabain, suggesting the requirement of an intact Na^+ gradient across the membrane.

The role of plasma binding proteins in the cellular uptake of T_3 and T_4 deserves special attention. In the monolayer system used by us, saturation of T_3 entry into rat liver cells could not be detected in protein-free medium. This failure can be explained in part on physical principles related to the unstirred water layer that surrounds the cells. Addition of 0.1% BSA to the medium obviates this effect by providing a buffer of protein-bound T_3 in the vicinity of the cells that is not readily depleted by the cellular uptake. However, apart from this, between 0.5 and 2% BSA also appeared to stimulate cellular entry of T_3, provided the free T_3 concentration in the medium was kept constant. The mechanism of this latter effect of BSA is unknown, but may involve an increased dissociation of T_3 from BSA induced by interaction of the complex with the cell surface. Such a mechanism has been proposed for the facilatory role of albumin in the hepatic uptake of fatty acids and certain drugs. Also in vivo studies have suggested a greater availability for albumin-bound T_4 and T_3 and perhaps even prealbumin-bound T_4 for tissue uptake than predicted from in vitro dissociation kinetics (Pardridge, 1981). Since such findings were also obtained with respect to transport of thyroid hormone through the blood-brain barrier, enhanced dissociation of albumin-bound T_3 and T_4 may also occur at the endothelial surface of capillaries. The possible role of factors in the microcirculation that lower the affinity of T_3 and T_4 for albumin remains speculative. For a more detailed discussion of the role of plasma proteins including thyroxine binding globulin in the uptake of thyroid hormone by tissues, the reader is referred to Ekins (1986).

The potential regulatory function of carrier-mediated uptake of iodothyronines into rat liver cells is strongly supported by studies using a monoclonal antibody against an epitope on the cell surface (Mol et al., 1986). This antibody did not only inhibit initial rates of uptake but also the cellular metabolism of hormone under steady-state conditions, indicating that the activity of the transport system determines the intracellular availability of iodothyronines. While non-specific effects of the antibody on the cell membrane were excluded by the demonstration of unimpeded Na^+,K^+-ATPase activity, transport rates of T_4, T_3, and rT_3 were affected to the same extent. These results are compatible with the existence of a common transport mechanism for different iodothyronines, and evidence was obtained that conjugates are taken up by the same pathway (Mol et al., 1986). However, it is not excluded that the antibody binds to a subunit that is shared by multiple transport systems.

Carrier-mediated uptake of thyroid hormone has also been observed in other cell types, among others in mouse and human fibroblasts (Docter et al., 1987), human hepatocytes (Krenning et al., unpublished), and rat pituitary tumor cells. Uptake was interpreted to take place by receptor-mediated endocytosis, as evidenced by the inhibitory effects of cytochalasin B, chloroquine, and monodansylcadaverine. It is of interest that the impairment of T_3 uptake induced by these compounds or ouabain resulted in a reduced occupancy of nuclear T_3 receptors, pointing to an important function of the transport system in the regulation of thyroid hormone bio-availability.

A wide variety of compounds of clinical interest has been shown to inhibit the transport of T_3 and T_4 into hepatocytes. Among these are iodinated substances such as radiocontrast agents and amiodarone, which are structural analogs of thyroid hormone and probably compete for binding to the transporter. Inhibitory effects have also been observed with propranolol, but in the hepatocyte monolayer system this was also associated with a decrease in cellular ATP. In contrast to starvation (Van der Heijden et al., 1986), propranolol induces a low serum T_3 level in humans in the presence of normal T_4 entry into the liver (Van der Heijden et al., 1988). Apparently, this effect of propranolol(-metabolites) is located at a step beyond uptake, e.g. enzymic deiodination of TH. Of special interest is that plasma of patients with severe non-thyroidal illness (NTI) contains excessive amounts of what seems to be a single factor that affects the interaction of thyroid hormone with different systems (Table 2). Uptake of T_4 by human hepatocarcinoma (Hep G2) cells has been studied from plasma spiked with radioactive hormone (Sarne and Refetoff, 1985). With plasma samples from healthy persons an excellent correlation was observed between cellular T_4 uptake and either the free T_4 concentration or the free T_4 index. However, in case of NTI patients uptake correlated closely with the free T_4 index, but poorly with free T_4 by dialysis, suggesting that T_3 resin uptake and T_4 cell uptake are affected to the same extent probably by the same factor. These studies have recently been extended by us using normal cultured rat hepatocytes, which enabled the effects of plasma factors on the hepatic metabolism to be determined. The results indicated that, if allowance was made for the variation in free hormone concentrations, metabolism of plasma T_3 and T_4 from patients with NTI was significantly lower than that from normal subjects. This was true for both iodide production and conjugate formation, suggesting inhibition of a step preceding these metabolic reactions, presumably exerted at the level of the cell membrane. Taken together, these findings provide strong evidence that changes in plasma membrane transport have important consequences for the cellular metabolism of thyroid hormone.

Regulation of Thyroid Hormone Expression

Thyroid hormone action on tissues is determined to a large extent by the occupancy of nuclear T_3 receptors, which in turn is a function of the available concentrations of free T_3 in the nucleus (Pardridge, 1981). It is not possible to give a general discussion of how the nuclear T_3 level is regulated, since in some tissues such as the central nervous system nuclear T_3 is provided largely by local conversion from T_4, while in other tissues like liver, kidney, and muscle, nuclear T_3 is mostly derived from plasma T_3 (Van der Heijden, 1986). Further, the architecture of the tissues has an important influence such that, owing to large fenestrae in the capillaries, liver interstitium is in rapid equilibrium with plasma. Therefore, the major barrier for uptake of thyroid hormone by the liver parenchymal cells is probably formed by the plasma membrane. On the other hand, in tissues with small or - as in the brain - absent fenestrae the capillary wall becomes a major obstacle, but little is known about the way iodothyronines traverse the endothelium. Consequently, most of the following discussion is focused on the handling of thyroid hormone by the liver.

With respect to thyroid hormone action in the liver it is important to distinguish the following aspects: (a) generation of T_3, (b) binding of T_3 to the nuclear receptor, and (c) degradation of T_3. Cellular uptake of plasma T_4 is the initial step necessary for this conversion to T_3 by the deiodinase located in the endoplasmic reticulum. Similarly, plasma T_3 must first of all traverse the cell membrane en route to the nuclear receptor and to sites where the hormone is metabolized, i.e. the cytoplasma for sulfation and the endoplasmic reticulum for glucuronidation and deiodination.

It has been suggested that T_3 enters fibroblasts and growth hormone-producing pituitary tumor cells by means of receptor-mediated endocytosis. This would implicate that T_3 is internalized in vesicles pinched off from the cell membrane, which ultimately fuse with lysosomes. It is not known how T_3 would leave this compartment and find its way to for instance the cell nucleus. No evidence exists either for or against the possibility that thyroid hormone is taken up into liver cells by receptor-mediated endocytosis. The experimental findings are also compatible with carrier-mediated uptake perhaps driven by a proper Na^+ gradient and resembling transport pathways for glucose and amino acids. In this case, iodothyronines would be in immediate contact with the cytoplasm which is abundant in low-affinity binding proteins that could be instrumental in passing on iodothyronines to the different organelles.

In order for T_3 to reach its main site of action it must penetrate the nuclear membrane. The latter contains pores that are sufficiently large to let fairly sized proteins move through. Nevertheless, based on differences between in situ receptor occupancy and in vitro affinity estimates it has been calculated that an even greater gradient may exist for T_3 across the nuclear membrane compared with the cell membrane (Oppenheimer and Schwartz, 1986). It seems difficult, however, to obtain direct experimental validation of the magnitude of such gradients.

Conclusion

In consideration of the importance of the liver for thyroid hormone turnover in euthyroid and perhaps even more so in hyperthyroid subjects, it is no less than expected that a dysfunction of this organ will result in abnormalities of thyroid hormone metabolism. In several clinical situations, such as starvation, systemic illness, and the use of certain drugs, production of plasma T_3 and clearance of plasma rT_3 have been shown to be decreased, and this phenomenon is commonly referred to as the low T_3 syndrome. Some of these conditions, i.e. fasting and illness, are associated with a reduced tissue uptake of iodothyronines. It is a logical assumption that the abnormal thyroid hormone parameters are the results of a defective plasma membrane transport mechanism, leading to a diminution in the availability of T_4 and rT_3 for intracellular deiodination. For other conditions, such as treatment with PTU or propranolol, it is likely that the defect is localized in the hepatic deiodination itself with resultant diminished rates of T_3 formation and rT_3

breakdown. Thus, irrespective of the actual mechanism, in many conditions associated with the low T_3 syndrome production of plasma T_3 and clearance of plasma rT_3 are impaired because of a diminished liver function.

References

Christensen HN, Hess B, Riggs TR (1954) Concentration of taurine, beta-alanine and triiodothyronine by ascites carcinoma cells. Cancer Res 14:124-127

Docter R, Krenning EP, Bernard HF, Hennemann G (1987) Active transport of iodothyronines into human cultured fibroblasts. J Clin Endocrinol Metab 65:624-628

Ekins R (1986) The free hormone concept. In: Hennemann G (ed) Thyroid hormone metabolism. Marcel Dekker Inc., New York, p 77-106

Krenning EP, Docter R (1986) Plasma membrane transport of thyroid hormone. In: Hennemann G (ed) Thyroid hormone metabolism. Marcel Dekker Inc., New York, p 107-131

Leonard JL, Visser TJ (1986) Biochemistry of deiodination. In: Hennemann G (ed) Thyroid hormone metabolism. Marcel Dekker Inc., New York, p 189-229

Mol JA, Krenning EP, Docter R, Rozing J, Hennemann G (1986) Inhibition of iodothyronine transport into rat liver cells by a monoclonal antibody. J Biol Chem 261:7640-7643

Oppenheimer JH, Schwartz HL (1986) Thyroid hormone action at the nuclear level. In: Hennemann G (ed) Thyroid hormone metabolism. Marcel Dekker Inc., New York, p 383-415

Pardridge WM (1981) Transport of protein-bound hormones into tissues in vivo. Endocrine Rev 2:103-123

Sarne DH, Refetoff S (1985) Measurements of thyroxine uptake from serum by cultured human hepatocytes as an index of thyroid status: reduced thyroxine uptake from serum of patients with nonthyroid illness. J Clin Endocrinol Metab 61:1046-1052

Van der Heijden JTM, Docter R, van Toor H, Wilson JHP, Hennemann G, Krenning EP (1986) Effects of caloric deprivation on thyroid hormone tissue uptake and generation of low-T_3 syndrome. Am J Physiol 251:E156-E163

Van der Heijden JTM, Krenning EP, van Toor H, Hennemann G, Docter R (1988) Three compartmental analysis of effects of d-propranolol and thyroid hormone kinetics. Am J Physiol, in press

Metabolism of Thyroid Hormone in Liver

T.J. Visser, M. Rutgers, W.W. de Herder, S.J. Eelkman Rooda and M.P. Hazenberg
Departments of Internal Medicine III and Medical Microbiology, Erasmus University Medical School, Rotterdam, The Netherlands

The thyroid gland produces predominantly the prohormone thyroxine (T_4) together with a small amount of the biologically active hormone 3,3',5-triiodothyronine (T_3). In normal humans and rats 80% of circulating T_3 is derived from outer ring deiodination (ORD) of T_4 in peripheral tissues. T_3 is further degraded by inner ring deiodination (IRD) to 3,3'-T_2. T_4 and T_3 are also metabolized by glucuronide (G) and sulfate (S) conjugation.

Different enzymes are involved with the peripheral deiodination of thyroid hormone. The type I deiodinase of liver is a non-selective enzyme that is capable of both IRD and ORD, and is inhibited by the drug propylthiouracil (PTU). IRD of T_3 (and subsequently ORD of 3,3'-T_2) by this enzyme is greatly facilitated by sulfate conjugation of the substrate. This enzyme is thought to be essential for the production of plasma T_3.

Incubation of 3'-[^{125}I]T_3 with rat hepatocytes results in the formation mainly of $^{125}I^-$ and T_3G. In the presence of PTU, I^- release is blocked and T_3S accumulates in the cultures. After i.v. injection of labelled T_3 to normal rats, mainly

Hepatic Transport in Organic Substances
Ed. by E. Petzinger, R. K.-H. Kinne, H. Sies

T_3G is found in bile, but in PTU-treated rats excretion of T_3S (and T_2S) is strongly augmented and T_3S also accumulates in plasma.

After i.g. administration of radioactive T_3G to rats, nonconjugated T_3 appears in plasma which is prevented by intestinal decontamination with antibiotics. This treatment eliminates obligately anaerobic bacteria which are capable of hydrolyzing iodothyronine conjugates. Resorption of administered T_3 is not impaired by decontamination.

After i.v. injection of labelled T_3G to normal and decontaminated rats, the conjugate is rapidly cleared from the circulation and 90% is excreted intact in the bile within 1 h. After 2.5 h, radioactivity (RA) reappears in plasma of normal rats in the form of T_3 (and I^-) which a) is similar in magnitude as the plasma RA after i.g. administration of T_3, and b) is diminished in decontaminated rats receiving i.v. T_3G.

The results indicate that conjugation of T_3 serves different purposes. Sulfation initiates the deiodinative breakdown of the molecule, while glucuronidation facilitates its biliary excretion. T_3G is hydrolyzed by intestinal bacteria and liberated T_3 is absorbed, indicating a significant enterohepatic circulation of the hormone.

Concluding Remarks

K.J. Ullrich
Max-Planck-Institut f. Biophysik
Kennedy-Allee 70
6000 Frankfurt 70, FRG

The conclusion remarks are those of a physiologist working on renal organic anion transport and watching carefully all presentations of this symposium. Dr. Berk divided in his talk the scientists working on organic anion transport in "lumpers" and "splitters". I think this is too exclusive, because many biologists start just with lumping results together. But finally, they are forced to split the results up into different categories: Nature in the molecular dimension of transport behaves more diverse than experimenters suppose.

Dr. Petzinger tried in his presentation to find common denominators between hepatic and renal handling of organic anions. There seems, indeed, to exist a parallelism between basolateral and luminal cell sides of liver cells on one hand, and small intestinal as well as kidney proximal tubule epithelial cells on the other hand: 1. In membrane marker enzymes (Table 1), and 2. in size of membrane proteins involved in bile acid transport (Table 2). In contrast, the Na^+ dependence of bile acid transport might change according to the direction into which bile acid transport occurs: In the liver from the blood into the canaliculi (=lumina), in the intestine and the kidney from the lumen into the blood. Another parallelism is given between basolateral transport processes in liver and kidney:

Hepatic Transport in Organic Substances
Ed. by E. Petzinger, R. K.-H. Kinne, H. Sies

Table 1 Marker enyzmes for plasma membrane in liver, small intestine, and kidney

1. Luminal Membranes

	Liver (canalicular)	Small intestine (brush border)	Kidney proximal tubules (brush border)
Alkaline phosphatase	+	+	+
γ-glutamyl transpeptidase	+	+	+
Leucine-arylamidase	+	+	+
sucrase	?	+	+
5'-nucleotidase	+	+	+

2. Basolateral Membranes

	Liver (sinusoidal)	Small intestine	Kidney proximal tubules
$(Na^{+}+K^{+})$-ATPase	+	+	+
Receptor-stimulated adenylate cyclase	+	+	+

In the liver three "multispecific" anion transport systems are proposed: 1. For bile acids and phallotoxins, 2. for bromphenolsulphthalein (BSP) and bilirubin, and 3. for long chain fatty acids and possibly T_4 and T_3. In contrast, for the transport of anions through the basolateral cell side in the proximal renal tubule, three distinct transport systems have been identified (Fig. 1). They interact with a wide variety of

Table 2 Comparison of bile acid transport systems in liver, intestine, and kidney

1. Liver	sinusoidal membrane	Na^+-dependent (49 kDa) Na^+-independent (54 kDa)**
	canalicular membrane	Na^+-independent (100 kDa)*
2. Intestine	brush border membrane	Na^+-dependent (99 kDa)*
	basolateral membrane	Na^+-independent (54 kDa?)**
3. Kidney	brush border membrane	Na^+-dependent (99 kDa)*
	basolateral membrane	Na^+-independent (54 kDa?)**

* The 99-100 kDa transporter from liver canalicular membrane and intestinal as well as renal brush border membranes interact with the same antibodies.

** The 54 kDa bile salt-binding proteins exist in the basolateral membranes of intestine, kidney, and liver. The proteins from liver and kidney have the same isoelectrical point and might be identical.

substrates and show overlapping specificities. The specificity does not reside in the configuration of a few atoms, but in strength, number, and distance of electronegative charges as well as in hydrophobic areas within the molecule. For the kidney, specific substrates are available, i.e. substrates which interact exclusively or preferentially with one of the three organic anion transport systems identified at the contraluminal cell side of the proximal tubule. Such a situation does not yet exist for the liver. Furthermore, the usual substrates to study liver organic anion transport are rather hydrophobic, i.e. bile acids, bilirubin, T_4, T_3, and hydrophobic molecules conjugated with glutathion, cystein, glucuronate, and sulfate. These substrates show a higher affinity

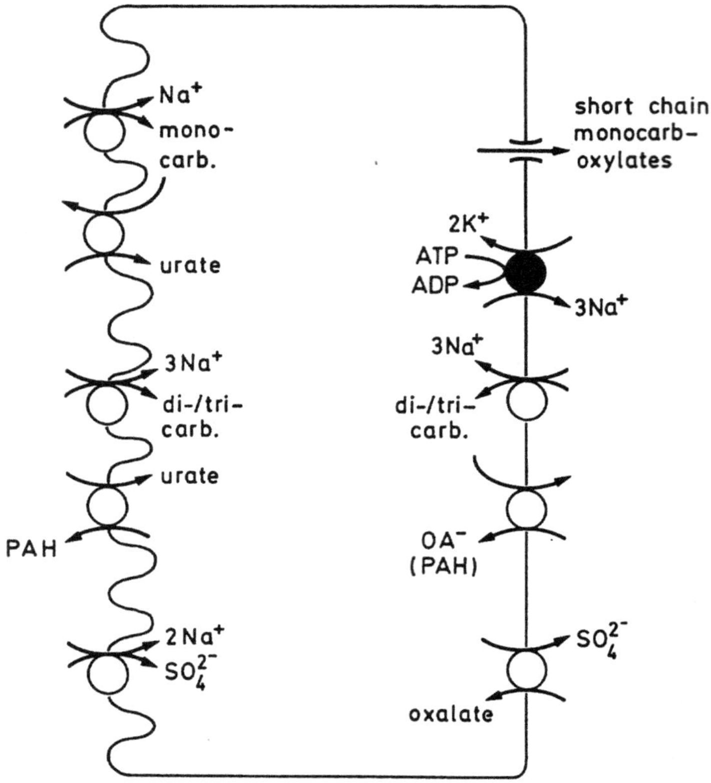

Fig. 1: Transport systems for organic anions in rat kidney proximal tubule cells. Lumen is on the left hand side; interstitium resp. blood on the right of the epithelial cell. Open circles denote transport systems, the filled circle an ATPase; ⊣⊢: channel; OA⁻: hydrophobic organic anions as, e.g., p-aminohippurate (PAH); Di-/tricarb.: short chain di- and tricarboxylates (citric acid cycle intermediates); Monocarb.: short chain monocarboxylates (hydrophilic).
(Burckhardt and Ullrich, unpublished).

to hepatic than to renal transport systems. In the kidney they interact preferentially with the PAH transporter, but often also with one or both of the other two organic anion transporters. The higher affinity for the hepatic transport systems may be the reason why affinity columns were used more successfully in the liver than in the kidney for isolation of transport proteins. The same holds for affinity labelling studies which are more advanced with hepatocyte membranes. A further improvement is shock freezing and irradiation in liquid nitrogen, which efficiently prevents diffusion artefacts during photoaffinity labelling, as shown by Dr. Kurz.

Several speakers reported about antibodies raised against the different transporters. It became evident that the chances to altering transport function and obtaining high affinity antibodies are higher with polyclonal than with monoclonal antibodies (P.J. Meier). On the contrary, monoclonal antibodies are the right tool to sense different epitopes of a transport molecule and their configuration change during substrate binding (D. Levy). Fortunately, the availability of antibodies against transport proteins opens the possibility for applying gene technology. Expression of mRNA which code for not yet identified transport proteins, in oocytes may help to clone the correspondent cDNA. Fractionation of mRNA before injection into the oocytes might be more successful, if the production and turnover of the respective transport protein is as high as in intestine or in regenerating tissue, or if the turnover can be stimulated by hormones or starvation (Kilberg)

It is noteworthy that the affinity labelling of sinusoidal membranes identified 46 kDa, 48-49 kDa, and 54 kDa proteins, the 49 kDa polypeptide being the most likely candidate for Na^{+}-dependent bile acid transport. The ontogenetic approach (F.J. Suchy) as well as the comparison between hepatocytes and hepatoma cells (D. Levy) favour the same protein. It is hoped that the amino acid sequence of these proteins will be soon available. Then, also the supposed relationship between hepatic transporter proteins and the similarly sized renal transporters can be further investigated. Finally, the reconstitution of the transport proteins into liposomes is obligatory to be sure that they function as translocators. I suppose that it is still a long way until point mutations in the transporter genes will provide information about the key amino acids and X-ray cristallography about the tertiary structure of the anion transporters.

Acknowledgement

I thank Dr. G. Burckhardt for helpful discussions and for compiling the tables.

Subject Index

www.ingramcontent.com/pod-product-compliance
Lightning Source LLC
Chambersburg PA
CBHW061153240926
51846CB00006B/1002
* 9 7 8 3 6 4 2 7 4 2 4 9 1 *